AF560136

Pvt. Ltd.

Sex Fixing

The Dawn of a New Era

B.S. Aulakh

Gregor Mendel Institute for Research in Genetics
144, N.J. Park, Baloke Road,
Ludhiana 141 001

2009

Studium Press (India) Pvt. Ltd.

Sex Fixing

The Dawn of a New Era

ISBN: 978-93-80012-17-9

Published by:

Studium Press (India) Pvt. Ltd.
4735 / 22, 2nd Floor, Prakash Deep Building
(Near Delhi Medical Association),
Ansari Road, Darya Ganj, New Delhi-110 002
Tel.: 23240257, 65150447; Fax: 91-11-23240273;
jngovil@studiumpress.com; shray@studiumpress.com

Printed at:

Salasar Imaging Systems
Delhi-110035 (India)

Contents

Part A. Preface		1-9
	1. About the Book	3
	2. About the Discovery	5
	3. Dawn of Hope	7
	4. Few Words by the Author	8
Part B. Introduction		11-56
	1. Sex Fixing: What the Experts Say	13
	2. A Massive Project	17
	3. A Treatise on Sex Fixing	23
	4. Historical Background	48
	5. Comparative Status of the Research	52
Part C. Theory and Basics		57-114
	1. Fertilization	59
	2. Sex Fixing	72
	3. Formulative Theory of Determination of Sex	97
	4. Prior Art	110
Part D. Reflections and Critique		115-163
	1. The Dawn of a New Era	117
	2. Need of the Hour	125
	3. Exploring New Vistas of Research in Pure and Applied Sciences	137
	4. Aulprofem: A Strong Case for a Nobel Prize in Medicine for India	145
	5. The Missing Cow: A Key to Success	154
Part E. Scope and Applications: A Bonanza of Benefits to Mankind		165-254
	1. Saving Lives – Fighting Death	167

2. Alleviating Poverty 177
3. A Biggest Environment Friendly 181
4. A Case for Rural and National Development 191
5. Saving the Female Gender: A Great Social Boon 200
6. Economic Crisis and the Way Out 205
7. Fighting Terrorism – A National Salvation 230
8. A Strong Case Against Narcotics and Addiction 237
9. Alleviating Poverty: A Humanitarian Cause 249
10. Delivering Designer Babies 251

Part F. Clinical Information 255-275
1. Clinical Information 257
2. Pharmacodynamics and Pharmacokinetics 266

Part G. Clinical Trials and Pharmacology and Toxicological Studies 277-339
1. Clinical Trials 279
2. Studies for Toxicology and Pharmacology 285

Part H. Test Dossiers 341-351
1. Method of Analysis 343
2. MSDS Data 347
3. International Safety and Hazard Regulation Data 349

Part I. Miscellaneous 353-363
1. Publications 355
2. Bits *n* Bytes 356

Part J. Charter of Fame 365-371
1. Brevet Canadien 367
2. U.S. Patent 368
3. New Zealand Patent 369
4. South African Patent 370
5. Australian Patent 371

Appendix I 372

Appendix II 377

Appendix III 378

Appendix IV 380

Subject Index 381-400

Food for Thought

Right before me there is vast ocean of discoveries...........and I am just a kid playing on the sea shore picking a few pebbles.........**Sir Isaac Newton**

Sincerely I want to admit that right from the day when Aulprofem was invented in 1992 and it has been long sixteen years after that; I am not able to understand that why the most kind Mother Nature chose me for the honor for which I don't find myself the rightful candidate in any way. There are big names and heavy weight scientists on the national as well as international scientific arena that could have been the better choice. I being a very small creature and a humble entity certainly do not deserve to be associated with such a great discovery like Aulprofem. It looks awkward, crude and non-adjusting when someone looks from the angle where I am placed. My whole being and existence is certainly totally indebted and grateful to the forces of nature that has decided and chosen me for this blessing and I have no words of thanks for them.

................certainly I have struggled and the scientific community has not accepted a research done by a lowly down to earth man of humble origins and apathy and neglect shown by the government and institutional organizations have done great harm to the cause of science but even then, based on the inner strength and the very nature of it; this discovery has survived and it gives me immense satisfaction and a feeling of inner accomplishment and joy and once again I find myself bowed in earnest prayer and thanksgiving to the most kind and benevolent Mother Nature.

B.S. Aulakh

About the Author

Popularly known as 'father of sex fixing', the author, is a professional drugs researcher. After qualifying in the pharmaceutical disciplines, he served a lecturer as well as a production chemist in many institutes and commercial organizations. It was sheer by chance, while working on simple drug chemicals that the discovery of Aulprofem happened. The intended objective, in no way was the discovery of a sex fixer formula for mammals or otherwise. To pursue his goal of carrying forward the drugs research in his own way and style, he founded the Gregor Mendel Institute for Research in Genetics; a privately owned and funded drugs research organization, at Ludhiana in India but the going was never so smooth for him. Despite running into financial hardships and social obstacles, the endeavor was eternally faced with conditions of near bankruptcy but slowly with success on international fronts with the clearing of the patents for this research in patents offices of many countries, including the United States, things eased out a lot. The other research projects that the author is currently associated with, include the discovery of a non-biological vaccine mimic, termed non-biological drug vaccine, a GIT drug and a gynecology drug.

B.S. Aulakh

E-mail : bsaulakhpost@rediffmail.com

PART - A

Preface

- About the Book
- About the Discovery
- Dawn of Hope
- Few Words by the Author

About the Book

This book is about the new and fascinating phenomenon of sex fixing. Sex fixing is a novel development in the field of biological sciences in general and reproductive physiology in particular. The organisms which are unisexual and dimorphs present a potential case for the practical application and commercial exploitation of this phenomenon, in their context. The kingdom *animalia* is the preferred field of study and exploring as regards this new technology. There has been recorded history of the existence and working benefits of this phenomenon in non chordates and chordates of lower phylogenic status. The ratites and certain reptiles exhibit this phenomenon and certain uses of this has been identified and explored. The kiwis, ostrich, certain lizards and some other reptiles exhibit this phenomenon. It has come to be known as temperature dependant sexual determination (TDSD). Even some patents have been generated in the United States and other countries' patent offices. The avians and mammals are advanced animals and do not exhibit TDSD. The new discovery of Aulprofem is the first such development in the field of sex fixing, in case of mammals. Being the first experience, there is ample scope that many of the aspects of this wonderful phenomenon may not be covered in full details but still we wanted to give our best.

The present book is most probably the first book on sex fixing, in mammals.

This book tries to cover the essence, scope and application of this phenomenon of sex fixing in higher chordates, particularly mammals. Fortunately enough, the first successful mammalian sex fixer has been discovered and making news and benefits in the scientific and commercial circles as well. This is an oral drug formula that is moving in the arenas associated with increase in dairy and livestock productivity under the name 'Aulprofem'. This drug produces female progenies in dairy animals and other mammals. The studies conducted so far and covered in this book, are on dairy animals; cows and buffs, pet animals, dogs, the economically useful animals like the goats, sheep, pigs, horses, etc.

The title of the book; "Sex Fixing: The dawn of a new era" is a misnomer and misleading to a certain extent for the present book deals with the subject of sex fixing particularly in mammals and it can not claim to be the first book on sex fixing as a whole because sex fixing as a phenomenon is not new in other animals such as non chordates and even lower chordates such as ratites and certain reptiles as mentioned above.

The subject of the present book is not the sex fixing in non chordates and lower chordates but it covers this phenomenon and its application in the most evolved family in kingdom *animalia*, the mammals. This focuses on the probable impact and outcome of such a new discovery and development on such animals including humans and the resultant changes in human society at large and argues in a unique criticizing and constructive way about the changes and advantages that are bound to follow with the application of this research on macro and mass levels in human society and thought. The various benefits that can accrue of this discovery to the mankind, is the subject as well as the debate point of this book and we have tried the best in our humble capacities to bring this aspect in the centre stage and under the analyzing magnifying spy glass at the same time. Being the foremost or one of the first books on the subject; many interesting angles of importance might not have been covered in this book and even some important discussions or debate points not even been touched but still in the present endeavor, we will try to our full capacity, to give as much information as we can or is at our disposal and knowledge. It will certainly not be surprising if within a few years more, there will be a lot of exploring on this subject throughout the world and the work described in the present book will be representing just a minuscule of the total work done at that time and which even now deserves to be should have been accomplished and covered. In any way and viewpoint, we are just standing on the beginning strip of a great scientific breakthrough and the journey has just started. Hope, it is most interesting and rewarding to one and all. Not only humans but also the other mammalian species; especially the ones already on the verge of being annihilated to extinction, should include in the grand list of beneficiaries form a mega scientific exploit. The hype is certainly there and is no hiding and can be easily felt. But still there are shortcomings and many things and outcomes that should have been a reality by now; are still sitting far away idling in the cozy lap of future. We have given the delivering that was in our small and humble hands and we feel satisfied to a certain extent that we have performed, may be to a smaller extent and we have maintained a sincere and committed stance in such trying to our best levels. We do not claim that it will be sufficient but it certainly is an honest attempt on the part of the author and the various people associated with him, this research and even the people associated in the printing, publishing and allied services needed in the present endeavor to make this work take the shape of a book and this way see the light of the day so that the message of research and progress reaches the maximum number of deserving and targeted people; and a word of thanks and obligation is extended from him to all such people and also to other well wishers and friends and still other people associated in any way and means for extending material and moral support, advice, criticism and making this effort, a practical feasibility and success.

B.S. Aulakh

About the Discovery

Aulprofem is a sex pre assignment and determination drug.

This is a small and compact answer but does not seem to suffice. To answer this fully, we need to have some more questions along with the answers to them. Let us give a try.

The sex fixing, what is it?

This question again can have a straight answer but it may not be easily palatable. It will be better if we invent two more questions and answer them first. These questions are as follows.

What is sex determination?

And secondly what is sex selection?

Let give them a try one by one.

Sex determination is the process of ascertaining the sex of the off-springs in zygote, embryo or fetus stage preferably in the womb, with the help of techniques like amniocentesis, ultra-sound scan etc.

And sex-selection is the process of selecting one particularly sexed (either male or female) zygotes, embryos, or fetuses and saving and keeping them and rejecting and destroying the other type.

A point here to note is that sex-selection is possible only when sex – determination has already been done.

The matter here becomes much of an unpleasant and tasteless one. Any reference or explanation to these questions, tries to point out the strange but true relationship between these words and the countless pitiable faces of the girl fetuses aborted, in male dominated societies of India, Pakistan and Bangladesh.

Only last year in a place called Paatran (see Appendix II), in Patiala district of Punjab in India there was a big story in press (Appendix III, Indian Express. Sep. 15, 2006) about the aborting of female fetuses in humans and dumping them into a well, in couples desirous of a male child and unfortunately carrying and detecting female embryos in the womb; in the premises of the nursing home, where this nefarious business was carried on. When that well was searched by the police sleuths; nearly 165 skeletons or remains of the fetuses were found in the well. This gory incident resembled the story line of a horror movie. The entire nation was sent into a shock mode when this story erupted in the media. The TV channels and programs that broadcast the in depth details about this incident, got high viewer ship ratings.

The morality of an entire nation and society, was under debate.

So, sex determination and sex selection have considerable moral as well as emotional overtones attached to them.

But scientifically speaking, emotions are not a commodity, worthy of trade. They do not command a price also. So we leave this humanitarian angle here only and proceed further in our discussion.

Sex pre assignment and determination is the process of deciding in advance as to which sexed zygotes will be conceived in future and conditions are so adjusted that only the particularly sexed zygotes and hence, off-springs are produced and even the possibility of conceiving and subsequent birth of the off-springs of the other sex type is altogether ruled out. This process is also called Sex-Fixing.

Another point to note here is that there is no place for the phenomena of sex determination or sex selection to occur if we have to go for the process of sex fixing. These two phenomena are not even needed in a usual course of sex fixing.

The discovery of Aulprofem belongs to the third category. Hence, it can also be called a sex fixer drug.

The Dawn of Hope

The mighty Sun of a new discovery dawns on the horizon and there is glitter all around the world of its magnificent, shining and blissful rays and they appear to spread the message of universal betterment and hope.

They seem to sermonize as under:

'This discovery of Aulprofem has in the real sense dawned an era of sex fixing in the world'.

The listeners seem to get interested. The sermon continues.

'The future of Biology in general and Reproductive Physiology in particular is definitely the Sex Fixing.

The effect of this research is bound to be such that not only there will be changes in the structure and functioning of future society but it will have profound effects on the ways the people think and even their behavior will be guided accordingly.

An existence of future society will even be difficult to be imagined without the presence of sex fixing phenomenon in the lives of the people. A day to day or the normal fabric of life will cease to carry a meaning without sex fixing.

The sex fixing will be there everywhere, knowing or unknowing, intentionally or without intention or in any other way in the lives, behavior, thinking, instincts or even in the ideologies of the people.

The new world definitely is going to prove to be an era of sex fixing and this phenomenon with its all prevalent presence and its ability to impress human thought and action, will definitely be one of the greatest forces that will constitute the strength of future world society.'

After this they pause for a moment, look at the world with a smile in their eyes and the people listen and the world seems to acknowledge in thanksgiving. The sermon ends here only; they spread their wings, appear to fly and just become invisible and disappear.

Where have they gone?

Who knows?

People whisper and gaze more deep for the answers, nod as if swallowing understanding and then go about in pursuit of their day to day chores.

Better we also follow them. But stop, let's try to be wise.

Let's give some reading to the book, turn some pages and try to uncover the mystery.

Perhaps we may find them again.

Few Words by the Discoverer

It is a truth that the discovery was a gift of Mother Nature to mankind and the discoverer made no particular contribution and effort in this. In fact, he does not even think him to be fit enough in any way of doing so. The very nature of this discovery is such that it opened new ways and vistas of success and trek in the forward directions that ultimately led and guided the discoverer himself. Today whatever has happened of the discoverer or the discovery, is the result of the very nature and inner strength of this research and no credit whatsoever should go to the discoverer.

There are few thoughts that need to be shared with the reader of this book. Even if somebody is of a spiritualistic nature or he is of a materialistic mind set, both ways the discussion leads to the one conclusion. There are certain forces of God or the supernatural that guide or lead the forces of nature to perform certain acts destined by God himself for this world and this discovery happened exactly in this way and the discoverer being a petty or worthless creature, played no appreciable role in all this. On the other hand in a materialistic thought, certain forces of nature try to work in such direction that the overall productivity and constructive efficiency of the world increases and man is a mere tool in this entire scenario and he is not the final force or destination. The position of the discoverer is such and such only.

There is no other thought and viewpoint.

All said and done, one thing is very real that we got a great visitor with us in the form of this discovery of Aulprofem. We welcome this or not; may depend on us but the survival of this discovery in no way is dependant on the liking or not of us all. The discovery has already been decided by the great forces of nature and it will definitely survive. So, let us welcome this benevolent and kind guest on our doors and allow it to bless and guide the entire mankind on its way to further betterment and development of the human society in still greater magnitudes. This is the demand of forces that guide and rule this world as well as is their decision and we being the truthful and worthy inhabitants of the magnificent creation of the Mother Nature; this beautiful world, should obey them.

The author being a small man and a tiny entity do not want to claim any credit or glory of this great research and he declares of having no appreciable role in the making of this worthwhile splendid discovery and does not even feel a right to call upon the citizens of the world to give consideration and importance to anything connected in any way with his name but only hopes in the best of his intentions and expectations that a great discovery done by the great forces of nature will in the long run prove much beneficial to all the great and worthy citizens of this world and prove a true milestone in the road to progress for the human generations of today and all tomorrows to come.

B.S. Aulakh

PART - B

Introduction

- Sex Fixing: What the Experts Say
- A Massive Project
- A Treatise on Sex Fixing
- Historical Background
- Comparative Status of the Research

Sex Fixing: An Important Field – What the Experts Say?

Ever since the advent of a wise thinking creature like man has occurred on earth, he has yearned for three things; first to conquer over death and mortality, second to be free from all disease and suffering and third to be able to master the choice of production of desired sexed off-springs. The fist two wishes can not be granted for reasons very obvious to a person expert in the field of thinking, practice and theory of scientific and philosophical arenas. But the nature has decided to satiate the third desire of the mankind; *i.e.* to produce the off-springs of desired sex. This has happened in the form of the success in the discovery of an *in vivo* mammalian female sex fixer, named Aulprofem (Aulakh, 2008). Aulprofem being the world's first mammalian sex fixer, produces female off-springs in dairy animals and other mammals. Even in a time period much before the actual happening of this discovery, this field of scientific research has attracted the attention of many a scientists and thinkers. The reason behind this may be that mammals are an important family and any success in manipulating sex ratios in mammals has much to do with the changing of the very fabric of social and economical aspect and essence of life. Today's world has seen considerable changes in the life of man. These days sex fixing is a dream preference; not only in the case of *Homo sapiens*, but also in the various other mammalian species that are connected with this most dominating organism on earth. The animals such as cattle, horses, dogs, sheep, camels, cats, pigs and the list is endless; all have shared and amalgamated their destinies with the man in the long journey of evolution and development of species and societies, due to their use and application by man for his own benefits; and there have been attempts to go for sex fixing for all of these species. It is a different question that there has not been reported much success till date but even then it has remained one of the most fascinating subjects ever to be tackled by the exploring and pioneering human mentality. Many scientists and social thinkers have concentrated their minds and thoughts over this and we will like to discuss a few of the examples in this brief write-up.

To and Chan (1944) has described the importance of sex fixing as 'The desire to control the sex of one's offspring is as old as recorded history. Most doctors dealing with pregnant or would be pregnant women have been asked this question one or more times whether there is a method which could determine the sex desired for a future child. Such methods have existed in folklore tales in many ancient cultures, such as Greek mythology, the Bible, the *Talmud* and indeed abundantly in Chinese scripts as well'.

'Dairy farm operations require the female of the species, cows, for milk production. Males of the species, bulls, for example are only important for breeding purposes. Thus, it is clear that any method that safely and effectively increases the production of female off-springs versus male off-springs would greatly benefit dairy cattle breeders and farmers and, therefore be very valuable. Such a method would allow more effective and efficient farm operation, save money and increase farm milk production'. These are the words to describe the importance of sex fixing in dairy cattle, preferably on the female side in the official patent specification description of U.S. patent granted to Zavos and Dawson (1991).

Hafez (1982) has pointed at the practical application of sexed semen technologies for the following purposes:

1. More female progeny from superior females for herd and flock replacements and increased milk, meat and pelt production.
2. More males for meat production from culled females and crossbreeding schemes, *e.g.*, dairy × beef crosses.
3. Ensuring male progeny as herd sires from top dam × sire crosses.
4. Ensuring appropriate progeny when progeny testing young bulls.
5. Avoiding intersexes in multiple births.
6. Gender selection in horses will provide more progeny for sale and more progeny for brood mare replacements.

In a recent review, the distinguished Edinburgh reproductive physiologist, Prof. R.V. Short (1982) has opined that regulation of primary sex ratio has remained a mirage on the horizon. Prof. R.H.F. Hunter (1982) has expressed his opinion thus 'A longstanding objective of the reproductive physiologists has been to separate X-bearing and Y-bearing spermatozoa so that insemination of the fractions would lead to the offspring of the known sex. There have been many claims of success, and certainly the sex ratio has periodically been shifted from approximately 50:50 males to females. However no methodology has yet yielded a predictable, repeatable and significant shift in the sex ratio, nor at present there are rational grounds for supposing that this can be achieved'.

Charles F. Westoff and Ronald R. Rindfuss (1974) has concluded in an article that if efficient technologies are evolved for sex fixing in case of human species; they would not have much impact on the overall sex ratios in the developed economies like U.S.A. In their words 'temporary effect would be a surplus of male births in the first couple of years. This would be followed by a wave of female births to achieve balance, and the oscillations would eventually damp out. Ultimately under conditions of sex predetermination, the sex ratio would be similar to the existing natural sex ratio at birth of 105 males to 100 females.......... In other developed countries, sex-control technology may have more of an effect on the sex ratio because of the much greater frequency of one child families. Indeed if fertility continues to decline in the United States and one child family becomes more common, an increase in the proportion of male births would then also be expected here. In the developing countries, the greater cultural emphasis on having sons suggests the possibility of a more radical impact on the sex ratio. But it is clear from the data that only the major impact in the United States would be on the first two births, with little effect on the overall sex ratio at birth'.

Bromwich (1989) gives a description on the working of human mind as regards the selection of sexed off-springs in these words,' A boy is desired by many couples as their first born, often because of inheritance issues, but in agrarian societies male infants are priced for their usefulness on the farm. And in others, because they bring wealth to the family when they marry'.

Sex fixing if practiced by uneducated and ignorant masses, can cause havoc in the natural balances in case of human populations. As estimated by Kumar (1985) nearly 2, 50,000 baby girls were aborted due to infanticide in India between 1979 and 1984 with the end result that there remained only 100 girls for every 111 boys; meaning very well that many men will be left without wives. But if such things happen, the Mother Nature will surely find a way to balance it to equilibrium as told by Egozcue (1993).

References

Aulakh, B.S. 2008. Sex assignment and determination in dairy animals. *Pashudhan* **34(4)**: 7.

Bromwich, P. 1989. The sex ratio, and the ways of manipulating it. *In*: Studd, J. *ed.* Progress in obstetrics and gynecology. Vol. VII. London: Churchill Livingstone, pp. 217-31.

Charles C. Westoff and Ronald R. Rindfuss. 1974. Sex preselection in the United States: Some implications. *Science* **184**: 633-36.

Egozcue, J. 1993. Sex selection: why not? *Human Reprod* **8**: 1777.

Hafez, E.S.E. 1982. Reproduction in farm animals. Reprint 5th edn. Lea and Fabiger, p. 499.

Hunter, R.H.F. 1982. Reproduction of farm animals. Longman. London, pp. 138-9.
Kumar, D. 1985. Should one be free to choose the sex of one's child? *J. Applied Philosephy* **2**: 197-204.
Short, R.V. 1982. Quoted by Hunter RHF in Reproduction of farm animals. Longman. London, pp. 138-9.
To, W.W.K. and Chan, Y.F. 1994. Gender preselection. *J. Hong Kong Med. Assoc.* **46**: 4.
Zavos, M. and Dawson, K.A. 1991. Method for X and Y spermatozoa separation. U.S. patent no. 4, 999, 283.

A Massive Project

James Watt.........a big name in science.

But there was also a time when he was a small boy in his childhood years like all other men in the world when they are in their early age.

One day he noticed a small thing.

In fact, a petty phenomenon.

The steam in the vessels was blowing the lids off, to find an exit from the vessels which were cooking the community food on the occasion of a festival. James Watt (see Appendix I) just happened to be there and it was no big news. Rather it was not a big happening either.

James Watt

(Glimpses of greatness from annals of history)

That is the reason that nobody noticed his presence there.

Who notices the small things in life?

But a big idea came and crept into the small mind of the young fellow. 'If the steam can blow the lids off, it can certainly push similar things with a force and this property can be put to some good use by devising a certain application of this pushing forceful virtue of the steam. At least some pushing and forcing instrument or machine can be developed by the use of this characteristic'.

This small idea did wonders in the later years and the world history is a witness that the steam engines developed on the principle of mechanical pushing thrust of steam, changed the very face of the civilized world. A very big industrial and automobile revolution was materialized with the application of this very small principle.

It is a fact that small things sometimes do miracles.

A small incident of discoloring of photographic films in the laboratory of Henry Becquerel (see Appendix I) led to the discovery of radio-activity and an insignificant looking development and observation in experimenting involving the cathode rays led Roentgen (see Appendix I) to recognize the outcome of a very high piercing radiation which later came to be known as X-rays.

Alexander Fleming
(Glimpses of greatness from annals of history)

The idea of the antibiotic property of the rotten moulds can form the basis of developing a whole new science of antibiotics and change the entire outlook and approach in the medicinal and therapeutic world; has been proved by the pioneering fruitful research in this field and as recorded in the annals of history and the name of Alexander Fleming (see Appendix I) has been immortalized in the process.

There are other examples as well.

The petty rain showers before the Waterloo changed the entire course of European and world history and the glorious career of one of the greatest professional military generals of all times, was ruined in defeat.

A small legislation imposing a small tax on the colonies in the British parliament set the cause for the American Revolution.

A little crafted letter by the chancellor Bismarck (see Appendix I), caused the fall of France.

A tiny idea in the brain of Louis Pasteur (see Appendix I), laid the new foundation for discoveries of many vaccine treatments against viral and bacterial diseases.

The history of smallness creating big results is very long.

Let us come out of the proven and golden pages of history into the present world of today.

A small thing has happened again.

We have a new and humble guest on earth in the form of the discovery of Aulprofem; the world's first mammalian sex fixer drug. It is also a small happening but with much bigger and fruitful developments and ramifications to occur in the immediate and distant future of human race.

Louis Pasteur
(Glimpses of greatness from annals of history)

It is a gigantic project by any standards!

Let's go a little deeper.

Aulprofem is a mammoth project, no doubt. It is a project that can cause a direct annual increase in the GDP of India to the tune of plus three lakh crores in rupees (one lakh is one tenth of a million and hundred lakhs form one crore) and a further indirect saving of materials and eatables worth nearly two lakh crores in rupees at the prices and worth value of rupee around 2007 AD. The total benefit and increase in annual GDP of India being nearly five lakh crores in rupees (nearly one hundred billion U.S. Dollars at an exchange rate of rupees fifty for one

U.S. dollar) or a plus sixteen percent in annual GDP increase can be a big revolution in the national economic development and betterment. Similar exploits can be expected in other economies of the world as well. This being the first mammalian sex fixer drug of the world is all set to open a new era of biological research and exploration. The importance of this valuable development in medical research is tremendous and it has given an altogether different view and look to the concept of sex assignment and determination. The concept of the designer baby will be the greatest benefit drawer from this. The heavyweight scientists like Ericsson, Meistrich, Corson, Gordon, Lindahl, Beernik, Barrat, Leger etc were all trying to give a new meaning to the concept of alteration of sex ratios in mammals but all failed miserably as the repeatability, verifiability and consistency criteria of the results of their scientific achievements concerned. This discovery based on its near 100% success rate to produce sexed offsprings in mammals, will definitely give a new meaning to the most basic and important biological phenomenon of sex determination which is virtually the foundation of all living creations in the universe. The great chromosomal theory of determination of sex will have to be changed even accordingly in the light of the successful derivations drawn from this magnificent scientific breakthrough.

There is another important angle. Better, we have a look.

The world is just on the verge of a food crisis of colossal dimensions. The hunger, starvation and famines are going to be the order of the day if the current trends of falling food grain productions and the failure of green revolutions due to decrease in hybrid vigor, are any indication of. The much touted second green revolution is nowhere in sight and considering its total dependence on genetic modifications, a concept that is never approved of by nature; there is certainly an inherent failure tag fixed to it even before its becoming a working reality. There is a sound reason for this. The nature has developed a genetic balance of its own choice in the course of long evolutionary period spanning hundreds of millions of years and it has always been a perfect one for any specific period of biological existence, depending upon the ways, nature has selected and evolved.

The world populations are growing. There is an ever increasing demand for food. Aulprofem can increase milk productions across the globe many folds to the present levels by increasing the numbers of high yielding milch varieties of dairy animals. Similarly meat and wool etc productions can also be increased by increasing the numbers of animals like goats, pigs, horses and sheep etc with the application of Aulprofem on these species. The situation at present is not good. We are faltering on other fronts as well. The pollution levels have risen to alarming heights. The green revolutions have already gifted the soils of progressive states like Punjab, Haryana, Andhra, Maharashtra, Western U.P. (see Appendix II states of India) etc with heavy doses of insecticide and pesticide poisoning. The foods, underground drinking water and even the atmospheric air has become loaded with toxicities of innumerable types. This has led to disease and suffering in animals and humans, of proportions and types unknown in history.

The animal wealth of India is 300 million dairy bovines. If only a 10% of them are taken for the application of Aulprofem for a short period of 10 years; then the annual increase in milk production will be nearly 160 million tonnes per annum based on a daily output per milch animal of just eight liters per day, giving about 400 grams of milk for every Indian per day in addition to saving nearly 70-80 million tonnes of food grains, 5 million tonnes of edible oils, 10 million tonnes of pulses and nearly 100 million tonnes of vegetables per annum. The total output of this milk if calculated in rupees @ rupees 20 per liter amounts to rupees three lakh and twenty thousand crores. The amount of money saved in the form of saving of food grains, pulses, edible oils and vegetables etc is rupees 80,000 crores, 20,000 crores, 25,000 crores and 100,000 crores when calculated @ rupees 10 per kilogram of grains, rupees 40 per kilogram of pulses, rupees 50 per liter of edible oil and rupees 10 per kilogram of the vegetables. The cumulative increase by the direct enhancement in milk production plus the saving in the form of food grains, pulses, edible oils and vegetables; comes out to be more than five lakh crores in rupees. The additional saving of the cooking gas, coal, kerosene, electricity etc used to cook such massive amounts of food is the added bonus on national economy, not calculated in this equation. Furthermore, it is a fact that the actual prices of these commodities of food grains, pulses, edible oils and vegetables etc are well above the prices put to calculating as given above. So, at actual prices the net saving and resultant enhancement in the annual national GDP both; are well understood.

This explains the mammothness of a gigantic project like Aulprofem!

The beneficing, the reverence, the acknowledging and the holiness are inter-related. The Sun is accorded a holy and pious status in the religious and mythological scriptures and books because it is benefiting the man in various ways and such is the treatment given to natural gifts to mankind like the ever benefiting Ganges and other countless rivers across the globe, the Himalayas and other many mountains and geographical boons like jungles, fertile valleys and so on. Anything that is benefiting man in any way is held in high esteem and respect by the mankind. The

Aulprofem is another gift by nature to mankind and it also deserve such a respectful, reverend and pious status to be accorded by man on it and it rightfully is the perfect candidate for the utmost care and attention to be showered upon it by the entire human race.

The discussion seems to prolong further.

The discussion of above colors and traits always tends to stretch with great elasticity and it is also not easy to sum it up so soon but we will prefer to pull a full stop here and proceed to halt and conclude. This may look hasty but we have a full course of study and description just opening up before us in the form of this research and the book and rightfully this prompts us to be concise here and close the discussion.

So to sum up; the Aulprofem is just going to be a milestone in biological research as well as proving and playing a savior force for mankind in feeding the ever increasing human populations, the world over and saving the world from toxicities of poisons of development like insecticides, pesticides etc and creating a clean, pollution free, hygienic and livable world.

References

Barrat, J. and Leger, D. 1979. Aveir des grossesses obstenues après stimulation de l' ovulation. A propos de 519 grossesses. *J. Gynecol. Obstet. Biol. Reprod. (Paris)* **8**: 332.

Corson, S.L. *et al.*, 1984. Sex selection by sperm separation and insemination. *Fertil. Steril.* **42**: 756.

Ericsson, R.J. *et al.*, 1973. Isolation of fractions rich in human Y sperm. *Nature* **246**: 421.

Gordon, M.J. 1958. The control of sex. *Scientific Am.* **199**: 87.

Lindahl, P.E. 1956. Counterstreaming centrifufation of bull spermatozoa. *Nature* **178**: 491.

Meistrich, M.L. 1982. Prspects of sexing mammalian sperm *in* R.P. Amann and G.E. Seidel, Jr. (*eds*.) Colorado University Associated Press.

A Treatise on Sex Fixing

The world is a bowl of woes.

This is a universal truth discovered by most of the wise men of all times. This is a different truth that most of such wise men are from religious background. Lord Buddha was one among them. He put forward his famous doctrine for acquiring eternal happiness. The underlining principle of his philosophy is that one should master the art to tame his desires as these are the desires of a man that make him unhappy when they get out of rightful proportions. So the key to happiness is that a man should learn, how to lead a contented life.

Thomas Alva Edison

(Glimpses of greatness from annals of history)

What should a man desire and what should he not?

There should be a limit on desires. But...........

Desires are unlimited, this is a universal truth.

Had man tamed his desire to fly and we would never have seen the flight of planes. And what about the eureka .... eureka ...... and ....

Had there been a satisfied Copernicus and we would never have known that a gospel in Bible can not order the Sun to stop midway in its journey around the earth.

And a Pasteur would have never invented the vaccine to fight the silent death snaring through the nostrils of the dogs, gone mad...............

The list is long and it is better a full stop be pulled.

There are some aberrations in the history of contentment.

One was named Thomas Alva Edison. He became restless and he made all the telephones to ring whenever a call came on them and in the process relieved all the telephone boys from the most boring duty of putting the telephone receivers to their ears and sitting in still mode of an astute pray person, engaged in deep worship. Another was George Stevenson who made the steam engines to move with kinetic motion and started the great revolution in the field of land transport. Charles Darwin, Isaac Newton, Albert Einstein, Marie Curie, Alexander Fleming, Ronald Ross (see Appendix I) and many others, also belong to the same breed.

Marie Curie
(Glimpses of greatness from annals of history)

Taming of desires and contentment seem to be good things and they have satisfied whole lots of generations after generations. They still continue to do so in countries like India. In India and other such places where people are sermonized to live lives full of big ethics and high morals by the self acclaimed men of God who themselves are epitomes of lust, luxury and greed and never step out of the cozy environs of their centrally air-conditioned *ashrams* and luxury cars and are held in the highest of the esteem by the lungi clad, semi naked and half fed frowning jokes in the name of humans. Such dirty and ugly jokes are not only the perfect pictures of self-contentment but also they are the fullest stories in themselves, molded in clay of complete miseries and laden with destitution and penury.

Such is the magic of contentment. It can keep the fabric of society intact and binding even in such drastic circumstances when it could have fallen prey to fear of being torn apart in pieces leading to chaos and resentment. So, contentment is a peace maker also.

But life is not a one sided story. It has the other side also. In fact it has many sides. Let us look at one of them.

The desires can be controlled. It is a beautiful derivation but sometimes it doesn't hold true. The desires can really fly and fly they do sometimes real high.

Isaac Newton
(Glimpses of greatness from annals of history)

Some men desire to be kings and history is a witness that they have become so. King Philippe of Macedonia had a son as he had other sons also and this son was called Alexander. The king had a very beautiful wife and so he had many other wives also and this wife happened to be the mother of this son Alexander. She had copulated with the king and incubated a son in her womb and then delivered it as a very cute and intelligent boy but physically he was not as strong as other boys of his age but very soon he got over his handicap and learnt the arts of war and statesmanship and started to dream and desire big. He got a teacher named Plato who taught him funny things like the ones of how to develop a great character, boldness and courage and above all the yearning to excel in life. His father also thought that his son was made for some bigger exploits but soon turned himself an obstacle in his way. The wife's desires grew and she reared a variety of snakes as pets in her private possession. One graceful day she conceived a beautiful conspiracy and stung with hatred driven strategy her own husband to death. Alexander became a king; which he never wanted to be by killing his father as he loved his father very much in spite of father's disliking for the son. Then afterwards in turn the son's desires grew and he led his big dreams and vast armies for the first time ever in history, to conquer the entire world.

Such is the power of dreams. The only condition being that a man should dare to dream and desire.

But every man is not Alexander. He is not Columbus either. He cannot dream so big that he leads his armies like Attila or Genghis Khan (see Appendix I) to most parts across the Asia and Europe and go exploring mega victories, one after the other.

Julius Caesar had a dream and so did Pompey had and it was the same dream that Caesar wanted his nephew Octavius to share with him and the courageous boy chose and opted to lead a life, full of challenges that ultimately made him the equivalent to the authority of God on earth.

So it is a proven fact that dreams can be real big sometimes. Only they should select the right candidates to grow their nurseries in the brains of such persons.

There is a story in America about a poor boy who had a strange dream that one day he would be the president of United States of America and the full world still remembers him by the name, Abraham Lincoln.

And so, there are some people who really dream and desire big.

This is a proven fact.

But most of the time it does not hold well.

There is a big dearth of people like Alexander and Pasteur and Darwin.

Those types of people are considered to be unpractical these days,

Unpractical and unwise.

Today's world prefers other type of people. The people who know the art of making wealth and fortunes. It does not matter what ways and means are acquired to achieve such ends. Such people are considered able and wise. Others, who are not so successful, also believe in this ideology. The biggest chunk of human population is composed of people with this mentality and mindset.

People have become very intelligent these days.

They just do not go doing things that have no worldly gains like adding materials and comforts to their life; at least of immediate nature.

Every man is busy. He has lot of work to do for his family and himself. He never gets time to do unnecessary things like the things aimed at the betterment of society and the making of great societies and nations. The things based upon bigger principles of humanity and high thought and cherished goals, aimed at making this world a better and livable one.

Everybody has grown wise.

Wise and busy as well.

So, nobody dreams now.

And if he dreams, he dreams about more practical things of life.

Like earning a job, making a house, rearing kids, pampering a girl friend and many more important things, falling into the same and related categories.

These are ordinary dreams. All are aimed to make life happy or at least they are supposed to do so.

But do they really achieve the goals to which they are intended and directed? Do they really make life happy? Happy and livable?

There is no satisfactory answer. It is better a full stop be pulled here.

How does a human mind work?

An endless voyage into the infinite world of fantasies and thinking and wishing.

There is a strange connection between thinking and wishes. All thoughts tend to grow wings of wishes.

If all wishes are granted!

Just a few examples.

Can we change the color of skies?

If only we could tame the tides of the oceans!

If only we could change the course of storms!!

If we could catch the stars on our palms or bring the cozy warm winter Sun into our bedrooms and feel its life-giving shine and glory in our veins!

These are the Sheikh Chilly wishes and can never be granted. But there is another variety. Just give a look.

If we could live forever; young and strong!

If we were ever healthy!

Again, these wishes are not in the jurisdiction of executive functioning of the forces of nature. So, She seems to be in no mood of granting them.

So, a man is destined to live a mortal life and that also without eternal health.

And properly so, every man is programmed to live and reproduce with a limited time frame. Reproduce and propagate his breed. This he does by producing the progenies. Progenies of either sex; male or female.

Another wish that he has reared from the times immemorial is as follows.

If one could have desired off springs! Desired off springs of the wanted sex. Wish a boy and beget it. Similarly wish a girl and have the wish granted.

This is also a big wish, So the Mother Nature has not satisfied it even after ten thousand years of human history of civilizations and countless efforts on research and development in science and technology.

But time has changed. There has been showering of measureless blessing and mercy.

And.......

This is the wish that the most kind Mother Nature has finally consented to grant. In fact she has already granted it.

The discovery of Aulprofem has been made.

Aulprofem is a drug that fixes the sex of the offsprings in the mammals. This is the first mammalian sex fixer ever reported in the entire history of humankind. This will have tremendous implications and benefits for the society and it will open before mankind the start of a new era in human development and betterment called the 'Era of Sex Fixing' in the entire world.

This was a brief account on sex fixing. And this seems to be sufficient also.

The story can end here but let us venture a little deeper.

What are the benefits of sex fixing to man, mankind and this universe? What was the need for it or more appropriately why the most kind and benevolent Mother Nature agreed and allowed it to happen. After all she never does anything unwanted and wasteful. So, the million dollar question is that why the Mother Nature gifted the mankind with this magnificent discovery of Aulprofem.

In fact the benefits are many. Many and countless.

It is truth that man is a selfish creature. Selfish and thankless.

And on the contrary the Mother Nature is so kind and merciful.

Here is a story.

Story of a worthy king, Ranjit Singh.

It was during the first years of nineteenth century that there ruled a kind and worthy King Ranjit Singh in north western parts of India. He was a just king and very popular with the people living in his kingdom. The people loved as well as respected him a lot. Once, he was having a stroll with his ministers and generals and a stone came from somewhere and hit him on his head. The blow was a hard hitting and it started to bleed. The king was hurt.

The soldiers ran from many directions and found the stone to be thrown by a teen aged boy who was busy picking mangoes from a mango tree and had not noticed the king's convoy coming that way. He was arrested and brought the next day before the king. The king recognized him and asked.

Why did you hit me?

I hit the tree and not you; was the answer.

Why did you hit the tree?

Because I wanted to have some mangoes.

This had an immediate effect.

The king was in thinking. If a tree when hit gives delicious and sweat fruit, then what should a just and kind hearted king do when he is hit. He thought and ordered that the boy be suitably rewarded. Everybody was surprised. But the king explained.

'If a tree which has nothing to do with this material world; can be so kind that it gifts the person hitting it with the most delicious and magnificent mangoes, then a king who is the savior and custodian of all people living in his kingdom, should be more kind and reward the person who has hit him unknowingly, with some sort of a reward. After all a king has to be better than a tree'.

Everybody listened and consented and praised the king.

The Mother Nature is just like Ranjit Singh, the worthy king.

The man is the senseless, ugliest creature that Mother Nature ever produced on earth. It has been continuously hitting the Mother Nature with whatever stone or weapon that it could lay its hands upon, right from the day it was born on this most beautiful planet of the entire universe. The Mother Nature produced near about ten million plant and equivalent number of animal species on this planet and all lived very obediently and amicably in cohesion with the earth, environment and other living organisms but this unworthy and thankless creation; the man, never lived in peace and harmony with himself and never let the other creatures of the world to peace. On top of that it has served a destructive act for hundreds and thousands of the plant and animal species to be extinct and disappear altogether from this beautiful world. It has been responsible for disturbing and destroying the total balance of *fauna* and *flora* in the world. It has created and spread pollution of countless types, all across the globe. Even the climates have changed. There is global warming. The snow on the poles is melting at an alarming rate. The ozone layer has been depleted. The experts are saying that the increased frequency with which natural calamities like earthquakes, tsunamis and tornadoes etc are happening; is also the result of the ill-doings of this selfish and evil creature.

But the Mother Nature is not like man. She is so kind and magnanimous that she still wants the goodness of this unworthy creature. She is still blessing this ungrateful entity. She has gifted him with the discovery of Aulprofem. Like a true mother, she always wants the welfare of his son, just like the mother of that Greek story.

The story has been sung in the folklores across the world since times immemorial.

There lived a young man and his mother in some Greek town near Capua. They lived happily until the young man met a wanton woman. She was very beautiful, so he fell in immediate love with her, at the first sight. The boy had some fortune; so the clever lady immediately offered him the comforts of body and beauty. The young man was infatuated with her and always spent maximum time in her company. The mother sensed something ominous and cautioned the boy but he never listened. One day, the lady told the young man to bring whatever valuables were in the house to her and he obediently did that. The other day she told him to throw his mother out of the house and the boy did that also. He loved his mother but more than that or even more than anything in the world; he had started loving that woman, worthy of the brothel. The mother had some precious ornaments on her body and the lady wanted to have that also. So, she ordered him to kill his mother and bring that jewellery along with the heart of his mother to her and the unknowing son killed his mother also. When he was on his way back to that woman with the ornaments and the heart of his mother; his foot slipped and he fell. Along with that the ornaments and the heart of the mother fell. When he was about to pick that from the earth; he felt it asking him, 'aren't you hurt my boy? You should know that your mother still loves you the best'.

The boy realized what he had done and repented and wept. Nothing else could be done.

The man is the true heir to that boy in the real sense. It has already readied the Mother Nature for her last ride on the death bed but the most kind and caring Mother Nature still wants the welfare of man. The man has done so harm to her but she still has decided to gift him with the power of Aulprofem, to benefit him in many ways than desired.

The truth.

The Aulprofem is a sex fixer drug. In fact it is the first sex fixer drug of the world and sex fixing is just going to be a great revolution in the history of mankind.

Another truth.

The world populations have grown tremendously in the last century. Almost all across the centuries; the world populations have remained fixed on a number of little less than one billion humans on earth in all the continents. But in the last one century the world populations have grown to more than six billion people across the globe. This has put tremendous pressure on the natural resources of the world. The jungles have got depleted, the mineral wealth has been exploited and the lands under cultivation have also shrunk.

This is a very bad scenario. The increasing populations need more and more food but the arable land has shrunk and is still in the process of shrinking. In fact, it has started to shrink faster. So, this is a big challenge for the rulers across the globe as well as for the policy planners and thinkers to think and chalk out programs and policies aimed to feed the ever bourgeoning world populations.

The sex fixing in dairy animals can increase the efficiencies of the cross breeding programs to almost double folds in the first step and then to many times multiplying in geometric progressions when the simple counting and calculating rules are applied. The production of two more exotic bred cows in the first step becomes the cause of production of four in the second place and with the same speed and success it becomes eight, sixteen, thirty two and similarly increasing with the more and more applications of such technology and milk production is increased accordingly with every such step. Similarly meat and wool etc productions can also be increased with the application of such a technology on meat producing animals like goats, pigs, horses and the wool producing animals like sheep, reindeers, yak and so on.

There is a world famous sermon in *Bhagavad-Gita*. The lord Krishna explains that whatever has happened in the past was good enough and it should be accepted with good soul and mind. Whatever is happening now, is also fair enough and we all should try our very best to make it the best performance of today with the best utilizing of our capabilities and capacities. An effort should always be made on our part to always stand by the most high and austere principles of humanitarian thought and behavior. If we learn from the past well and do our duties in the right and just manner in the todays of ours; then there is no reason to believe that the good things will not happen in the future. A man should always accept and understand his rightful duties and deliver them with the best of the belief that they will be aimed only for the betterment and welfare of the world and mankind. There should not be envy, jealousy, treachery, deception or any other foul play in our thoughts and actions and we should go on doing and delivering the good and right deeds and actions endlessly without wishing even for any recognition and reward; *in lieu* of the humanitarian contribution that we make towards making this society and world a better and livable place.

His sermon can be put in simple words like this.

'Accept whatever has happened in the past as good only and mindful of your rightful and just humanitarian duties and responsibilities, go on doing only the good in present and without expecting any benefit of conferring of honor and recognition on you for this act of good doing; leave everything else to God and rest assured that all that happens in future, will be good only'.

This is a very good advice that the entire human race should learn and accept by heart and brain otherwise from few hundred or thousand years from now the history of our magnificently inhabited planet will definitely record that there once lived an unruly and savagely creature called man who did a great wreckage and harm on this planet and created all types of nuisance like pollution, sabotage of environments and disturbed the natural balances of existence in this biological eco system and ultimately won the wrath of nature and was wiped like anything from existence by the ultimate survival forces of nature and done to extinction from this most conforming, adjusting and livable niche called earth and in this way the worthy and competent Mother Nature was able to preserve and secure the existence of life, in this beautiful world.

Man has done badly in the past. Man continues doing so in the present. Then how can it be digested that the deliverance in the future will be good as regards this claimingly perfect but most non adjusting creation of nature, called man.

Nature created life on this planet and She has preserved it by endless endeavors and means. In doing so in the entire history of existence and evolution; nature has evolved certain mechanisms with utmost calibration and perfect craftsmanship in such a way that every species of plant, animal or prokaryotic nature is an excellent example of the best skill in the art of the nature. The most striking is the rules and conditions for the natural selection of the survival capacities of the fittest of the species and the breeds of organisms. A single species takes about millions of years to evolve. The phyllogenetic histories of many animals like dogs, horses, giraffes and man are well established and known. Then how come that a single species called man has already become the destruction force to make thousands and hundreds of thousands of such painstakingly evolved species by nature, extinct.

Will the great forces of nature accept this indignation and humiliation of the mightiest power in the universe called Mother Nature?

This is a question that should attract the attention of one and all otherwise the doom is the writing on the wall and everybody should read it otherwise it will be too late very soon.

The nature has started displaying signs of its anguish and displeasure. There have been instances of the fury of the nature like tsunami of recent years, the earthquakes that have started to happen with increased frequencies and vigor. The examples of them were shown in Latur, Bhuj and Islamabad. The cyclones and hurricanes in and around the Atlantic are some more examples of such an outrage.

There are people who are called astrologers. A man by name Nostradamus has lived in this world or is supposed to have lived. There are many stories about him. Many of them seem to resemble propaganda and appear to be concocted myths and lies. There are well known arts like palmistry, *reiki*, prophesying etc. It is a different fact that no right thinking person should believe in them. There is another thing called intuitions and many literary exploits and silver screen stories have been based on them. Even they make movies on the subject revolving around tarot cards reading and it has become one of the flourishing professions in the world these days like the damned *vaastu-shastra* that only is aimed at pushing superstitions and hollow thinking in the minds of the gullible neo educated.

We do not believe in either of the above mentioned arts nor do we endorse upon the people to believe in such nonsense things. But there is certainly one thing in which we are very much sure about that future of a man or a material can be known and if available data and facts are critically collected and evaluated; then it is not difficult to foretell the future as regards a substance or even a man.

The future of the entire human race seems to be extinction from this world of life.

The available data and material evidence if piled and understood in the right way and manner, point only in direction and that direction is not of the progress or flourishing of man. But on the contrary that direction seems to be heading towards the utter devastation and disappearance of this most highly evolved breed of the organic live matter on earth.

The *Bhagavad-Gita* has a lesson for the entire human inhabitants of the world.

'Either live the life like human beings and care and love your fellow beings, the nature and its wonderful different creations. Develop humanity and compassion in your hearts and enlighten your inner thought and feeling to be free of savagely greed and lust for power and supremacy over your fellow beings and the forces of nature or be prepared to perish. Even if you make no preparedness for this, you will not be spared the event and before you know anything, the entire human race on earth will be an organic species gone on the notorious band wagon of extinction'.

Dinosaurs lived on this planet for dozens of millions of years. Insects were in existence even before the dinosaurs and they still are some of the most successful animal species. All other animals are equally successful and existing in harmony with themselves, with other animals, plants and with the great forces and rules of nature. They are equally important and dear to the Mother Nature. All plants are equally compatible and obedient to the forces that created them and that nourish and survive them. Even then when the circumstances demanded and there was time for it, the great forces of nature just annihilated the formidable dinosaurs, the mammoths and the dagger toothed tigers. Even the big families of ferns and gymnosperms went along with them. The forces of nature do the annihilation job with the extreme degree of skill and deftness just like the 'Legion' crack commandoes (see Appendix III) of General Franco's division, in the war of Spain. No mercy or attachment is shown in any of such exercises.

Those were the loyal and faithful sons of nature and they perished. Natural selection always works with strict maintaining of the professional standards and degrees of skill. No concessions are granted to anybody concerned. Only the 'survival of the fittest' is the running and the success criteria. All other candidates and species just are thrown in the historical paleontological dust bin of organic evolution. Some are even not so lucky enough to adorn this disposal destination and are wiped out of the memory lane of nature's permanently lost hard discs of the crushed computers of natures' office of biological records.

There is a *sloka* of great *Sufi* saint, Baba Farid (see Appendix I) which runs like as under:

O Farid, when you go and see the bell hanging out of the gates. It is beaten regularly without any fault of it. Then what will become of you, you being a sinner.

So, those were the loyal and obedient sons of the Mother Nature and they perished as the great surviving rules of the nature found them just not fitting and appropriate. They were selected in one part of time calendar of existence and when the time passed and circumstances changed; they were mercilessly shown the way off the living pastures.

But man is different. He is a proven sinner. He has spread dirt of carbon emissions. He has littered the bigger tracts of natural habitats with the concrete debris lading of the pure and naïve earth as pointed by the great Lev Tolstoy (see Appendix I). But the voice of wise men is never heard by this hopeless creature. Today the very survival of this planet is in danger due to this creature only. Many species of animals and plants have perished due to the misdeeds of man and it is also true that many have developed unwanted and deadly mutations, cancers and are on the track to extinction.

How can the most kind and merciful Mother Nature tolerate this slaughter of species and how long?

The scientists have declared that the world is on a fast track to destruction and that the dooms' day is not far off. But the most important thing the scientists and theorists forget to note is that this wonderful and magnificent world will never end. It will go on and on forever. Even the species that the Mother Nature has produced with great toil and care will continue to exist. There may be random annihilations but the greatest deletion from the arena of life will be the species of *Homo sapiens septens* and nobody will ever mourn this subtraction from the sum total of life forms in this world. Actually nobody will have time to do so. The struggle for life and existence is always so intense and consuming and the bowl of life is so full and mesmerizing that it appeals and attracts and before a creature comes to know of anything, the act of play is over and the next episode of life unfolds an entirely different scene.

This is life and this is all about life.

The truth may appear bitter and crude but this is the truth after all.

The earth will live on and so will the world and the universe and great forces of nature will continue to create, control and guide the various fabrics of life and life forms and this story is never ending just like the spirit of the great sermon in *Bhagavad-Gita*.

'The change is the rule of the nature'.

This is the essence of existence and the summary of all wisdom and intellect in the world.

In fact nobody is more intelligent than the Mother Nature.

And nobody is more competent also.

We may develop illusions but the truth cannot be hidden for long and this unfortunately is the ultimate truth of life.

The sex fixing can control pollution and help improve the environments.

This is clear like daylight and one does not need to elaborate much on the subject. Here is a little explaining.

The biggest source of the pollution is from the three sectors; industrial, automobile and agricultural poison chemicals. Industry and automobile emissions are a matter of concern from a long time but recently agricultural poison trade transacting has also taken a front seat in pollution circles. This becomes more important in the sense that major agricultural produce are directly consumed by people and even the agricultural waste material like hay and straw etc are eaten by the dairy and meat animals and the residual matter if any of these chemicals and poisons can easily find its way into the biological systems of these animals and this way finds the entry into the bodies of the humans feeding on the milk or meat of such animals. The people have resorted to the use of insecticides, pesticides, and fertilizers to such alarming levels that these chemicals' presence and accumulation in the grains, fodders, feeds etc have gone much above the danger levels. This has been a major causative factor for many diseases and ailments in the humans feeding on such foods and animal products from animals which feed on such agricultural products. The diseases such as cancer, asthma, skin diseases, allergies, liver and renal complications, optic and neurological disorders like LGBS and many more have come to light in recent years and their occurrence and spread to such alarming levels is the matter for grave concern.

The Aulprofem can be a powerful tool in fighting such cancerous poisons. It can increase milk productions to great levels and this is such a food that is by and large free of such poisoning if the milch animals are not fed upon harmful chemicals and poisons. By doing so we can decrease our dependence on other foods produced by agricultural outputs and hence the poisoning levels and other precedence of occurrence in the foods can be minimized. The fodder crops usually are free of regular fertilizer and pesticidal etc usage. Hence the use of such posisoning techniques is not resorted to in case of cultivation of fodder crops that are eaten by the dairy and meat animals. Even if such fertilizers and herbicides are used, they are done so in minimum quantities as the use of herbicide or weedicide is useless in a fodder crop and even the urea and phosphate or sulphate etc fertilizers are not needed to be applied in a usual fodder cultivation.

So, the sex fixing offers a better option to human race to get rid of the poisoning in food materials and the animal products meant for cultivation and usage by humans.

Gurbat is the Urdu word for poverty.

Gurbat is the nursery of terrorism. These are the words of the Pakistan's Prime Minister, Mr. Gilani. We may not agree with him but the people who have seen the nurseries of terrorism and witnessed the procedures and modes which breed and cultivate terrorism, will definitely tend to agree with Mr. Gilani; if not fully well but then also to a greater extent.

Zakaat is the money donated by the Muslims in the name of God and to be used for religious purposes and for the welfare of Muslim religion and people, especially the poor. This money donated in big amounts in rich Muslim countries of western Asia, finds its way to the poor countries of India, Pakistan, Bangladesh, African countries and so on. Lots of religious and charitable bodies in these countries thrive on this money. They run various organizations and activities like *madrasas* and religious awareness and propagation programs etc. These include activities and teachings that tell and elaborate the points of Islam's superiority over the other religions of the world and in this way urge these people to bring maximum other people under the shelter of Islam and the prophet.

This is not bad in any sense when viewed from a religious angle. Everybody in the world has a right to spread and propagate his religion.

But why?

Why a person wants to spread his religion when there are other religions in the world. Are they inferior to a particular religion and because somebody is having faith in certain religion and he wants to bring as many people from other religions under the fold of his religion and in the ranks of his religion mates; as he can. The motive seems to be a better one only. He wants the welfare of those men by bringing them on the way of the God and his prophet only with the lofty and pure ideas in his mind.

But are all other religions truly bad?

A person becomes wiser by listening to such a philosophy or at least he thinks that he has become so. He starts to love all those people who believe in his religion and with a strange feeling, starts to distinguish himself from the people believing in other religions. You talk to such a person and he will try to go to lengthy details that his religion is superior to or is at least better than the other religions. He in a strange way starts looking down upon the persons who do not believe in his religion and are gone astray from the path of God in his views. He thinks that those people have gone even away from the path of righteousness and humanitarian thought. He either learns to hate or feel mercy on such people who do not belong to the category of him or his own people. By his own people, he means the people believing in his own religion.

Even if he feels mercy on such people; this mercy is very soon converted into a feeling of helplessness or the deep urge to bring those people of other religions under the fold of his own religion and he longs for the immediate betterment of such people by making them believe in his own religion. This eagerness soon gets converted into a sort of desperation when he finds that those people are not just yielding and they prefer to stick to their old religious beliefs. He falls into such a dilemma where he wants the welfare and betterment of others and those unruly and foolish people just do not try to understand the lofty thinking and high ideology on the part of him to save those people from the curse and blasphemy of God by remaining in their old religions and beliefs. And here comes the point when the person concerned starts becoming desperate and he starts developing a strong eagerness to bring those people on the path of God and righteousness. He does not digest the inability or the folly of such people of not conceding to come to the path of their own betterment and salvation. He thinks that those people are doing the severe mistake of sticking to the path of sin that will definitely lead them to the gates of hell and torment.

He wants to save them.

The non listening fools! He thinks.

He even feels pity for them. His inner soul compels him to go and save them. But they are non understanding fools and just not paying any heed or consideration.

Such unruly fools! He thinks of them. They are really disgusting. Disgusting and hateful. He thinks.

This is the time for the feeling of mercy and the urge in him for the betterment of those people; to get converted into a feeling of hatred for those people in the ranks of the other religions.

He no more feels mercy on them. He no more rears soft and good feelings for them. On the contrary he starts to hate them and sometimes even starts to hate them strongly.

This is the underlining principle of the *krieg* of one religion with the others.

And all religions are same in this basic concept; so interestingly common in all of them.

In fact so, this is the basis for the *krieg* of all religions with one another.

Not surprisingly, this is the reason that all religions teach and propagate hate these days.

This hate is the mother of many problems.

This makes a person a zealous propagator of his own thoughts and exhorts upon him to force his ideology and religion upon others and not accept anything belonging to the belief and understanding of the people from other religions.

This is the stage when he starts believing in his religion to the greatest levels and thinks that it is only his religion which is the ultimate salvager of mankind.

He becomes a fanatic.

Such people always want one religion in the world and that religion is the religion of their own. They can go to any extent to bring other people to the fold of their own religion. Even they can be violent also. Such violent people may even take to the path of forceful extremism to bring other people to the fold of their own religion or to prove the supremacy of their religion over the other people and religions. This is the stage when they turn to the ideology and actions of terrorism.

Such people have become full fledged terrorists by the time they reach this stage of mental preparedness and conditioning.

This is a brief dissection of the entire anatomy of religious intolerance and terrorism based on religion.

And this is a basic feature with all the religions of the world and not a feature restricted only to Islam.

All religious terrorists compulsorily undertake this basic course of study before they become full-grown terrorists.

This was the philosophy that prompted the Christian invaders and missionaries in the new world of America and the blind continent of Africa to propagate such preaching and bathing the entire continents in the blood of the innocent *Inca, Maya, Zulu* and countless other such people, annihilating the entire civilizations in the name of God and His son at one time and same were the lofty and holy sentiments of the Nazis that were conceived and applied with all the wish and pureness of religious betterment and upliftment of the people who were *Zionists* and were thus away from the love and affection of the son of the God and so they were rightfully done to deserving death in the concentration camps and boiling gas chambers in Germany and the Europe.

This is the essence of the policy of the religious intolerance and its practical execution leads it to a situation called Nazism.

Such is the philosophy of Taliban and Mujahedeen these days.

But here is a relief.

A negative philosophy attracts the negative people the most.

There are people who are big failures in life and they have nothing particular to achieve or prove in life. Once, they are given an opportunity; they are drawn head down into this. The well placed and successful people are often not keen to join such movements. Usually there are frustrated, unemployed and idle fellows in the society who look at such an opportunity as the God sent and they just go and grab it. Otherwise also, they have nothing particular to perform. These people are not ordinarily well to do and they belong to the lower strata of life.

It is basically the poor and lower class people who are more attracted to such philosophy and thought. It is a well obvious truth that what else they stand to perform or gain in life? Very few people in upper and middle classes rush to such destructive destinations.

Even in lower and upper middle classes, the young men and other people who are attracted to such ideologies are a minority in these classes and also they are people who are out of the usual struggle for existence in capitalist societies and are either unemployed or with their business' or fortunes' gone ruined. Few are shrewd politicians and religious propagandists who find career opportunities in this and they build their successful political and financial careers on the basis of such ideologies of hate.

These people form the minuscule core of leadership that generates and develops the much needed leadership initiative in such movements. Many shining names and careers are carved in the process.

Hitler is one example. Khomeini is another and the list is very long. Bin-Laden, Bhindranwale, Bal Thackeray, Slobodon Milosevic (see Appendix I) and the list is never ending.

These people made big names and attained successful positions in their lives and are some of the most talked about persons of human history; soaked in the blood of countless innocents that fell prey to the designs of the vicious hate campaigns and propagations of these successful men.

But the followers of them are molded of some different and inferior clay. They do not have the acumen and caliber to rightfully digest and understand the real game of these master planners. They come forward in the search of some missed chance in life so that they also can prove at least, some thing in life but usually they constitute the material of the sacrifice.

Fortunately these poor people have little choices left in their lives and they end up easy takers of such bogus ideologies and hollow interpretations of religion and religion based philosophy.

If only the poverty can be eliminated; such people who run and organize such nefarious movements; can be defeated in their evil designs aimed at the destruction of normal human life on earth.

And Aulprofem is a known poverty fighter.

The application of this technique on dairying can increase the milk production across the globe to many times the present values and this is the money that will be generated at the grass roots levels of villages and far flung areas even in the remotest and poorest belts of the maximum number of poor countries in the world. These are the usual poverty strongholds of the nations and these are the places that produce easy recruitments for the deadly army of terrorism. If prosperity flows in these areas and employment is generated there, thereby decreasing the number of people in the garb of idling and worklessnes, the proportion of people getting attracted to the folds of terrorism will definitely become lesser.

And devoid of recruitment, terrorism will die a natural death.

So, sex fixing becomes a strong case against terrorism.

Based on the same lines, the Aulprofem will be a strong case against addiction and intoxication menace that has spread across almost all the nations in the world. The cocaine grown in Latin American countries and golden crop of poppy produced in Burma and Afghanistan are the basically poverty produced scourges. The poor person, who is empty bellied and fighting with misery and shortages in life, usually does not want to listen to the ethical and lofty talk. Before he can know anything worthwhile; the compulsions of ordinary human needs and the lure of money and glamour attached with the sinful life, pull him deep into the world of intoxicants' trade. Once, he is in; there is usually no way leading to the outside.

Such is the story of countless Latin American and south eastern Asian youth who fall easy prey to the evil designs of the international drug cartels and mafias.

Again in suburbs of United States, Canada and Europe, the unemployed young men are the easy targets of the drug racket runners and they recruit them with the lure of money and glamour and before a new entrant can understand anything; the all exits from hell are just sealed immediately.

The Aulprofem being a prosperity generator and the sworn enemy of poverty and unemployment, will employ these people before they are attracted to the sinful, easy looking and deceptive ways of life in these countries. So here we have a strong case against addiction and intoxication throughout the world in the form of sex fixing technology that will definitely act a big salvager of humanity.

There are many benefits that sex fixing has in store for the mankind other than being a great feeder, employment generator, poverty fighter and social and moral emancipator.

It can open brand new chapters in the biological and drug research arenas.

The sex fixing is a new research and a research field as well. This is a different story that many heavy weight scientists have tried luck in this field but success by and large eluded them. This is the main reason to why this field of research was not established as a separate and independent research domain. This overall, had a devaluating impact on the status of this field as a full fledged and recognized subject of research. But the discovery of Aulprofem and its success in manipulating mammalian sex ratio is competent enough to attract the attention of the world scientific community and more research in this field will definitely prove it to be one of the most rewarding and benefitting research domains aimed at the welfare of the entire mankind.

One more point.

Many new fields of study will open, with this invention playing the foundation stone and basis for them.

The most important phenomenon in biological sciences is the fusion of a sperm with that of an ovum and the process of it is still shrouded in mystery. This discovery of Aulprofem will definitely prove a great tool in understanding this biggest secret of biological sciences. Not only this; if this happens, this will go a long way in understanding the exact mechanism by which the life evolved on this planet, some billions of years ago. Even many questions revolving around the mystery of evolution will get solved. This attaches even more importance to this discovery.

There is much stress laid on the techniques of embryo transfer, *in vitro* fertilization (IVF), embryo labeling etc. Many of them will lose relevance in the face of a bigger achievement in the form of Aulprofem that manipulates sex ratios *in vivo* with such a simple methodology and with a greater success rate. This will annihilate the basis for the embryo labeling in the expensive and extremely low success rate techniques like IVF, commonly known as the test tube baby techniques. The embryo transfer technology will then lose its relevance and application value. This no doubt, will be a great advantage of such a mega discovery like Aulprofem.

Sex fixing in dairy and other animals, will open the way for its application on human species as well. Of course, this will have social and moral overtones and implications attached with this but already the girl child is a much neglected lot in many countries and societies of the world. Under such circumstances, a science based success in favor of the weaker and fairer sex will definitely prove an emboldening and strengthening point for the ultimate emancipation and victory of femalehood in a world dominated by men folks. The womanhood will attain greater acceptance and respect with such a science based development. This will be a great moral victory attached to a scientific achievement.

There is a saying in Bible that if you have attained light of life and your brother is still in the darkness; then you lose your right to be in light and you remain still surrounded in the darkness.

The scientific discovery of a colossal standing and pedestal definitely finds its due place in the world of achievements and success, in natural due course of time.

There is no stopping the Ganges or the Amazon from flowing right into the sea. A mega research gets respect and recognition; there is no refuting this point. And a big scientific achievement definitely plays its part in helping and serving the mankind, is also true and binding.

This has happened with all the great discoveries and inventions done so far and will continue to happen this way in the distant futures to come as well. The science has taken the job of horse wagons and bullock carts and it has also replaced the countless men working in big dirty crowds under unhygienic conditions of the crowded cloth weaving mills, weaving a few hundred meters of cloth per day with a couple of sophisticated computer controlled knitting machines weaving many thousand meters of cloth per hour with the employment of a few men in very pleasant and healthy air conditioned environments. The medicinal science has rid the mankind of intense disease and suffering and still continues to do so. The physics, the chemistry, electronics and numerous other fields of science has served the mankind in their respective domains and has made the lives of humans much easier. The airplanes, the air-conditions, the television, the mobiles and internet; all have been helpful in the development of modern science based civilization of progress and prosperity. The press and media have contributed to the raising of moral and intellectual levels of people, across the globe.

The rich or poor, the elite or commoners, all have benefitted from the progress based on science. But still there are researches that are aimed to benefit only a small fraction of the total humanity and this small fraction is usually the people from the upper crust of life. They are either rulers or the privileged class like the politicians, bureaucrats, business men or the extra ordinary achievers like world class sportsmen, film stars, painters, writers and other such people. The benefits from such discoveries do not reach the people on the lower strata of life. These developments usually fall in the category of inventing products and tools, aimed at making life comfortable and luxurious. There are some people who own private planes, personal cruise liners, super luxury bungalows, fleets of limousines and can afford the latest in medical treatments but the earth is full of such people also who still live in the living conditions resembling the ones to that of the medieval times. Such people often find it difficult to buy a bicycle, live in rented, shabby slum dwellings and when they fall sick; they have no other option except either to receive the third class obsolete medications and treatments in the poorly equipped and badly maintained government hospitals or prefer to die. The treatment in a government hospital is synonymous with the letting to die for a person in exact meaning of the word. So, the service of science is dependent on a person's social standing and position in the society. The same scientific research serves the two different persons; one rich and another poor, in two different ways. Many times it serves only the rich and the poor is never cared by it.

This is true that greater numbers of people fall in the category of masses lying in downward lower strata of life and a small proportion of people usually fall in the rich and privileged categories. So, the great discoveries which are meant to benefit the entire humanity, turn in the long run benefitting a very few upper class people only. This is an irony of life as the greatest talents on earth; the scientists prefer to make discoveries meant to benefitting the whole mankind but their toil and talents go waste in the sense that they end up being benefactors of a very small section of populace.

This is a different story that some researches are aimed to benefit the general masses and this virtue is inherent in their very nature. The discoveries in the domain of green revolution, were in this domain of scientific achievements and they ended up increasing food grain productions to greater levels thereby ending up feeding the common masses on the verge of malnutrition and starvation. The discovery of DDT is another discovery which benefitted the common people and saved countless valuable human lives from death from diseases like malaria and typhus. The medical discoveries in big magnitudes usually fall in this category. It is a pity that in an era of consumerism, the doctors have become a most morally corrupt group of people who has plotted and colluded with themselves and with the pharmaceutical, laboratory equipment and test chemical manufacturing multinational companies to hijack the medical treatment, totally out of the reach of the common people especially in the poor countries of the world. Even a small fracture of a femur or humerous bone costs so much that an ordinary person can not even dream to be able to pay for; even if he piles the earning of his whole lifetime in such countries. Many people prefer to remain untreated themselves or keep their near and dear ones away from the safety of medical treatments and opt for the agonizing slow deaths that surely are already waiting for them to clinch their deadly claws over these people. People die in times when they can easily be treated and cured. This is a strange story of the victory of science and the related failure of the human civilization and society. Electricity is such another discovery which has made the life of humans much easy and livable but it is still an apathy that nearly 52% of households in India are still devoid of an electricity connection.

Aulprofem on the other hand is such a discovery that automatically will be aimed at the welfare of maximum number of humans irrespective of what they are; rich or poor. This is not a virtue of this discovery but on the other hand, a factor inherent in the nature of it. The welfare of humanity is the nature of Aulprofem. This particular trait of Aulprofem makes it a peculiar benefactor of mankind. The increase in milk production, the increase in meat production, the increases in wool production and related benefits, cannot be stopped at the upper levels and stratum of society and they will automatically percolate with great force to the down under, thereby reaching the maximum number of common masses who are the most suitable and legitimate candidates for such benefits. This is the quality of the discovery of Aulprofem and this is the achievement and victory point of it also that makes it a particular tool of human welfare and betterment.

So, the Aulprofem will attain the light of the life and also it will spread the light of human betterment to the maximum number of common masses in the world down the lines of rich and elite existence and right to the lowest sections of the human society that have remained so and are still in the garb of extreme resourcelessness and penury and finding it most difficult to make both ends meet. The light of the God's mercy will definitely fall on such people and this will be the greatest service from a mega discovery that is born and will definitely bring the whole mankind in the light of prosperity and development.

The sermon in the Bible is definitely preaching this and there is something extremely holy attached with it. Let us all hope that message of God prevails to one and all and a discovery in the lines of His will, spreads the light of hope and goodness, everywhere.

Robert Browning...............a great poet.

He sang like this once.

The snail is on the thorn......

The God is in the Heaven........

And all is well with the world.

But we have the eyes and we can see for ourselves that the thorn has torn the belly of the snail and the God has taken flight from the Heavens and the world is standing with the ghost of extreme hunger and starvation, staring in its face with the deadly and frightening eyes.

In fact, the world is already in a state of semi starvation. Many nations are already in the grip of severe food grains shortages. The agricultural productions have considerably fallen and are still going down further. The first green revolution has failed. In fact, it has betrayed the man. The second green revolution is nowhere in sight. The arable lands are decreasing..........................

There are many other ugly angles looking much similar to the ones above.

Then how can the God rest in Heaven?

And nothing is well with the world.

But..........

There's a twist.

The Aulprofem is born.

And there will be plenty of milk and meat and wool and.........many a things.

There will be no hunger .... and there will be prosperity.

And world will no longer remain a bowl of woes.

And there will be happiness.

The thorn will look beautiful.

The God will be smiling.

And everything definitely............will be fine with the world.

Historical Background

Prophet said in Bukhari sharief, there will be more females........

So, there are historical mentions of sex fixing perspectives and possibilities in human societies of past with reference to future.

To mention a few more examples.

The great king Augustus (see Appendix I) wanted to have a male heir to his throne but he was unlucky not to have one and that was the biggest failure he attributed to himself, in his otherwise very successful career as a statesman and a king. Until he breathed his last as a sad man who could not provide Rome a male heir to the throne, he reared a sort of guilt consciousness. Had he begotten a son who could be emperor after him, he would have died a very happy man. The builder of Russian empire, Peter, the great (see Appendix I), wanted to have a son who could be heir to the throne after the accidental deaths of his two sons but none of his wives could deliver the wishful. The king Janak of Mithila (see Appendix I) in India, known for his adherence to high principles and morals was blessed with a female baby by the Mother Earth as told in the Indian mythology. King Henry VI (see Appendix I) also longed to have a son and the un-begetting of a son from his first wife led Napoleon Bonaparte (see Appendix I) to marry a second time with the Austrian princess. The wish of the great Mughal emperor Akbar (see Appendix I) for a son is world famous and the most powerful man on earth of that time had to go bare footed to the shrine of Hazrat Salimuddin Chisti at Fatehpur (see Appendix I). The entire history of mankind is full of innumerable events of the wishes of people of variable rankings in society, ranging from kings to general masses to have the off-springs of desired sex. There have been times when people had yearned for female off-springs also. The people in Finland have a history of the desire for the female child. The countries in Indo-China still have the tradition of many people wishing for a female child even now. So, the mankind has a long history of the existence of the wishes of the people of all ages to have the progeny of desired sex and in this way, it is the simplest conclusion of the presence of the desire for sex fixing in human societies of all types and of all times.

Societies in Asia, Africa, Europe and other places all have exhibited the desire to have off-springs of desired sex in almost all ages of recorded history. The people in medieval times particularly after the rule of Muslim sultans and emperors set in, in India, became infatuated with the thought of having one or more sons in their families. The craving for one or more sons resulted in a sort of society that became inimical to the presence of fair sex in families and *vice versa* and the end result was that societies became dominated by men. This led to a sort of discrimination against the female sex. The girls were forced to face all sorts of prohibitions and social barriers. A similar condition existed in Afghanistan even in a time at the start of 21st century under the rule of Taliban (see Appendix III). The people of Iran have also had such an experience under Khomeini (see Appendix I) and his clan. If such things can happen even today, then we can very well imagine what could have been the plight of women folks under such rulers in the medieval times. Their rights as honorable citizens came under questionable standing. The social environment was such that they were not treated as humans but in certain cases, worse than animals. They were deprived of the right to be educated, they were forced to remain locked inside the boundaries of their homes for the most of the time, they could not participate in social and political functions, they had to bear each and every dictation and decision taken by men folk and thrust upon them and the situation was such that they even did not have the right to protest. The start of evil practices like *sati*, dowry etc; are the products of those cruel times. The burning alive of a widow after the death of her husband was an indicator of the extent of the degradation; the society had suffered in those times. Some other evils that the society had cultured, included the practice of polygamy practiced by men, the tradition of the marriage of girls at very tender age to men of considerably advanced ages; sometimes of double, triple or even more the age of girls and many more related evils like selling of girls for providing physical comforts to men or binding with them to make families. The practice of marriage of girls at ages around thirteen, fourteen or even less is still widely practiced in the states of Bihar, Madhya Pradesh, Rajasthan, Orissa (see Appendix II, states of India) and others in India even these days. We can very well imagine that under such conditions, whenever there is a birth of a new child in the families; the natural reaction of every human mind is the preference for a male child. So, the desire for sex fixing, may be it for a male child; has always existed and still exists in the thinking of every human. Even the female ones *i.e.* mothers, daughters and grandmothers etc., don't want a female child as the new entrant in their families. If such is the condition these days, then what could have been the situation in medieval times which are remembered in history as black times till today?

Right from the beginning of the great British democracy, they have had only one female Prime Minister, Margaret Thatcher (see Appendix I). Even the more vociferous democracy in the United States, can not still boast of having even a single female incumbent in the White House in the entire course of their existence and development as a role model of a sort for other democracies in the world. The great Soviet Union never had a female secretary to the powerful Communist Party of the Soviet Union or even the female head of the Polit Bureau. The Chinese communist party and the government, also have similar traditions.

The social structure and organization of a society always decides its behavior. So, the history has witnessed many a sad happenings due to the unholy dominance of society by men. It was a good thing that a scientific breakthrough like sex fixing was not reported in those times; otherwise it could have proved very, very detrimental to the status or even the existence of women folks in societies of those times. But the truth remains that the people would have opted and rather gone for a choice for sex fixing to decide the off-springs of male sex, probably resulting in the great harm to the status and even the existence of contemporary human society. The nature decides many a things and probably this might have been the reason that such a scientific discovery could not see the light of the day, because the society was not mature enough to use it in a proper and fruitful way. The deciding of nature for the discovery of Aulprofem today in the 21st century in the age of information technology and development of society to higher levels of excellence in social and mental behaviors; probably may be the reason for the timing of the evolvement of such a technology as of now.

The simple truth about sex fixing is that it is such a development that has the potential to change the very structure of societies and even the thinking of the people on much larger scales. The fact that this discovery has occurred on earth now, should pave the way for greater progress and prosperity of mankind and it should not become a tool to harm of any kind, to the man. The mental and thinking levels of humanity by and large have come up to those where it can be hoped that there will only be the benefits and not losses for the human race. It is a good thing that Aulprofem produces only the female sexed babies as of now. So, there is not a danger of its misuse in its application on the human society as could have been if it were a technology aimed and developed to produce male offsprings. On the contrary there can be more benefits and that also tremendous ones in the form of mega increases in the milk, meat and wool etc production with the application of Aulprofem in cattle, horses, camels, sheep etc animals and the humanity can be liberated from the impending danger of hunger, starvation and famines that have already started to stare the mankind in its face with the gradual fall in the food productions all across the globe with the failure of the much touted green revolutions. So, historically speaking also, the Mother Nature has very well timed the discovering of the world's **greatest benefactor and savior research** for humanity, in biological sciences and this at least, seems to be a very pleasant surprise for today and all tomorrows to come.

Comparative Status of the Research to Existing Technologies

Compare and contrast.

Are two words, two methods, two principles?

But in a discussion that ought to have some logical standing.

They should have been better classified as uncalled guests.

Uncalled and unwanted.

But they are omnipresent. They creep up and are applied everywhere.

Applied and savored with a universal touch.

Just like salt and potatoes.

To be part of all societies and cultures.

They find the greatest use in discussing and developing thought and philosophy and of course science. They help in assigning and understanding the real place and status of a material entity, development or thought to the ones, already existing.

Lets discuss these things regarding Aulprofem as follows:

The technologies in competition to the Aulprofem medication are basically sexed semen technologies. They are basically *in vitro* technologies whereas Aulprofem is an *in vivo* technology. So, a comparison is basically not appropriate in this case but even then when compared on the basis of relative virtues and vices, they stand nowhere as compared to Aulprofem as discussed below under following headings:

1. Drawbacks of sexed semen techniques
2. Advantages of Aulprofem over sexed semen techniques

1. Drawbacks of sexed semen techniques

The production of sexed semen techniques are still in the mire of controversial standing as commented by Hunter, "Spermatozoa are haploid cells possessing only half the usual number of chromosomes

The superior *in vivo* alternative

for the species, and there is no evidence that haploid, as distinct from diploid cells, express their genetic constitution in surface characteristics. This being the case, it is difficult to imagine, how chemical, electrical or bio-physical separation of X or Y spermatozoa might be obtained" (Hunter, R.H.F., 1982, Reproduction of Farm Animals. Longman. London, pp. 138-139).

Further in the absence of any testing parameters to ascertain the accuracy of percentage separation of X or Y spermatozoa, it is difficult to believe the truth of existence of separated X or Y spermatozoa as pointed by Hafez, "Experimental attempts have been hampered by the lack of laboratory tests to evaluate the degree of sperm separation" (Hafez, E.S.E., 1987. Reproduction in Farm Animals. 5th edn. Reprint. Lea and Fabiger. Philadelphia, p. 501).

Some other drawbacks of the sexed semen techniques are as under:

1. Poor conception rate
2. Congenital defects
3. Expensiveness

4. Difficulty in transportation and handling

5. Controversial standing

The production of sexed semen and its resultant handling and storage is a specialized job. This involves the application of cryo-preservation techniques and apparatus. Any failure in maintaining the required freezing temperatures, adversely affects the survival of the spermatozoa and the resultant fall in conception rates. Even with the usual maintaining of the temperatures and conditions such as the preservation and culture media etc, the desired conception rates are not maintained. In case of the semen subjected to separation of X and Y spermatozoa, the fall in vitality and viability of the sperms happens. There may be many reasons to it, the detailed scope and need to elaborate on which does not come in the domain of this write up but even then a mention of the few such ones as the decreased concentration of the semen due to subjection of them to sperm separation techniques, the fatigue and related exhaustion quotient attached with such exercises, the subjecting of the sperm to un natural *in vitro* environment and chemicals, media etc, may be some of the reasons for the poor performing health of the sperms.

Of the other defects, the congenital deformities are a big reality with the spermatozoa treated with unnatural methodologies and there is no hiding of the fact these days that any maneuver that exposes the micro cellular level live entities to unnatural conditions, produces certain damages or deformities in them and such sperms if allowed to unite with the usual ovum will transport these defects to the resultant zygotes also; there is no hiding of this fact. So, the increased or inherent danger exists, of the congenital defects in the progeny resulted from the union of such sperms with the ovum.

IVF (*in vitro* fertilization) is a complicated and expensive technique, there is no refuting it and there are practical problems and hardships associated with it. The formation of sexed zygotes as a result of the IVF technologies, is a further practical hindrance. The application of usual cryo preservation techniques and further laboratory procedures and manuals, adds to the impracticality of this technology. The transport of such sexed embryos is a further handling headache and the poor standing on successful delivery of the desired sexed offspring is also a question mark in such exercises. Usually more than one embryo is transplanted in the genital tracts of the recipient mothers who may carry multiple twins in their wombs. Under such conditions, the very question of getting sexed offsprings is under suspicion. Also the instances of miscarriage or unnatural abortions are more in such mothers. By and large, due to practical problems and complicacies, such techniques are still mired in controversial standings.

2. Advantages of Aulprofem over sexed semen techniques

Almost all drawbacks of presently used sexed semen techniques, can be overcome with the application of Aulprofem and superiority of this technique over other techniques is obvious as discussed under following headings:

1. Un-affected conception rate
2. Congenitally compatible
3. Economically affordable
4. Easy transport and handling
5. Effectiveness
6. Stability and optimum storage conditions

Aulprofem is a natural *in vivo* methodology. Hence, much of the usual drawbacks associated with unnatural *in vitro* technologies are not associated with it. This being a natural procedure, does not involve any unnatural aspiration of the ova from the genital organs of the mothers into the artificial conditions of the culture media and man made climatic imitation of the natural environs of the biological systems of inside the mother's body. So, the usual drawbacks of artificial and unnatural man made techniques are not found in this methodology. The conception rates are not affected. This is congenitally compatible. This is economically affordable to the farmers also as with the additional spending of a few hundred rupees; he gets a desired female calf with an exotic breed stock. There is no cryo-preservation technique involved and handling and related transport problems are not existent in such methodology. This is a simple formulation of organic and other compounds which is conveniently packed in transportable single dose liquid preparation and can be stored in just ordinary temperatures. This has optimum storage conditions and being a natural technology, is as effective as the otherwise natural recourses of insemination and fertilization processes. The conception rates as well the natural reproduction capacities of the genital organs and the animals are not affected in any way.

The comparative status of Aulprofem as a sex fixing technology can be better understood by considering that this methodology yields the required outcome and results of a usual sex fixing choice with the much enhanced success rate of near absolute manipulation in the mammalian sex ratio with the least of the efforts. The only thing that a person has to do is to administer an oral drug formulation at one particular time and leave the rest to the drug to do the wonders and deliver the rightful.

There is a story of a gooseling in a duck yard.

Once upon a time, there lived many a duck in a duck yard. There were many ducklings also. One particular time, nobody knew how it happened but there a strange ugly little bird was born. No duck wanted to own it as her own. Even the duck which had hatched it from a healthy egg, did not have an inclination to rear or raise it. The little one was very upset and depressed but it also grew with the flock of duck kids and time kept on flying. A stage reached when all the little birds reached young age and were on the threshold of the adulthood. The strange bird by then, had grown in awkward proportions and it was an unpleasant sight, away from group. Nobody found it suitable to mix with or talk to it. The poor bird was very sad. He always tried to maintain a safe distance from the flock. Everybody despised him and stayed away from him. But on a fine evening, a big flock of the swans flew into that territory. The chief of the swans happened to look at the poor ugly duckling and he immediately recognized it as one of their own. He called the poor bird towards him with love and affection and took it along to the swans' group and praised the little bird. He declared to one and all that the little bird was the most beautiful swan throughout the world and further added that the swans would adopt it as their rightful prince. Everybody was surprised and the flock of ducks felt sorry for the treatment, they had showered on the little gooseling. The gooseling was very happy and he flew away with the swans.

The Aulprofem is an entirely different technique to the existent ones and it has no comparison or contrast points to be considered and applied in a usual compare and contrast tactic. This is like a true gooseling, found accidentally in the yard of ducks full with duck progenies in greater numbers but it is different and does not resemble with any of the others. This has benefits also that other methodologies can not be expected to have. This is far, far better and more practical, cheaper, easy to use and safe and without any side effect or usual drawbacks of the sexed semen technologies.

This in a way can be compared only with itself and certainly has no comparison with the others. This is the true and precise concluding of the discussion in this matter.

PART - C

Theory and Basics

- Fertilization
- Sex Fixing
- Formulative Theory of Determination of Sex
- Prior Art

Fertilization:
The Fusion of Sperm with the Ovum

The most impressive event in the entire world of life and creation is the binding and fusion of sperm with the ovum (**Figs 1, 2, 3**). In spite of the fact that numerous scientists across the globe have worked on the subject, very little is known about this still. Masaru Okabe a Japanese scientist (Inoue *et al.*, 2005) has described thus, 'representing the 60 trillion cells that build a human body, a sperm and an egg meet, recognize each other and fuse with each other to form a new generation of life'.

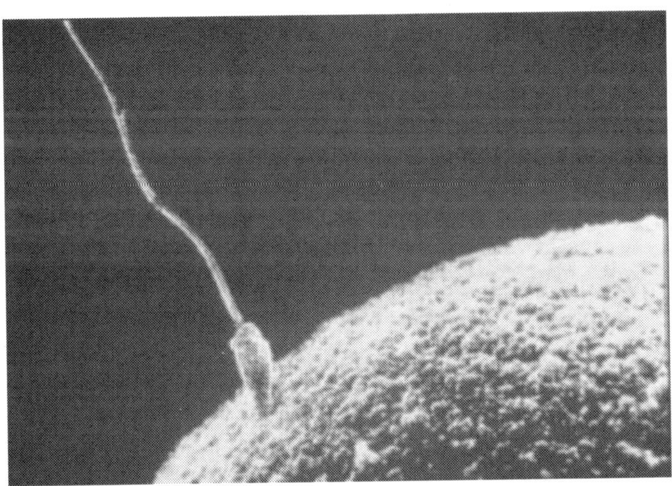

Fig 1. Shows the union of a mammalian sperm with an ovum

This is an apt and beautiful description in simplest of the words about the process of fertilization of male and female sex cells in humans.

Basically, sperm egg fusion being a cell-cell membrane fusion is as other fusion events, a two step mechanism. The first step involves the identification and attachment of the two cells, sperm and ovum and in the second step the physical merger of the plasma membrane lipids takes place. We try to give a brief account of this whole process in the following write up.

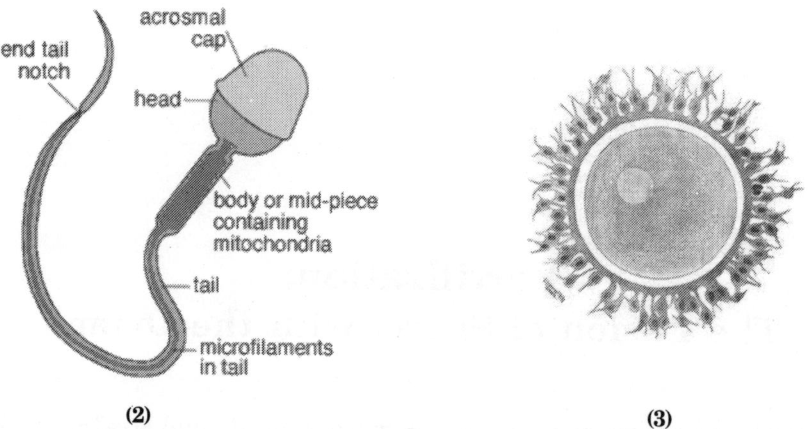

Figs 2 & 3 Showing a mature mammalian sperm (2) and an ovum (3)

Introduction

The fusion of two cells, an egg and a sperm is a rare event in sharp contrast to the cell division **(Figs 4 & 5)** which is very frequent and prominent during the entire development of live organisms. Of course, there are some other fusion processes in the living world. The focus of study has to investigate the fusion processes taking place in living world like the virus-cell fusion **(Fig 6)**, intracellular vesicle fusion **(Fig 7)** (that is so crucial for protein trafficking and exocytosis) and the cell-cell fusion.

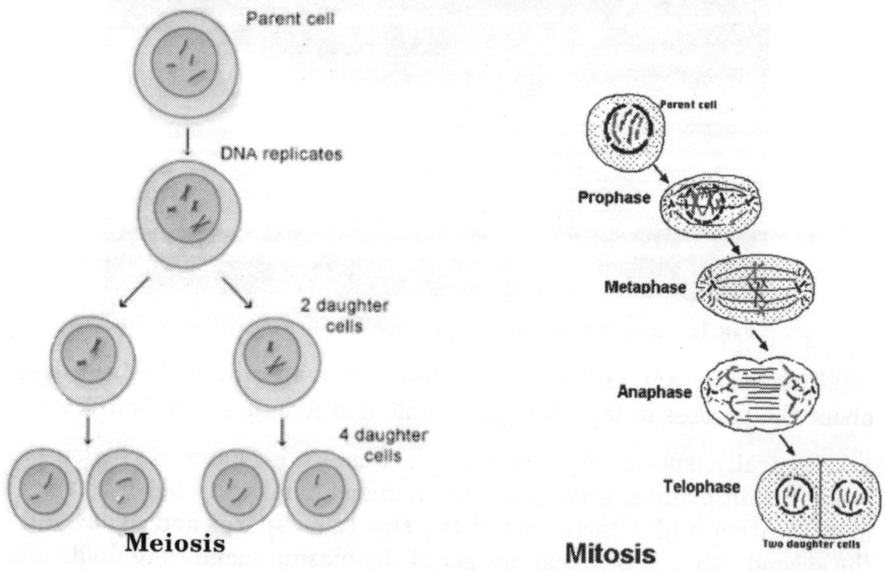

Figs. 4 & 5. Show the meiosis (4) and mitosis (5) processes during division of cells

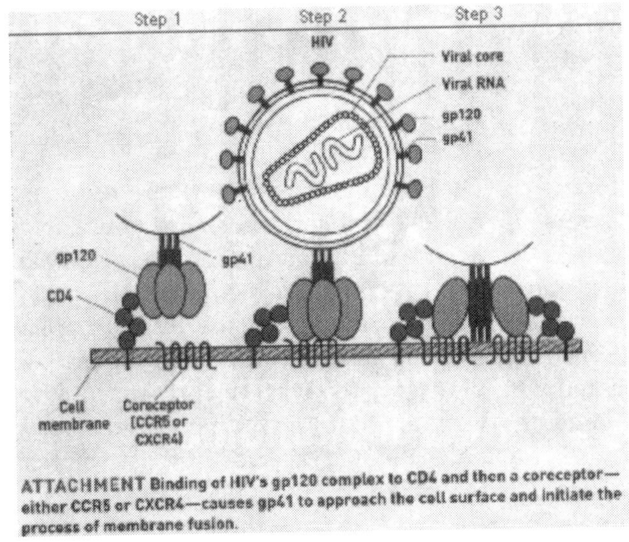

Fig 6. Shows the attachment and binding of a virus with the host cell

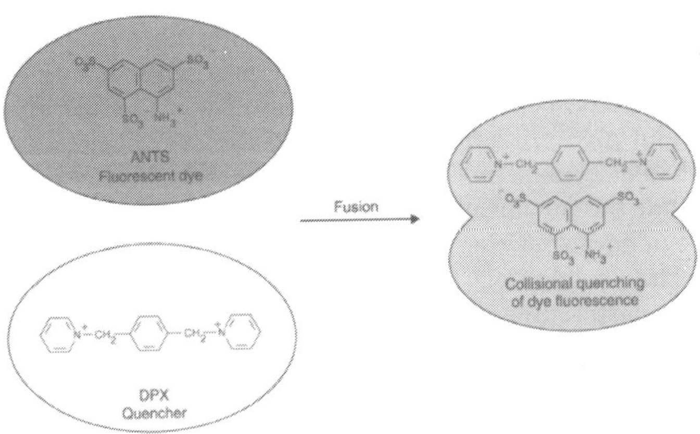

Fig 7. Shows the theoretical procedure of the intracellular vesicle fusion

Although little is known about the cell-cell fusion, virus-cell fusion and intracellular vesicle fusion are better understood (Jahn *et al.*, 2003). So, in order to study the first, the other two are emphasized and it is a fact that many similarities do occur as regards the mechanistic concepts in the last two and the first *i.e.* cell-cell fusion. Many candidate molecules have been targeted like tetraspanins like CD_9, glycosylphosphatidylinositol-anchored proteins, ADAM family proteins like fertilins and many more that appear to participate in the processes involved in the usual fusion of a cell with the other. The various steps involved in this process are as under.

Sperm movement and penetration of cumulus cell complex

After a sperm leaves the male reproductive organs and enters the female reproductive tract, it still has to travel a long way and overcome many obstacles before it can reach the destined destination, the ovum. Of the 30, 000,000 human sperms ejaculated during coitus, only about 200 reach the site of fertilization in case of human fertilization. This explains on an average about the complexity and hardships in the way of a usual mammalian fertilization. For comparison, the Russian misadventure of the famous French emperor and general Napoleon Bonaparte (see Appendix I), can be quoted which was started with nearly two hundred thousand soldiers and ended up with just fourteen of them returning alive, back in France. The hardships in the way of sperms are just colossal and out of imaginable limits. The sperm has to undergo a process called capacitation before being able to unite and fuse with the ovum. This takes about 5-6 hours in case of humans. Capacitation is triggered by the bicarbonate ions in the vagina which enter the body of the sperm and directly activate soluble adenylyl cyclase enzyme in the cytosol. This action produces cyclic AMP in the cytoplasm which helps to initiate changes associated with capacitation. The main changes that take place during capacitation include the increased metabolism and motility of the sperm, altering of the lipid and glycoprotein composition of the sperm plasma membrane and marked decrease in the membrane potential to a more negative value so that the membrane becomes hyper polarized as a result of this. There is evidence that chemical signals released by the follicle cells that surround the ovulated egg, attract the sperm to the egg. But the nature of the chemo attractant molecules is not known. Once, it finds the egg, it migrates through the layer of the follicle cells that surround the ovum and it has to penetrate this layer to finally get attached to the outer coat of the cell, called zona pellucida.

When the sperm reaches the egg more correctly called oocyte at this stage because it has not yet completed meiosis, it comes across the mass of cells called cumulus oophorus surrounding the egg. These cells are ovulated along with the egg. Sperm has to swim across these cells to reach the egg and in the process dissolves the extra cellular matrix that holds these cells together. The sperm carry with them in the plasma membrane and in the acrosome the enzyme called hyaluronidase that digests this extra cellular matrix and helps the sperm to reach the zona.

The attachment with the zona pellucida

The zona pellucida of mammalian eggs is composed mainly of three glycoprotein layers, ZP_1, ZP_2, and the ZP_3 **(Fig 8)**. These glycoprotein

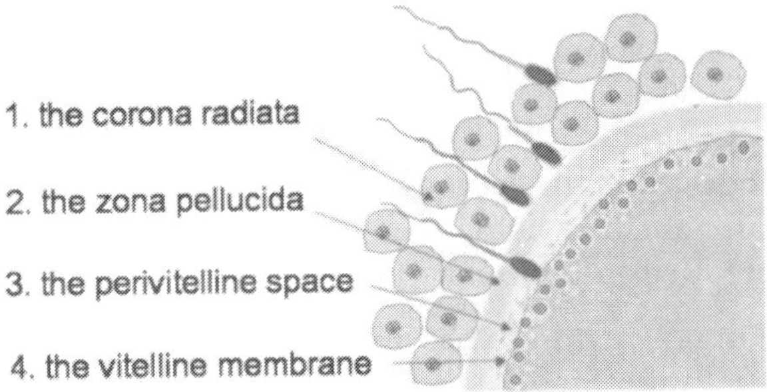

Fig 8. Shows the binding and penetration of a sperm through zona pellucida

layers are exclusively grown by the developing oocyte. The meeting point for the mature gametes is the oviduct and sperm has to cross the barrier of zona pellucida after it has crossed the barrier of cumulus. It is shown that acrosome intact mouse sperm initially binds to the carbohydrate region of ZP_3. It is the ZP_3 which is responsible for the species specific binding of the sperm with the ova. Several proteins on the sperm surface bind with specific O-linked oligosaccharides on the ZP_3 which are the supposed receptors on the zona. The complete identity and function of these carbohydrate molecules is not known. On binding to the zona, the sperm undergoes a process called the acrosome reaction **(Fig 9)**.

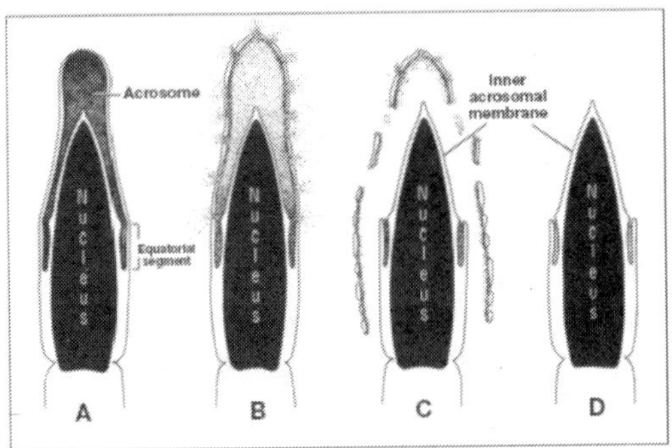

Fig 9. Shows acrosome reaction in a sperm

The acrosome reaction

The acrosome reaction is marked with the release of the contents of the acrosome by intracellular exocytosis. In the mouse sperm at least it is established that the trigger for the acrosome reaction is the ZP_3 in zona which induces an influx of calcium ions into the sperm cytosol. This in turn initiates exocytosis. The acrosome reaction is very important for the fertilization and it exposes various hydrolytic enzymes that act a help in the sperm tunneling through the zona pellucida. This also exposes other proteins on the surface of the sperm that bind with ZP_2 to establish a sort of secondary grip which helps the sperm to maintain a tight hold when the sperm burrows through the zona. The acrosome reaction also exposes certain other proteins that help in the fusion of the sperm with the ovum after it has burrowed in the zona.

During binding to the zona, the sperm maintains its motility although a reduced one. This provides the additional mechanical force while digging through the zona.

Fusion of the gametes

When sperm has burrowed through the zona, it comes to lie in the perivitelline space, lying in between the plasma membrane and inner boundary of the zona of the ovum and starts the process of binding and then fusion with the plasma membrane. A specific portion of the gamete having a unique protein population, lipid organization or composition or membrane morphology is the topological selected part for fusion (Stein *et al.*, 2005). The sperm plasma membrane overlying the acrosome that does not take part in the acrosome reaction is termed as the equatorial region, and the fusion process is generally associated in this region of the sperm (Yanagimachi, 1988). The oocyte plasma membrane has also two differential regions, microvillar free region overlying the meiotic spindle and the microvillar rich region covering the rest of the oocyte surface. Sperm oocyte fusion occurs predominantly (Johnson *et al.*, 1975) or exclusively (Ebensperger and Barros, 1984) in the microvillar rich region. There seem to act many mechanisms that help in these processes. There are different candidate molecules that seem to help in these processes. One among them is the surface protein on sperm called fertilin. There is another protein as adhesion partner on the egg membrane called CD_9. This is a member of the tetraspanin protein family and is abundantly expressed on the egg surface. There are clusters of the microvilli on the egg surface which cluster around the sperm to ensure that it is held firmly so that fusion process is facilitated. The plasma membrane of the sperm is drawn in continuity with the plasma membrane on the tips of the microvilli of the egg and their meeting points are digested with the fusion proteins or the surface proteins and the sperm fuses with the ovum. After fusion the entire cytoplasm of the sperm is drawn, head first into the body of the ovum along with the microvilli which are also resorbed into the egg. It is supposed that the extra cellular domain of the α-subunit of the fertilin contains a hydrophobic region that resembles the fusogenic region of the viral fusion proteins. When the cytoplasms of both the gametes have fused, the fusion is said to be complete.

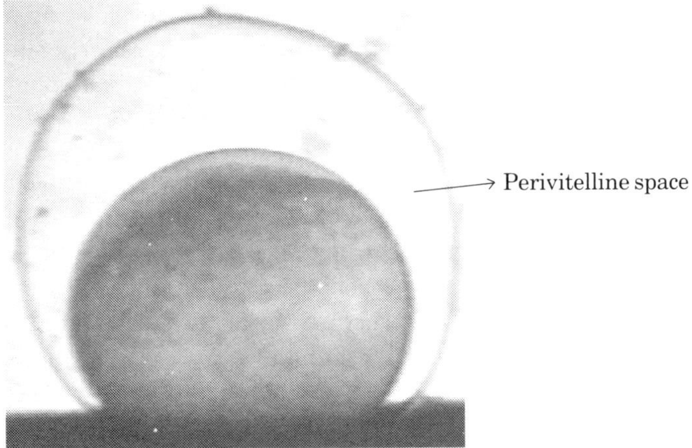

Fig 10. Showing the perivitelline space in an ovum

The egg cortical reaction

There are chances that more than one sperm may enter the perivitelline space **(Fig 10)** after crossing the zona and be able to fuse with it. Then a stage is reached called polyspermy. This is a rare condition. This may result in the formation of multipolar or extra mitotic spindles. This leads to the faulty segregation of chromosomes during cell division and non diploid cells are thus produced. This sends a particular signal and further development is stopped. But in normal cases, there operates a mechanism to stop such development. A rapid depolarization of egg plasma membrane which is caused by the fusion of the first sperm

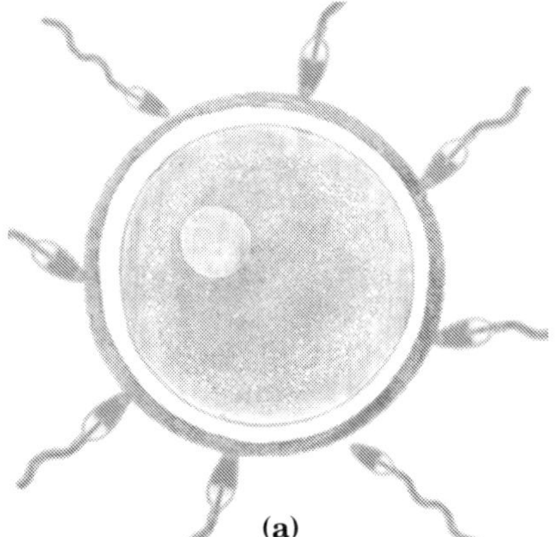

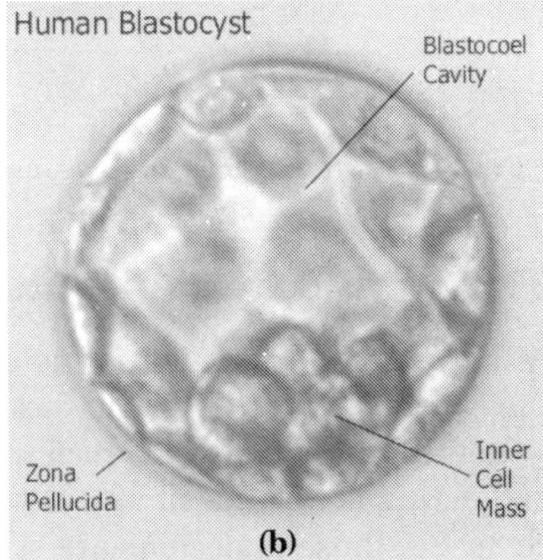

Figs 11a, b. Shows cortical reaction (a) after the completion of cortical reaction and (b) the zygote starts the further development

prevents the further sperm from fusing and acts as a primary block to polyspermy. But the membrane potential reverts soon back to normal after fertilization. Hence a secondary block to polyspermy is provided by the cortical reaction **(Fig 11)**. When the sperm fuses with egg plasma membrane, it causes a local calcium ion increase. This increase spreads in the cell like a wave and is soon followed by prolonged calcium ion oscillations, probably caused by a protein that is introduced into the egg by the sperm. The calcium ion oscillations cause the cortical granules to shed their contents by exocytosis which include certain enzymes that change the nature and structure of zona pellucida and it becomes hardened. Now no more sperm can bind and fuse with it and this acts as a permanent block to polyspermy. The fused gametes now give rise to a single diploid cell called **zygote (Fig 12)** and this is the completion of the process of fertilization.

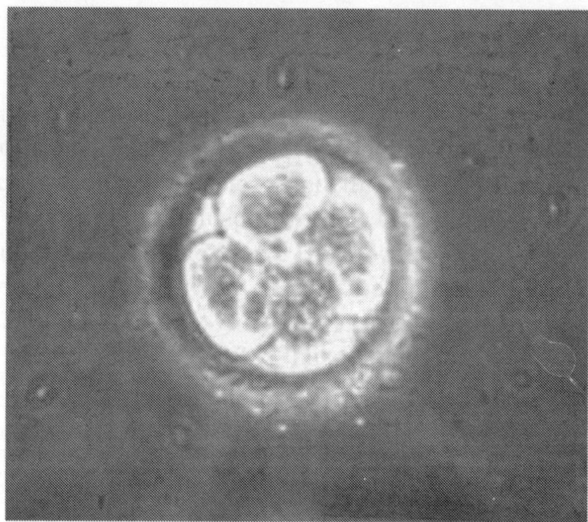

Fig 12. Shows a zygote that has just started further development into an embryo

The candidate molecules in the fusion of gametes

A discussion on fertilization is not complete if a mention is not made of the various molecules that act as facilitators of this one of the most wonderful acts of nature. Much work has been reported on these supporting and acting molecules in recent years but the final picture has not become clear. So, these molecules are still considered the candidate molecules as their candidature of being a party in the process of fertilization is still under investigation. We present a brief account of these candidate molecules down below.

The attachment and binding molecules

It is very difficult to ascertain a particular compound strictly to a particular category such as attachment, binding or fusion as an attachment or binding function defect impairs fusion process also and the level of research in this field has not reached the satisfactory levels so that we know for sure the exact mechanism of sperm-egg fusion. An agent or molecule for attachment can also be a molecule for binding or fusion. Even then, we will try to study them suitably as told below.

In virus-cell fusion, the carbohydrate moieties or cell adhesion domains on proteins or other molecules in the plasma membrane are the attachment points for fusion (Bomsel and Alfson, 2003). Such reactions are much complex and a variety of molecules take part in them. In fruit fly *Drosophila* myoblast fusion **(Fig 13)**, four immunoglobulin *(Ig)* super family members,

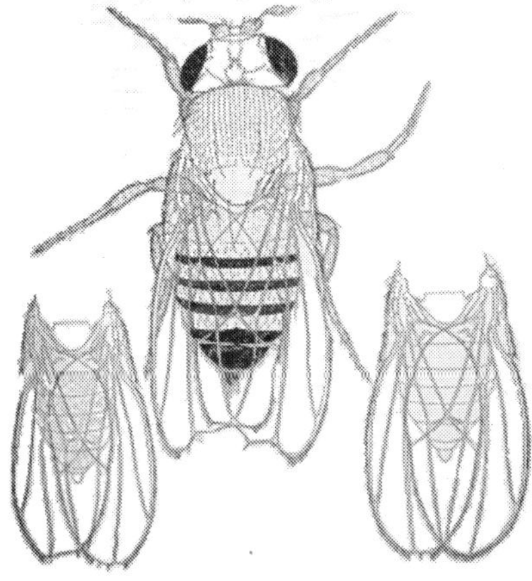

Fig 13. Shows the fruit fly *Drosophila pseudo-obscura* which attracted the attention of scientists after the first gene mutation was reported in it

Sns, Hbs, Duf and *Rst* (Taylor, 2002) take part as cell-cell fusion domains. In alga *Chlamydomonas*, antibody *fus1* and in *Caenorhabditis elegans*, sperm protein SPE-9 containing the epidermal growth factor, EGF, is the attachment domains in gamete fusion.

Glycosylphosphatidylinositol (GPI) anchored proteins are a class of extra cellular glycoprotein attached to membranes by a lipid anchor. The study using the enzyme, phosphatidylinositol specific phospholipase that cleaves this anchor, hampers the gamete fusion (Coonrad *et al.*, 1999). Oocyte specific stop of Pig-A, another enzyme acting in biosynthesis of this anchor, also impairs the gamete fusion (Alfieri *et al.*, 2003). Another

GPI anchored protein is the CD_{55} whose targeted deletions do not result in the impairment in fertility, thus eliminating a potential role for this moiety (Alfieri *et al.*, 2003).

The other attachment molecules are ADAM family proteins. Until now some 39 ADAM family proteins have been identified and about half of them are either testis specific or testis enriched, indicating that they have important sperm functions (Stein *et al.*, 2005). ADAM-1 and ADAM-2 were first discovered in a monoclonal antibody screen for the guinea pig sperm proteins. The best studied ADAM's are ADAM-2 known as fertilin-beta and ADAM-3 known as cyritestin. Initial studies in immuno action studies with peptides from the disintegrins loop of these two ADAM's, potentially inhibited the sperm-oocyte fusion *in vitro*. ADAM-2 deficit mice sperms show defective transport through the female reproductive tract. ADAM-2 and ADAM-3 gene knockout mice exhibit defective ZP binding (Cho *et al.*, 1998). ADAM-3 is released at the time of acrosome reaction and thus may not have a role in plasma membranes binding (Kim *et al.*, 2004). These proteins need to be investigated further in order to ascertain their exact roles as binding or fusion agents. The new evidence is emerging which establishes this category of compounds to be effective fusion agents for sperm-egg fusion.

Binding and fusion molecules

The binding and fusion molecules occur both on the sperm and the ovum. The ADAM's also seem to fall under this category. They definitely have binding roles for sperm-egg binding and a molecular dissection of this compound fertilin or ADAM-2 might provide a better understanding of its direct role, if any in fusion (Stein *et al.*, 2005).

Tetraspanins are the egg proteins that are called tetraspanins because they have four membrane spanning segments. CD9 is one among them. Tetraspanins are supposed to mediate several types of protein interactions including primary reactions with the Ig super family, membrane anchored growth factors and secondary reactions with other members of the tetraspanin family. Many of these proteins and these reactions form the 'tetraspanin-web' **(Figs 14a & 14b)** (Boucheix and Rubinstein, 2001). There are tetraspanin enriched domains where large numbers of tetraspanins regulate critical cellular functions by their cooperative action with themselves and other proteins, in these specialized domains.

The two most worked out members are CD_9 and CD_{81}. CD_{81} is a co-receptor for the hepatitis C virus (Cormier *et al.*, 2004) and both CD9 and CD 81 are implicated in the mammalian myoblast fusion (Schwander *et al.*, 2003) and monophage/macrophage fusion in mice (Takeda *et al.*, 2003). The mice with deficiency in CD_9 are normal and healthy females and

Fig 14. Hypothetical model for tetraspanins. Tetraspanin web compounds are still a mystery for the scientists

after mating many sperms are found in the perivitelline cavity, indicating that earlier sperm-oocyte interactions take place normally but the failure of attachment or fusion occurs at the plasma membranes fusion levels (Kaji *et al.*, 2000). The CD_9 exhibits an orientation in the membrane, characteristic of a tetraspanin where N and C termini are intra cytoplasmic and two extra cellular loops, one small EC_1 and other large EC_2 project into the extra cellular space.

Other possible candidate molecules are integrins but gene knockout experiments in mouse have not shown any decrease in fertilization rates, thus ruling out a role for these compounds in the fertilization (He *et al.*, 2003). Recently the possible role of integrins mediated signaling in oocyte activation has been reported that can stimulate the protein kinase C and cortical granule exocytosis **(Fig 15)** in cortical reaction in mouse oocytes (Tatone and Carbone, 2006).

Summary

Mammalian fertilization is a process that begins when a sperm binds in a species specific manner to the zona pellucida of the egg and this binding induces acrosome reaction in the sperm. This releases the contents of its acrosomal vesicle, exposing enzymes and proteins that help the sperm to burrow in the zona (Fig 1) and make its way to the egg plasma membrane in order to fuse with it. The binding and fusion of the plasma membranes of the sperm and egg take place and the cytoplasms of the both meet in the formation of a single cell structure. This fusion initiates calcium ion signals and the egg undergoes cortical reaction to release cortical granules and enzymes that alter and harden the zona to prevent any further entry

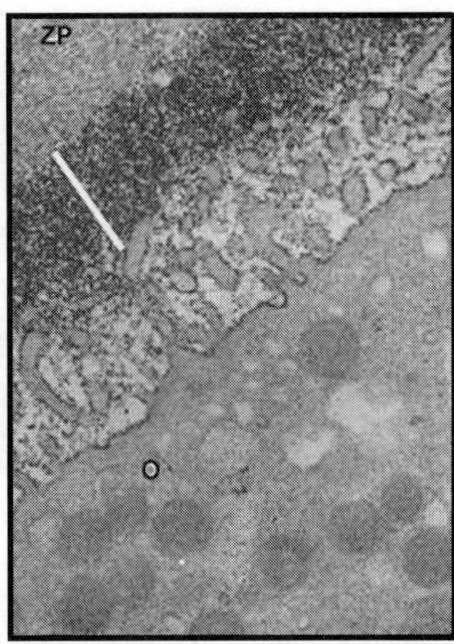

Fig 15. Fertilized ovum shows cortical granule exocytosis

of a sperm to avoid polyspermy. The calcium ion signals also start the development of the zygote. The haploid pronuclei come together and get organized on a single mitotic spindle and a new diploid cell called zygote comes into being and this becomes the foundation, on which a new life finds its essence and beginning. During all this process, many proteins, carbohydrates, enzymes and other molecules act as attachment, binding and fusion agents and facilitate this whole magnificent process to the finish. The exact roles for many of these candidate molecules have been worked out completely and for many others it is still the subject of pioneer investigative research.

References

Alfieri, J.A. *et al.*, 2003. Infertility in female mice with an oocyte specific knockout of GPI-anchored proteins. *J. cell Sci.* **116**: 2149-55.

Bomsel, M. and Alfsen, A. 2003. Entry of viruses through epithelial barrier: pathogenic trickery. *Nat. Rev. Mol. Biol.* **4**: 57-68.

Boucheix, C. and Rubinstein, E. 2001. Tetraspanins. *Cell Mol. Life. Sci.* **58**: 1189-05.

Cho, C. *et al.*, 1998. Fertilization defects in sperm from mice lacking fertilin beta. *Science* **281**: 1857-59.

Coonrad, S.A. *et al.*, 1999. Treatment of mouse oocyte with PI-PLC releases 70-kDa (pI5) and 35 to 45-kDa (pI5.5) protein clusters from the egg surface and inhibits sperm oolema binding and fusion. *Dev. Boil.* **207**: 334-49.

Cormier, E.G. *et al.*, 2004. CD9 is an entry co-receptor for hepatitis C virus. *Proc. Natl. acad. Sci. USA.* **101**: 7270-74.

Ebensperger, C. and Barros, C. 1984. Changes in hamster oocyte surface from the germinal vesicle stage to ovulation. *Gamete Res.* **9**: 387-97.

He, Z.Y. *et al.*, 2003. None of the integrins known to be present on the mouse egg for to be ADAM receptors are essential for the sperm-egg binding and fusion. *Dev. Biol.* **254**: 226-37.

Inoue, N. *et al.*, 2005. The immuno globulin super family protein Izumo is required for the sperm to fuse with eggs. *Nature* **10**: 434 (7030): 234-38.

Jahn, R. *et al.*, 2003. Membrane fusion. *Cell* **112**: 519-33.

Johnson, M.H. 1975. Mosaicism in organ concanavalin A receptors on surface membrane of mouse egg. *Nature* **257**: 321-22.

Kaji, K. *et al.*, 2000. The gamete fusion process is defective in eggs of CD9 deficient mice. *Nat. Genet.* **24**: 279-82.

Kim, E. *et al.*, 2004. Synthesis, processing and sub cellular localization of mouse ADAM 3 during spermatogenesis and epididymal sperm transport. *J. Reprod. Dev.* **50**: 571-78.

Schwander, M. *et al.*, 2003. Betal integrins regulate myoblast fusion and scaromere assembly. *Dev. Cell.* **4**: 673-85.

Stein *et al.*, 2005. Sperm-egg fusion: events at the plasma membrane. *Journal of Cell Science* **117(26)**: 6269.

Takeda, Y. 2003. Tetraspanins CD9 and CD81 function to prevent the fusion of monoclonal phagocytes. *J. Cell Biol.* **161**: 945-956.

Tatone, C. and Carbone, M.C. 2006. Possible involvement of integrin mediated signaling in oocyte activation: evidence that a cyclic RGD-containing peptide can stimulate protein kinase C and cortical granule exocytosis in mouse oocytes. *Reprod. Biol. Endocrin.* **04**: 48.

Taylor, M.V. 2002. Muscle differentiation: how two cells become one. *Curr. Biol.* **13**: R 964-R966.

Yanagimachi, R. 1988. Sperm-egg fusion. *In* Current topics in membranes and transport. Vol. 32. *Eds.* F. Bronner. Academic Press. San-Diego, CA, pp. 3-43.

The Sex Fixing

Sex pre selection is the attempt to influence the sex of a child before fertilization (To and Chan, 1994).

The fertilization is the most impressive event in the life of the organisms multiplying by way of sexual reproduction. The gametes which are haploid in nature, fuse to form a zygote which is diploid and contains the chromosomes from both the sexual organs *i.e.* male and female. In organisms which are hermaphrodites, both the sex organs are located

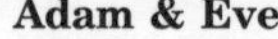

Adam & Eve

Kamadev

(Glimpses of excellence from archives of creativity)

on the body of one organism like in annelids **(Fig 16)**, coelenterates **(Fig 17)** or echinodermates. It is a different question that the male gametes from one organism are transferred to the female gamete of the other organism by cross fertilization or male gametes of one organism fertilize its own female gametes by a process called self fertilization. Such exercises are common in plant kingdom.

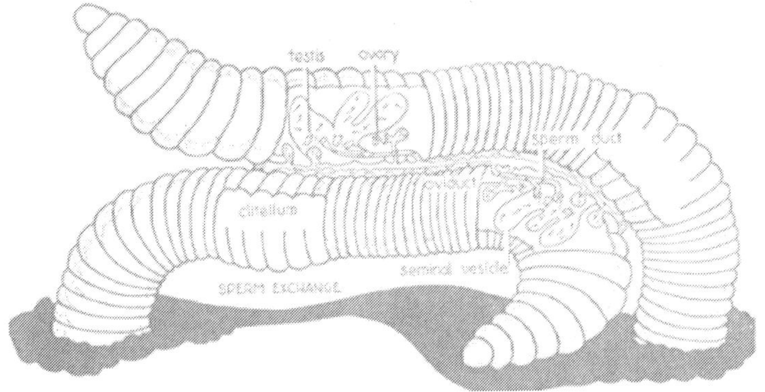

Fig 16. Cross Fertilization in earth worms

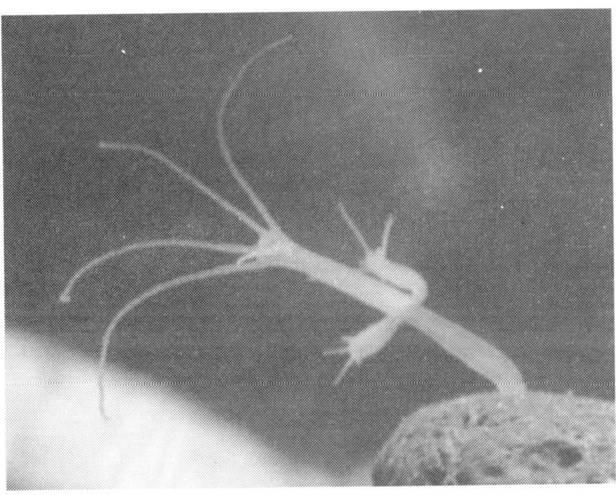

Fig 17. *Hydra* propagates by sexual as well as asexual modes of reproduction

The nature prefers cross fertilization and it has devised various methodologies by way of which the cross fertilization takes place in most of the cases in both the kingdoms for examples the copulation in hermaphrodite earthworms in annelids and copulatory union in cockroaches in arthropods in kingdom *animalia* and cross fertilization by way of pollination vector processes in most of the angiosperms in plant kingdom. Even in *Spirogyra* **(Fig 18)** and *Chlamydomonas* **(Fig 19)** cross fertilization takes place by specialized processes. In *Chlamydomonas* the gametes are termed male and female depending upon their size and motility. In other thallophytes like fungi, such cross fertilization also takes place.

The sex fixing process is a meaningless phenomenon in bisexual organisms but in case of unisexual organisms this is a very important process and is guided by various mechanisms to produce male and female offsprings in almost equal numbers. The nature and its worthy agent, the natural selection has perfected this art of production of male and female progenies in fifty to fifty ratios with a remarkable success rate. With slight deviations not withstanding, the natural primary sex ratio of fifty to fifty male to female offsprings, holds good in most of the cases. The sex ratios and the deviations from them are dealt with in a different chapter in this book and are not the subject of deeper study in this

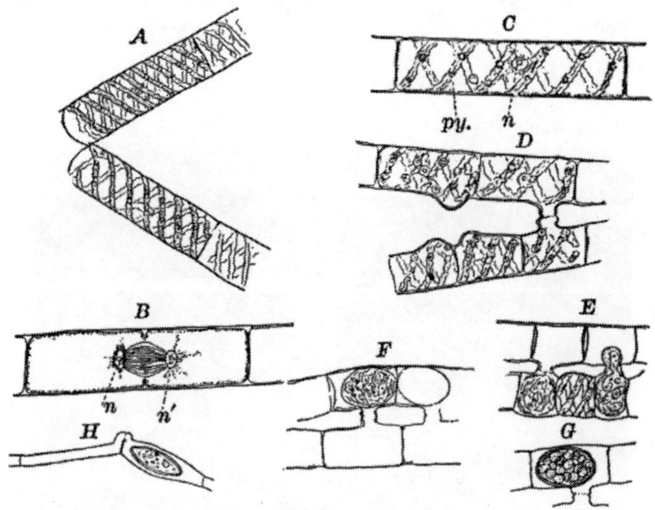

Fig 18. Sexual Reproduction and cross fertilization in *Spyrogyra*

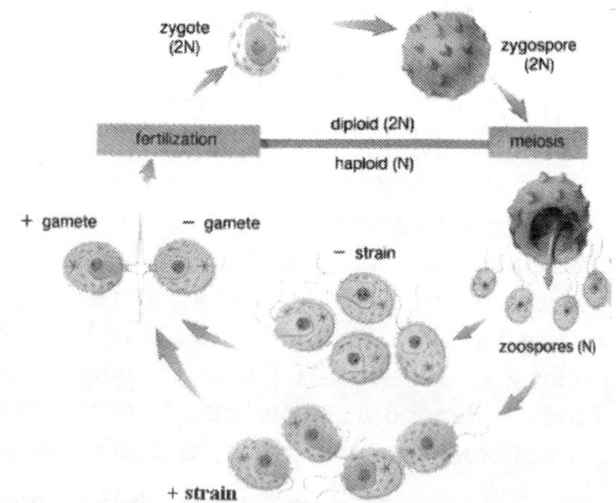

Fig 19. *Chlamydomonas* display sexual as well as asexual modes of reproduction

chapter. So, we leave this discussion here and move to the topic of sex fixing.

The subject of sex fixing has attracted attention of the breeders in both the kingdoms of plant as well as animal ones. The papaya tree is a unisexual tree and male and female trees are distinct from each other. There have been many attempts to produce female trees in more numbers in case of this species. Bull *et al.* (1980) and Crews *et al.* (1983) have patented techniques to produce sexed offsprings in lower chordates in ratites and lizards in case of animal kingdom and there are many other attempts to achieve this in non chordates such as arthropods, echinodermates and other non chordates. In 1966 Madeline Charnier (Crews *et al.,* 1993) discovered the phenomenon of temperature dependant sexual determination (TDSD) in African rainbow lizard, *Agama agama* **(Figs 20 & 21)**. The techniques followed by workers like Crews *et al.* (1993) is a modification of the TDSD technique in the sense that it adopts the use of certain chemicals in lizards and certain reptiles to achieve the

Figs 20a,b,c. Shows the lower reptiles such as lizards (a) and ratites, such as Kiwis (b) and Ostrich (c). which display the phenomenon TDSD

Fig 21. Reproduction in lower chordates display phenomenon of TDSD as deciding the sexuality of the offsprings

goal of sex fixing and has bagged international recognitions and even patents from the patent offices including the U.S. patent office. Compounds like estradiol-17 beta and dihydrotestosterone are employed to achieve this goal of begetting sexed offsprings in above technique and also the same technology has been extended in case of being applicable in ratites like kiwis, ostriches etc (Crews *et al.*, 1995). The reason behind this TDSD and estradiol-17beta and dihydrotestosterone dependant sex fixing is cited as the existence of unorganized chromatin in these species. Hence the sex of the offspring can be altered with TDSD and the use of compounds listed above.

In case of higher chordates and especially in mammals, the sex chromatin is much organized. Hence, the methodologies employed in case of sex fixing of lower chordates, can not be applied. But in spite of many obstacles in way of fixing the sex of offsprings in mammals; lot many scientists have worked on this subject. It is a different question that success has by and large eluded them and a predictable, repeatable and significant shift in the sex ratio has not been achieved (Hunter, 1982) and the regulation of primary sex ratio has remained a mirage on the horizon (Short, 1982) but even then the sex fixing in the mammals is considered to be the most sought after and preferred subject of the reproductive physiologists and biologists the world over and this field has fascinated and attracted the attention of the scientists.

Down the history lane

The fertilization is the union of an X chromosome containing oocyte and a sperm containing either an X or a Y chromosome giving rise to the diploid zygote containing an XX or XY chromosomal configuration. This automatically leads to the conclusion that the ovum being one and always with the consistent and compulsory presence of the X chromosome; the burden of deciding the sex of the offspring lies on the sperm. If a sperm with an X chromosome unites with the ovum, then it results in the formation of an XX or female zygote leading to the conception and subsequent birth of a female offspring and if the sperm has a Y chromosome, then it leads to the conception and birth of a male offspring **(Figs 22 & 23)**. This has led the scientists to focus on the sperm to explore possibilities of producing sexed offsprings. Furthermore the scientists have been working to demonstrate the differentiation in morphology, physiology, biochemistry or immunology of the X or Y chromosome bearing spermatozoa **(Fig 24)**.

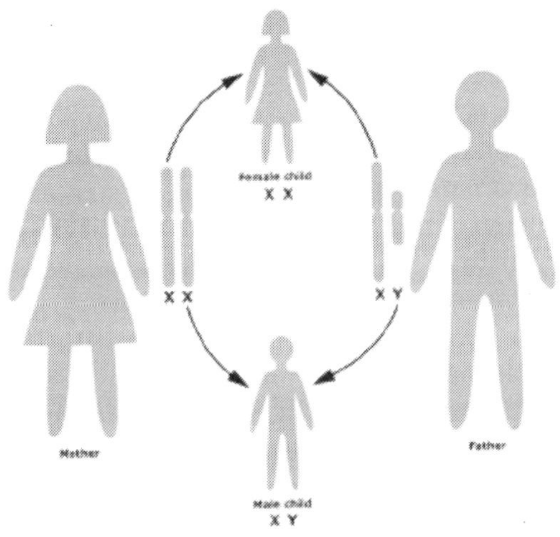

Fig 22. Diagrammatic depiction of sex determination process in humans

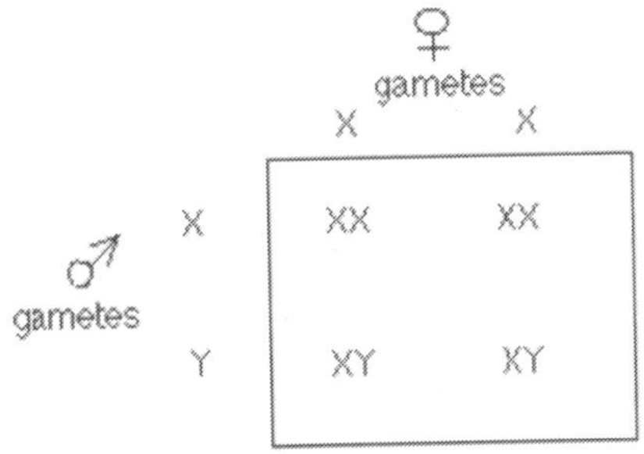

Fig 23. Shows diagrammatic depiction of determination of sex, resulting in equal number of XX (Female) and XY (Male) zygotes

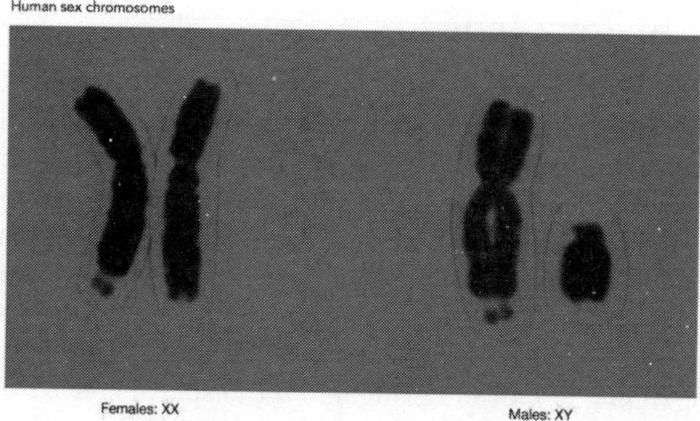

Fig 24. The chromosomal basis of sexuality as XX (female) and XY (male)

It has been shown in certain studies that the Y chromosome is smaller in size and greater in motility to that of the X chromosome. But the content of the DNA or the chromatin matter in a chromosome is a very negligible amount as compared to the total cytological contents and mass of the sperm plus the seminal plasma and environment around the sperm mass. In bull semen, the compact bull sperms mass tends to increase three times its actual volume if placed in a hypo osmotic medium and tends to shrink if placed in a hyper osmotic medium. The presence of an X or Y chromosome can cause a difference in size and shape of the sperm, in weight, density, motility (type and velocity), surface charge, surface biochemistry or internal biochemistry (Foote, 1982). It is however not well established that all the X chromosome bearing spermatozoa are more in size and mass to that of the spermatozoa with Y chromosomes. Even the motility difference is based on the F-body staining technique but is corroborated in some clinical studies. The difference in charge between X and Y sperms has been the foundation of many a separation techniques reported so far but is not corroborated in many studies in the field. Albumin column separation technology as developed by Ericcson and Glass (1982) has been used commercially. Among other techniques used so far, a small description is given as under:

Lawrence in 1941 and McPhee and Eaton in 1942 did review the various factors such as diet, stress, season and -pH etc that could have an effect on the manipulation of the sex ratio but it is a proven fact that no conclusive evidence was found in careful scientific evaluation of these methodologies. Lindahl in 1956 reported on the technique of counter streaming centrifugation technique in case of bull spermatozoa to separate fractions rich in X and Y sperms. Way back in 1939, a Russian scientist Schroeder had claimed success in separating X and Y bearing spermatozoa in case of rabbits with 80% success. The female producing sperm were attracted towards the anode and male producing sperms were attracted towards the cathode. Gordon in 1958 used this technique in case of rabbit and bull semen and has claimed success. Shettles in 1960 discovered two morphological forms of the sperms; the large containing X chromosome and the smaller Y chromosome containing sperm. He proposed that being smaller in size, the Y sperm may move fast. But Diasio and Glass (1971) showed that this is not holding good. Schilling *et al.* in 1971 used the technique of sedimentation of immobilized sperm on certain media like skimmed milk powder, glycine, sodium citrate, glycerol and egg yolk etc and found about 70% of female offsprings with the sperm that had traveled the greatest distance and more male offsprings with the semen that had traveled the least and was still in the top layers. Bhattacharya reported on the same technique in 1966 and found no shift in the sex ratio. Meistrich in 1982 reported on a technique based on velocity sedimentation. Again in 1982, Meistrich reported on centrifugation sedimentation. Bangham in 1961 reported on a technique based on electrophoretic separation of sperms. Later on Nevo *et al.* in 1961 and Hafs and Boyd in 1971 reported on this technique but the technique was not found to be effective in later investigations. Bhattacharya *et al.* (1966) worked once again on convection countercurrent galvanization principle and reported 85 to 90 percent success to separate both X and Y spermatozoa. Sherbet *et al.* in 1972 and Moore and Hibbitt in 1975 reported on sperm separation, based on iso-electric focusing using columns with fluid stabilized in differential density gradients with different –pH values of the fluid in different density gradients. Bennet and Boyce in 1973 worked on a technique based on immuno separation of mouse sperm with the use of an antisera containing Y linked histo-compatibility antigen. This produced 45.4 % males compared with 53.45 for controls with artificial insemination and naturally conceived births. But in all the exercise nearly 70-80% sperm were killed. Pinkel and Gledhill in 1982 reported on a principle based on flow sorting by DNA content. Y-sorting is 72-80% successful but low sorting rate and considerable viability loss of sperms, is the disadvantage. Johnson *et al.* (1993) reported on this technique of flow cytometry and reported enrichment of X fractions up to 82% and Y fractions up to 75%. Steeno *et al.* in 1975 reported 70% success in separating X sperm in certain fractions of the filtrate when sperm were placed over top of Sephadex G50 column and left to swell overnight. Schilling *et al.* used this technique in 1978 and found the methodology of no use. Corson *et al.* in 1984 used this technique and was able to separate 65-85% X sperm. Out off the 17 infants that were born of this exercise; 4 were males and 13 females. Bryant *et al.* in 1984 got a U.S. patent using a technique based on Sephadex column separation using a single male specific antibody or two antibodies using a protein substratum. Later on Zavos and Dawson in 1991 improved on this technology and used essentially two antibodies instead of one and a non protein substratum. The first antibody is a male specific monoclonal antibody and the second antibody is covalently bound to an immuno absorbent protein free substrate such as agarose beads which binds freely to the first antibody. This increases the efficiency of the exercise and has also bagged a U.S. patent. More recently Beernik *et al.* in 1993, reported over Ericsson technique with a total of 1034 results; the sex ratio modified for production of male births up to 71-76%. The same article also advocated a derivative technique enhancing the female birth ratio to a 69%. Even the authors concerned, did not offer any explanation for the shift from a male oriented sex ratio to a female oriented one in this exercise. Barrat and Leger in 1979 reported 8.7% lowering of sex ratio using clomiphene citrate and/or gonadotrophins. Beernik in 1984 reported single live births of 791 males and 777 females using above methodology. Sampson showed in 1983 that multiple births display a marked skewness towards male births by a margin of 7-10%. The use of clomiphene citrate in these cases reverts the sex ratio back to 50:50 normal.

But a coin always has two sides. A study by Vidal *et al.* in 1993 showed the efficiency of Sephadex based techniques to be null and void and found the sex ratio to hover around 50:50 normal natural values. In another study a little earlier; Cellsoft, New York in 1988 (Carson, 1988), failed to notice a bimodal distribution of sperm velocities; a basic assumption of Ericsson methods. This and the later study by Vidal *et al.* raised doubts about the Ericsson technique. Lobel *et al.* in 1993 also confirmed above doubtful outcomes.

Summary

In the above and the foregoing discussion in chapter dealing with fertilization, many things become obvious. The first and foremost is that till now there is not a foolproof, predictable, repeatable and viable technology for producing a significant shift in the sex ratio in mammals (Hunter, 1982). Even as opined by professor Short (1982), the manipulation in mammalian sex ratio, still remains a mirage on the horizon. There may be practical problems to attaining this. One among them is the lack of laboratory tests to evaluate the degree of sperm separation (Hafez, 1982). The things have remained so from a considerable long time. Even in forties of the twentieth century and earlier; many people who worked on this subject, reported success and it seemed that the ultimate goal of mankind to have desired sexed babies in mammals; pets and humans was either achieved or the good news was just around the corner. But as the time passed, nothing fruitful came out. Despite this fact many people claimed success and publicity but still the conclusive evidence was not found in careful scientific investigation of their work and claims (McPhee and Eaton, 1942). Even now, many workers are continuously reporting on the subject and many of them have certainly been successful in manipulating mammalian sex ratio in their pioneering research work and experiments, may be for random and not consistent and longer intervals of time but sex ratio has periodically been displaced from unity (Hunter, 1982). Only time will tell whether these success trials and tales are due to chance or whether control of sex is to become possible by any of these or some other means still to be explored (Salisbury and Van De Mark, 1961).

Mechanism of action

An important point to note about sperm and oocyte fusion is that it is a cell-cell membrane fusion event taking place by two steps; the attachment of two membranes and then he subsequent physical merger of the plasma membrane lipids (Stein *et al.*, 2005). The membrane fusion in animal systems has been found to occur basically in three contexts: 1. Virus-cell fusion (Fig 4) 2. Intracellular vesicle fusion (Fig 5) necessary for protein movement and kinetics and also for exocytosis and 3. Cell-cell fusion. There is scant knowledge still about the cell-cell fusion but the virus-cell and intracellular vesicle fusion are comparatively much better understood (Jahn *et al.*, 2003). So, these two mechanisms which are better understood; can throw a light on understanding the true cell-cell membrane fusion process. Any membrane fusion event undergoes three basic steps namely; the membrane recognition and attachment is the first step and is achieved through a protein-protein mediated or protein-carbohydrate mediated binding of the two membranes. This step involves target specificity that is achieved by the involvement of fixed or particular compounds of protein

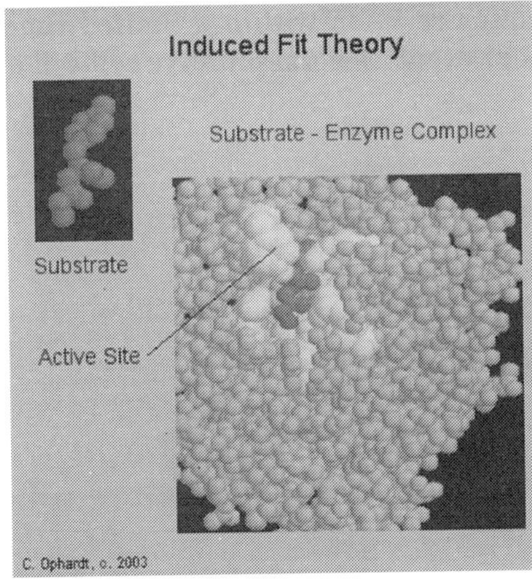

Fig 25. Induced fit theory displaying the working of a lock and key mechanism in bio-chemical reactions and processes

or carbohydrate nature taking part in this process. Second step is of the membranes appositioning. The activity of selective proteins for fusion, brings the two membranes even closer to each other. In many cases this is achieved by fusion proteins spanning the inter membrane space and physically linking the two membranes through protein-lipid or protein-protein interactions. This is just like specialized nuts and bolting in a mechanical engineering operation like a lock and key mechanism **(Fig 25)**. One or a variety of operations brings about an irreversible conformational change in the protein structure and positioning and the fusion protein folds back on itself. This shape change brings about a hinge like motion that draws the two membrane inserted ends of the protein very close together, pulling the membranes with it (Smith and Helenius, 2004). The third step involves the lipid mixing. Once the membranes are very close, lipid mixing occurs between the proximal membrane leaflets and then the distal leaflets, leading to cytoplasmic digesting of the two membranes and then the actual happening of the cytoplasmic continuity between the two cells (Jahn and Grubmuller, 2002). It is a point of debate whether proteins are necessary for this step because liposomes are known to fuse without proteins. It has been noted in some viral fusions that a particular fusion protein actively disrupts the organization· of the lipid in the membrane; thus facilitating the fusion event involving lipids leading to membranes fusion (Tamm *et al.*, 2003).

Attachment proteins are responsible for the initial interaction between two membranes. In virus-cell fusion, carbohydrate moieties or cell adhesion domains on proteins or other molecules in the plasma membrane; mediate virus-receptor interactions (Bomsel and Alfsen, 2003; Dimitrov, 2004). It is thought that in many cases multiple proteins or compounds participate in a sort of a complex reaction within a limited time frame in a single virus-cell attachment event. Several putative cell-cell attachment proteins

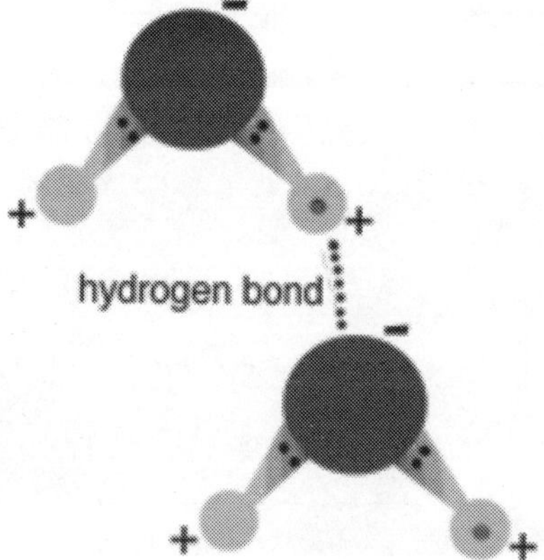

Fig 26. The actual binding and reactions in biochemical processes are preceded with simple machanisms like hydrogen bonds

share these characteristics. A better studied example is the four immunoglobulin *(Ig)* super family members involved in *Drosophila* myoblast fusion namely; *Sns*, *Hbs*, *Duf* and *Rst* (Taylor, 2002) (Fig 13). Each of these proteins has several *Ig* like domains which are well defined cell-cell

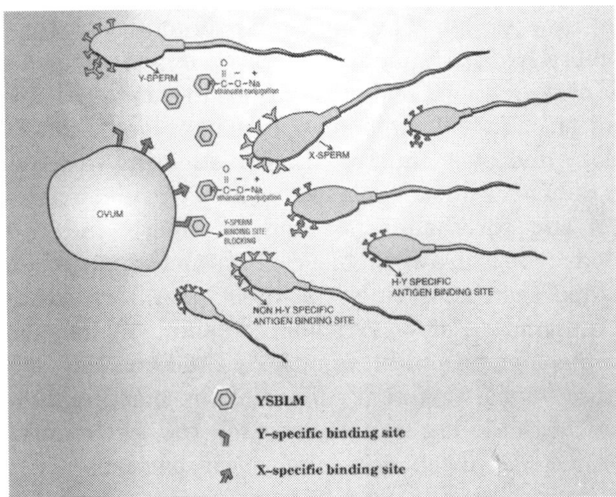

Fig 27. Shows the hypothetical inducement of YSBLM and blocking of Y-specific binding sites on ovum

adhesion domains and it is likely that these proteins function in co-operation in case of attachment process taking place *in vivo*.

Fusion proteins directly facilitate the mixing of two membrane bilayers. In virus-cell fusion, a fusion protein typically contains a single transmembrane domain and a fusion peptide which is a sequence of some 10-30 residues that form an amphiphilic domain at the N-terminus (class I) or within the protein (class II) that is crucial for fusion (Chernomordik and Kozlov, 2003; Jahn *et al.*, 2003) just like hydrogen bonds **(Fig 26)**. It is a different question that putative cell-cell fusion proteins are diverse in structure and only a few match the portrait of a canonical viral fusion protein but the presence of putative fusion domains or in simple words the receptors on the egg surface and the respective binding ligands on the body of the sperm oolema is a well established fact. A simple hypothetical interpretation of the mechanism of selectivity of fusion of X or Y sperm with the ovum is demonstrated in **Fig 27**.

The sex chromosomes are of two types; X and Y, and so are the sperms based on the point whether they contain an X or Y sex chromosome. They are accordingly placed in two categories and called X and Y sperms also. The putative fusion protein domains on the oolema of the ovum or even the protein carbohydrate attachment domains on the zona pellucida may be different in case of X and Y sperms.

The available knowledge about the exact mechanism and methodology of the sperm-egg binding and fusion is far from being complete and it will be no surprise if it happens to be such for a couple of coming decades or even a century or so. The molecular elucidation technologies and spectroscopic advances aimed at working out structural arrangements and atomic organizations and orientations in compounds in the vicinity of organic and biochemical chemistry; have improved a lot. Even in clinical pharmacology and biochemistry, the investigating research has traveled from the old days of bucket science to the molecular science. There were days when the forefathers of endocrinology and applied clinical pharmacology were drawing blood from the veins of the animals in bucketfuls and then analyzing it in their laboratories and reporting the results as the amount of a particular compound or harmone present per liter of the plasma, serum, blood or body fluid etc. Now we have come to an era where only a few micro liters of the blood or other body fluids pipetted out of the living systems can tell the whole details and stories about the molecular levels and composition patterns. We are now living in an era that is so drastically different from the time that was in, only a couple of decades back. And to top to that so much research work is being conducted all over the world to know and elucidate the exact molecular mechanism and pathways involved in the process of exact fusion of a sperm with an ovum. In recent years alone, a lot of valuable knowledge and data has been generated by workers like Kaji (2000), Inouye (2005), Jahn *et al.,* 2003, Primakoff and Myles (2000), Cho (1998) and many more. We certainly know far better and more accurate than what we knew a few years back. So, it can be hoped that in a few years from now or at least a few decades from now, the picture will be much clearer. There is a reason to be optimistic and expecting good outcomes. But the complete workout of the comprehensive details about the fertilization process and mechanism, will take some time. Considering the present levels, it will definitely take few more decades and even a century and it will not be a surprising outcome. We, at the maximum want to have as much knowledge that can enable us to work out the molecular elucidations, additions or deletions etc that when applied in the form of certain treatments, remedies or other applications; can start, interrupt, restart or guide the process and mechanism of the sperm-egg fusion along the expected lines. In short, we ought to have sufficient knowledge of this process that can make us able and equipped to have medication and drug therapies to entirely direct and control the fertilization mechanism along the desired courses of outcome of having to produce sexed offsprings as and how wished. This will be nothing less than developing effective and working drug therapies for begetting, denying or designer ordering a baby with the wanted sex.

And this will definitely take some time.

But a bird in hand is worth two in the bush. The available data and information on the subject being insufficient and lacking; poses problems but even then we can hypothesize to a certain extent and come forward with a working explanation in the light of whatever knowledge and findings, we have on the subject.

A detailed account of the available knowledge about physiology of the sperm-oocyte meeting, binding and fusion processes is given in the chapter related to fertilization and a detailed account of the possible candidate molecules is given thereof. So, trying as much as we can to avoid repetition, we now keep ourselves centered on developments that have actually taken place the world over regarding this subject and try to figure out the exact mechanism that may be involved to fix the sex of mammalian offsprings.

There has been much study in the last years about the nature of the possible candidate molecules that act as mediators in the attachment, binding and fusion processes of the sperm-egg fusion. The most appropriate among them are the ADAM family members like fertilin beta and cyritestin etc. The other molecules are tetraspanins like CD_9 (Hemler, 2003), CD_{81} (Cormier *et al.*, 2004), glycosylphosphatidylinositol anchored proteins like CD_{55} (Coonrad *et al.*, 1999), integrins (Tatone and Carbone, 2006; He *et al.*, 1998), disintegrins (Primakoff and Myles, 2000), fertilins, cyritestin (Cho *et al.*,1998) etc on the process of gamete fusion. These compounds are the actual moieties that mediate or take part in the reactions that do occur during the process of sperm binding and fusing with the ovum. These researches have clearly established the receptor sites on oocyte and their respective ligands on the surface of sperms. In fact, these are the compounds that individually or collectively or by a well orchestrated co-operation with one another, form the putative protein-carbohydrate, protein-lipid or protein-protein attachment or fusion domains or the receptor points on the surface of the zona pellucida or the oolema of the oocyte and on the other hand these very compounds have certain members that are assigned the task of acting as binding ligands on the bodies of the sperms which in turn go and attach with these attachment and binding domains. But the bigger questions are far from being answered. The 'tetraspanin-web' is still a big mystery (Boucheix and Rubinstein, 2001). The role of other candidate molecules the likes of which are mentioned above, needs a lot more investigating. Recently there have been attempts to differentially bind X and Y bearing spermatozoa using H-Y antigen antibody interactions using protein and non protein substrata such as agarose beads and U.S. patents granted for them (Zavos and Dawson, 1991; Bryant *et al.*, 1984). These developments clearly demonstrate the existence of differential binding sites for X and Y sperms and this difference can be exploited in developing a viable technology for sex fixing in mammals, either *in vivo* or *in vitro*. The present technology of the discovery of Aulprofem is an *in vivo* technique and it can be hypothesized that there are differential receptor domains on the body of the ovum and their corresponding ligands on the bodies of the X and Y bearing sperms **(Fig 27)**. A receptor site for X sperm does not receive the binding ligand of Y sperm as it does not correspond to its binding morphology and physiology and vice versa. The ADAM family members like fertilin beta and cyritestin are the possible attachment and binding ligands and tetraspanins like CD_9 and CD_{81} etc are the putative attachment and fusion domains on the body of the ovum along with GPI-anchored proteins which lend a helping and co-operative hand throughout these processes.

The attachment and binding ligands may be of different constitution as regards the X and Y sperms. So are the attachment and fusion receptor domains on the body of the ovum. If the binding ligands or the receptor domains can be occupied or reacted with certain moieties that can engage them before the actual binding and fusion with the receptor domains; then they are no longer free or effective enough to react with these receptor domains. If the binding ligands of one type of sperm for example X are occupied; then the binding ligands of the other sperm type *i.e.* Y being free; will go and bind with their respective receptor domains. This will make the happening of a male/XY zygote thus ruling out chances of a female/XX one. The same picture can be true of the attachment and fusion domains on the body of the ovum. They can also be X or Y specific. And if one of them are occupied or reacted before the actual binding or fusion; then the only chances are left for the other type to be free and able to take part in the actual binding and fusion process.

The material of present research produces certain Y sperm binding ligand mimics (YSBLM) moieties in the living system and also in the genital system of the female animals undergoing this treatment in such a way that these YSBLM moieties differentially bind with receptor sites meant for Y-sperms on the body of ovum thus occupying these sites and inhibiting their subsequent binding and/or fusion with true binding ligands on the bodies of the Y sperms and this way over rule the chances of an XY zygote being formed. This leaves all the chances of an X sperm to come and combine freely with the X sperm specific receptor sites on the ovum and form an XX zygote. So, it can be inferred that the most appropriate candidate molecules that act as receptors on the oocyte like CD9 and binding ligands on the sperm like fertilin beta are also X and Y sperm specific. So we can have tetraspanins like CD9x and CD9y and fertilins like fertilin beta-x and fertilin beta-y. The importance of this technology becomes even more marked with the view that it will not only provide an insight into the differential but at the same time equivalent in competence behaving and working of X and Y bearing sperms in the process of fertilization but will also definitely provide more valuable data and knowledge on the role played by the candidate molecules in understanding fully well the exact mechanism of gamete fusion in further research on this subject.

The immunological reaction

The Aulprofem produces an immunological reaction in the body of the animal ingesting it, is well digesting but given the fact that the drug produces an effect that is sufficient to produce a bigger outcome as the production of sexed progenies; the importance of this reaction becomes even more pronounced and valuable. To study this effect of the Aulprofem, we have to study the nature of this sex fixing activity especially under practical conditions. We have to study the various parameters and effects of this preparation on the body of the animal and the animal behavior plus its actual effect on the production of sexed progenies and the pattern of the action expressed in this entire exercise. We have a certain feedback from the trials in the field conditions and studies. We will discuss some of them as follows:

1. **The production of sexed progenies:** The Aulprofem produces sexed progenies; is not a hidden fact and in fact this is the basis and cause for the writing of this book either. There have been given many clinical studies in the other chapters of this book which elaborate and highlight this action of the Aulprofem and we need not go in further details of it as of now.
2. **The amount of the preparation and expiry criteria:** The preparation needs a certain concentration threshold to achieve the desired effect *i.e.* the production of sexed progenies. The product ceases to produce the desired effect below this concentration and becomes a blank performer. This fact is furthermore compounded and corroborated by the expiry criteria. The product is basically a short expiry material and stops the delivering of wanted sexed offsprings after the term of the expiry period, which is less than an year after the manufacture date; having been consumed. When applied and studied in the field conditions to produce sexed progenies, the outcomes are interesting and specific as regards the effectiveness and action mode of the preparation and discussed as given just a while from now.
3. **The temperature effect and sensitivity:** The product is not stable at higher temperatures. In fact it contains a lot of ethanoate content

which is volatile in nature and tends to fly by converting itself into vapors with the result that the overall concentration of the product is decreased. Therefore it is advised to store the preparation at temperatures below 25 degree Celsius. The clinical co-relation of this condition presents certain interesting outcomes in the practical application conditions and discussed as given just a down below.

4. **The action reversal:** In application involving the expired or the older samples of the product, the sex fixing effect ceases to exist. The same is the outcome with samples stored at higher temperatures or the samples containing lesser amounts of the active materials in the product. A very striking feature of the product has come to light with the stopping altogether of the action of sex fixing activity in samples stored at and above temperatures of 35 degree Celsius and also with the samples stored for more than one year from the date of manufacture. Actually this was the reason that this property of the product came to be known when actually used in the field conditions under higher temperatures of certain districts of Rajasthan and Haryana (see Appendix II, state of India) and also with the samples used with more than one year period after the manufacture. A further striking development was the outcome in the form of production of more male progenies instead of female ones with the Aulprofem preparation which was stored at temperatures at or above 40 degree Celsius and with more than one and half year after manufacture. In some studies the number of male sexed progenies went as above as sixty to sixty five percent. A detailed account of these studies will be given in another write up probably in some other book or research article for reasons including that as of now the data are not bigger enough and suitably compiled but the trend is certainly obvious and we have just enough data to notice this. Probably more studies are needed to reach a more accurate conclusion.

The clinical co-relation and immunological derivations of the above manifestations in the practical and field conditions are very interesting and show certain particular action mechanisms of the usual drug action modules as described below:

The action threshold

There is a minimum required concentration of the drug composition that is required to produce the desired effect. This is just like the pain threshold of the NSAIDs (non-steroidal anti-inflammatory drugs) or the active minimum level of antibiotics, sulphonamides, anti-malarial, anti-carcinogenic, tranquilizers or so on. The action threshold is minimum concentration in body fluids like blood plasma etc that is required to produce the desired therapeutic effects for a particular drug. This is like a crossing window, the height of which determines the cross over action. The height of the effort or the jump above and over the window height; determines the success of the cross over attempt. If the height of the jump is lesser than the cross over window, then there is no possibility of the jump being successful. Similarly the composition of the drug molecules should be equal or more than the minimum required for the production of the required therapeutic action and there is certainly a minimum level of concentration of the drug molecules required to produce that given therapeutic effect. Suppose a minimum 500 mg of the drug capsules are required to produce a particular say 0.0005 mg/ml composition level of a particular drug molecules in the body fluids, then the given 500 mg dose has to be given and there is no escape from this point. Any dose less than this given dose amount will not produce the required concentration level *i.e.* 0.0005 mg/ml in the body fluids and the overall effect will be null and void.

The Aulprofem also has to be given under similar conditionalities. The expired samples decrease the composition concentration and so does the products stored at temperatures higher than the optimum ones due to the evaporation of the active ingredients from the preparation.

The immune response

The application of the drug in the body of the organism induces an immune response to counter and undo the probable or perceived harmful actions of this application on the living system. The application of any material; may be it is food or may be it is drug or may be it is anything, initiates a certain response in the living system of the organism; is a biological truth. There is no refuting this fact. So, the application of the active ingredients that form the preparation Aulprofem set in motion a certain immune response is for certain. The YSBLM (Y sperm binding ligand mimics) are produced in the living system due to this. They are the result of the some primary effect or the immune response of the living system to the application of the material or they are the result of the some secondary response to the effect of this material already produced. After all the natural living system is very well stabilized in a sort of buffering or a variety of chemical, physical, biochemical or any other equilibrium. The –pH, chemicals, biochemicals, the ions and electrolytes equilibria, the fats, proteins compositions and coefficients, the fluid compositions and concentrations etc; all are well maintained in the living systems. A slight modification or alteration of this equilibrium point sends signals in the living system. These signals may be strong or weak depending upon the circumstances. These signals are well received and the action to undo the effects of the alteration or change in the equilibrium point is initiated immediately and without a delay of any type. This action may be small or big depending upon the circumstances involved. This is all about the natural functioning of the living systems. There is not much particular or specific in the present case. The usual immune response produces YSBLM in the living system by any of the antibody production modes as told above involving the immune system.

The YSBLM

The YSBLM are the Y specific antibodies or moieties produced in the living system as a result of the administration of the material of the drug Aulprofem *i.e.* sodium ethanoate and ethanoic acid; to the living organism which is the female animal in this case. They mimic the structural stereochemistry outlooks of the binding ligaments present on the bodies of the Y sperms. These antibodies or moieties are best produced in the living system in a short time period of few minutes to few hours depending upon the circumstances in the living system after the administration of the material to the organism. The state of the physical health of the animal at the time of such administration plus the behavior as well as mental health may also play their roles in this entire process. The presence of these antibodies or moieties in the living system of the organism may remain in effective concentrations for quite a couple of hours after their formation. The YSBLM are of protein, lipid or carbohydrate nature; may be the subject of further exploring on the subject. They may bind to the O linked oligosaccharides, tetraspanins, integrins or any other compounds that form the composition or structure of the putative receptor domains on the ovum for the sperm. Even they may be of mixed nature. The compounds of the drug fall under the domain of predominant organic composition; hence the nature of these YSBLM moieties and their chemical substitutions may also be such. The material under the present study may have certain other companions in the gallery of chemicals in the sense that these moieties may be produced by the administration of some other compounds that are not known to produce them at present. In spite to these moieties, there may not be overruled the chances of certain other moieties that may have the property to block the receptor sites for the X bearing sperms. Nothing particular can be concluded as of now. Only the future research has an answer for this.

The multilayered immune response

The production of YSBLM is the result of the primary or the secondary or subsequent immune response. This is not the primary question here but the fact that it is produced is the subject of discussion. The production of this entity in the living system once actuated, causes a shift from the natural biochemical equilibrium in the living body system existing prior to it. Now the system has to revert back to the original one existing earlier or a little while ago. The system initiates the action for this. The immune response once again is the principal player and the mechanism. The natural processes taking place in such mechanisms are perfected by the nature in such completeness in the long course of evolutionary developments spanning over tens or hundreds of millions of years in the entire phylogenic history of life and living organisms.

So, the system once again swings into action. This time these are the YSBLM as the main target of its correcting exercise. They have to go if the system has to function normally and it is a fact, the system has to revert back to the truly original and old live system and equilibrium.

YSBLM below action threshold

If the system produces YSBLM in concentrations that are not sufficient to block all Y sperm binding and fusion sites or if the application of the material is not sufficient enough to be able to actuate the starting of the production of YSBLM in the living system; then a very interesting situation may emerge. The immune system has to swing into action and the immune response has to initiate and carry on the correcting exercise. The immune response takes into target, the YSBLM in low concentrations and rectifies the effect of their occurrence by annihilating them out of existence. This cleansing exercise may remove also some other occupants of the receptor sites for Y sperms on the body of the ovum, thereby making these receptors more competent and efficient to receive the Y sperms, just a little later. Even they may be more enthusiastic and active. This, as a result will tilt the primary sex ratio comparatively in favor of the male sex. The lower the material, the lesser is the amount of YSBLM formed and there may be a stage when the amount of the material may be as lower as to cause a negligible and inappreciable production of the YSBLM, with the effect of the binding of the receptor sites also being non appreciable. This may lead to a situation when the binding of the receptor sites for the Y sperm with the YSBLM is negligible but sufficient to initiate the immune response to wipe away these YSBLM later on, out of existence and also removing more such occupation, if present, of the receptors for the Y sperm binding sites with similar other binding moieties. The result will again be propping up of the Y sperm receptor sites and molecules. This will end up producing further male progenies. The end result in such circumstances will be the production of more male progenies in the lower dosage regimens of the material. The lower the dose, the more chances are for the production of more male dominated sex ratio. This may also have a maximum limit, of how much a dose can be lowered to produce such an effect.

There may be a situation when both the binding and fusion elements for the Y and X sperms may be different but located on the same receptor site on the body of the ovum. A single receptor site for this purpose may have receiving molecules for X and Y sperms in more (or perhaps equal) numbers. A receptor site in such conditions may be a cluster of X and Y sperm binding molecules.

There may be another interesting situation when the occupancy of one type of receptor points, say Y, may prove a sort of enhancement in activity for the other type of receptor points, say X, just like a see saw mechanism. Both the receptors located in same receptor site may alternate the activity and depression in action modules with each other. If one type of receptors say for receiving X sperms is enhanced in activity, it depresses the activity of receptors for the sperms of other type say Y and *vice versa.* The sudden increase in the living system of the material of the drug may produce an action in the system of the production of the YSBLM and this very reaction resulting in the engagement of the Y sperm blocking sites on the body of the ovum may stimulate certain molecules that increase the reactivity of the X sperm binding or fusion sites and *vice versa.* Under such conditions the immune response generated elimination of the YSBLM, will eventually end up lowering in activity of the X sperm binding receptors thereby increasing the binding and fusion efficiency of the Y spermatozoa. Further, if there are more chances of the Y sperm to perform, then its activity to perform may produce certain stimulating molecules and if this activity is less; then the amount of stimulating molecules will be less. In conditions of Y sperm binding sites being occupied; there are chances that the stimulation for these sites to perform is also less due to non secretion of such stimulating compounds; with the resultant overall lethargy in action due to the non performance of these binding or fusion sites. The reason may be the lesser production of biochemical stimulant molecules. Under same conditions; there may be chances that the binding and fusion sites for the X sperm, may get more excited due to non competition and the production of certain stimulating chemicals from the X sperm receptor binding and fusion sites and compounds.

A bull gets attracted to a cow only when she is physiologically fit to receive him because she at this time is emitting certain odorants that act as stimulants for the bull to react. Similar conditions may be acting on the micro level for pulling the Y and X bearing sperms to the ovum and in the process pulling them to their respective binding sites. If the binding sites for type of sperm are occupied; then there is no question that this binding site is emitting those chemicals to pull that particular sperm towards it with increased activity and energy and in the process, it will no longer emit those activity stimulant molecules. The activity of such a sperm will get a dip and it will no longer be offered an allurement to direct itself to that binding domain. The binding sites for the other type of the sperm will be continuously offering such charm and food for attraction for the sperm and pulling it towards him. These activity enhancers may be of homo or hetero nature. The unoccupied receptor site on the body of the ovum may be a stimulating factor for the sperm and for itself also as well. It may be secreting certain chemicals to stimulate the sperm but at the same time it may be secreting certain chemicals to keep its own stimulation upbeat. Same may be the case with the sperm. How a bull finds a cow, how a sperm finds an ovum; there may run a strange similarity and co-relation in behavior and activity between the two. The stimulation of smell plus the energy of the adrenaline in a bull and the stimulation of the chemo-attractant plus the energy of the mitochondria or power center in the sperm. Similar is the case with the receptor site on the body of the ovum. It attracts the sperm with a similar mechanism that a cow uses to attract a bull. The attraction and bonus of stimulation for binding and fusion are offered and acted, to get the ultimate fusion process done by the receptor sites that are free and not occupied. An occupied site will lose all the operational activity involved in the bribing attraction mechanism and will end up a non performing entity but the case is different with the receptor site that is free. Same may be the case with the binding ligament sites on the bodies of the X or Y bearing sperms with similar mechanisms and principles at work.

The YSBLM in overwhelming concentrations

When YSBLM are in overwhelming concentrations, they block all the available receptor sites for the Y bearing spermatozoa and thereby stop the combination of Y sperms with the ovum, leaving all chances for binding or fusion to the X bearing sperms, leading to the formation of an XX zygote, giving birth to the female offspring. The concentration of the YSBLM, may even be higher than the required for blocking all the available receptor sites for Y sperms. In such circumstances there may be excessive YSBLM in the body fluids even after the occupancy of all existing receptor sites for Y sperms. This excessive concentration may be just excessive or even multifold of the required one; depending upon the amount of the material administered and the state of the normal living body and immune systems of the organism.

The YSBLM in lower concentrations

The presence of the YSBLM will invite an immune response from the system, is for true. So, their concentration in lower levels will create conditions for the insufficient blocking of the receptor sites for binding and fusion of the Y sperm. The immune response will make their elimination faster from the scene and in spite to that it will ensure the concentration in the living system, of the moieties that act in cons to the action of YSBLM moieties. The receptor sites for the Y sperm will become more vitalized, renovated and effective after this exercise with the end result that the primary sex ratio will tend to tilt the other way; that is in favor of the male sex after this entire exercise. The see saw mechanism may tend to explain this shift in sex expression in the offsprings.

References

Barrat and Leger. 1979. *J. Gynecol. Obstet. Biol. Reprod.* (Paris). **8**: 332.

Beernik, F. *et al.*, 1984. Presented at the annual meeting of American Fertility Society.

Beernik, F. *et al.*, 1993. Sex pre selection through albumin separation of sperm. *Fertile. Steril.* **59**: 382-86.

Bhattacharya, B.C. *et al.* (1966). An Attempt to determine the sex ratio of calves by artificial insemination with rpermatozoz separated by redimentation. *Nature.* **211**: 863.

Bomsel, M. and Alfsen, A. 2003. Entry of viruses through epithelial barrier: pathogenic trickery. *Nat. Rev. Mol. Biol.* **4**: 57-68.

Boucheix, C. and Rubinstein, E. 2001. Tetraspanins. *Cell Mol. Life. Sci.* **58**: 1189-1205.

Bull, J.J. 1980. *Qrtly. Rev Biol.* **53**: 3-21.

Bryant, F. *et al.*, 1984. U.S. patent no. 4, 448, 767.

Carson, S.A. 1988. Sex selection: the ultimate in family planning. *Fertil. Steril.* **50**: 16-9.

Chernomordik, L.V. and Kozlov, M.M. 2003. Protein-lipid interplay in fusion and fission of biological membranes. *Annu. Rev. Biochem.* **72**: 175-207.

Cho, C. *et al.*, 1998. Fertilization defects in sperm from mice lacking fertilin beta. *Science* **281**: 1857-1859.

Coonrad, S.A. *et al.*, 1999. Treatment of mouse oocyte with PI-PLC releases 70-kDa (pI5) and 35 to 45-kDa (pI5.5) protein clusters from the egg surface and inhibits sperm oolema binding and fusion. *Dev. Boil.* **207**: 334-49.

Cormier, E.G. *et al.*, 2004. CD9 is an entry co-receptor for hepatitis C virus. *Proc. Natl. Acad. Sci. USA.* **101**: 7270-74.

Corson *et al.*, 1984. *Fertil. Steril.* **42**: 756.

Crews *et al.*, 1993. U.S. patent no. 5201280.

Crews *et al.*, 1995. U.S. patent no. 5377618.

Dimitrov, D.S. 2004. Virus entry: molecular mechanisms and biomedical applications. *Nat. Rev. Microbiol.* **2**: 109-122.

Diasio, R.B., Glass, R.H. 1971. Effects of –pH on the migration of X and Y sperm. *Fertil. Steril* **22**: 303-7.

Ericsson, R.J. 1973. *Nature* **246**: 421.

Ericsson, R.J. and Glass, R.H. 1982. Functional differences between sperm bearing the X or Y chromosome. *In* Prospects for sexing mammalian sperm. RP Amann and GE Seidel. Jr. *Eds.* Colorado University Associated Press.

Gledhill, B.L. 1982. *Fertil. Steril.* **40(5)**: 572-574.

Gordon, M.J. 1958. The control of sex. *Scientific Am.* **199**: 87-94.

Hafez, E.S.E. 1982. Reproduction in farm animals. Reprint 5th edn. Lea and Fabiger, p. 499.

Hafs, H.D. and Boyd, L.J. 1971. Galvaniç separation of X- and Y-Chromosome bearing sperm. In sex Ratio at Birth Prospects for control. C.A. kiddy and H.D. Hafs (eds.) American society for Animal science.

He, Z.Y. *et al.*, 2003. None of the integrins known to be present on the mouse egg for to be ADAM receptors are essential for the spermegg binding and fusion. *Dev. Biol.* **254**: 226-37.

Hemler, M.E. 2003. Tetraspanin proteins mediate cellular penetration, invasion and fusion events and define a novel type of membrane micro domain. *Annu. Rev. Cell Dev. Biol.* **19**: 397-422.

Hunter, R.H.F. 1982. Reproduction of farm animals. Longman. London, pp. 138-39.

Inoue, No *et al.*, 2005. The immunoglobulin super family psotein Izumo is required for the sperm to fuse with eggs. *Nature* **10**: 434

Jahn, R. and Grubmuller, H. 2002. Membrane fusion. *Curr. Opin. Cell Biol.* **14**: 488-95.

Jahn, R. *et al.*, 2003. Membrane fusion. *Cell* **112**: 519-33.

Johnson, L.A. *et al.*, 1993. Gender pre selection in humans? Flow cytometric separation of X and Y sperm for the prevention of X linked diseases. *Human Reprod* **8**: 1733-39.

Kaji, K. *et al.*, 2000. The gametefusion process is defective in eggs of CD9 deficient mice. *Nat. Genet.* **24**: 279-82.

Lawrence, P.S. 1941. The sex ratio, fertility and ancestral longevity. *Quart. Rev. Biol.* **16**: 35-79.

Lindahl, P.E. 1956. Counter streaming centrifugation of bull spermatozoa. *Nature* **178**: 491-92.

Lobel, S.M. *et al.*, 1993. The sex ratio of normal and manipulated human sperm quantiated by the polymerase chain reaction. *Fertil. Steril.* **59**: 387-92.

McPhee, H.C. and Eaton, O.N. 1942. Experimental attempts to modify the sex ratio. *J. Hered.* **33**: 429-33.

Meistrich, M.L. 1982. Prspects of sexing mammalian sperm in R.P. Amann and G.E. Seidel, Jr. (*eds.*) Colorado University Associated Press.

Mitra and Chowdhury, 1990. Abstr. *In* Animal Breed. Abstrs. **58(4)**: 2354.

Moore, H.D.M. and Hibbit K.G. 1975. Isoelectric focussing of boar rpermatozoa. *J. Reprod. Fertil.* **44**: 320-332.

Nevo A.C. *et al.*, 1961. Electrophoretic focussing of boar rpermatozoa. *Exp. Cell Res.* **23**: 69-83.

Primakoff, P. and Myles, D.G. 2000. The ADAM gene family: surface proteins with adhesion and protease activity. *Trends Genet.* **16**: 83-87.

Sampson *et al.*, 1983. *Fertil. Steril.* **40**: 481.

Salisbury G.W. and Van De Mark, N.Z., 1961. Physiology of reproduction and artificial insemination in cattle. WH Freeman & Co. San Francisco, p. 95.

Schroeder, V. 1941. Physico-chemical methods of sex regulation of progeny of mammals. Russian contributions. 1939 Genetics Congr. Am. Documentation Inst. Doc. 1565. Abstr. *In* J. Hered. **32**: 248.

Schilling, E. *et al.*, 1975. Failure to separate human X and Y chromosome bearing spermatozoa by Sephadex gel filtration. *Andrologia* **10**: 215.

Shettles, L.B. 1970. Factors influencing sex ratio. *Int. J. Gynecol. Obstet.* **8**: 643-48.

Sherbet, G.V. *et al.* 1972. Characterization of ionogenic groups and estimation of the net negative electrical charge on the surface of cells using natural pH gradients. *Exp. cell Res.* **70**: 113-123.

Short, R.V. 1982. As quoted in Reproduction of farm animals by RHF Hunter. Longman. London, pp. 138-39.

Smith, A.E. and Helenius, A. 2004. How viruses enter animal cells. *Science* **304**: 237-42.

Stein *et al.*, 2005. Sperm-egg fusion: events at the plasma membrane. *Journal of Cell Science* **117**: 26: 6269.

Steeno, O. *et al.*, 1975. Separation of X and Y bearing human spermatozoa with sephadex gel infiltration method. *Andrologia.* **7**: 95-100.

Tamm, L.K. 2003. Membrane fusion: a structural perspective on the interplay of lipids and proteins. *Curr. Opin. Struct. Biol.* **13**: 453-66.

Tatone, C. and Carbone, M.C. 2006. Possible involvement of integrin mediated signaling in oocyte activation: evidence that a cyclic RGD- containing peptide can stimulate protein kinase C and cortical granule exocytosis in mouse oocytes. *Reprod. Biol. Endocrin.* **4**: 48.

Taylor, M.V. 2002. Muscle differentiation: how two cells become one. *Curr. Biol.* **13**: R 964-R966.

To, W.W.K. and Chan, Y.F. 1994. Gender pre selection. *J. Hong Kong Med. Assoc.* **46**: 4.

Vidal, F. *et al.*, 1993. Sephadex filtration and human albumin gradients do not select spermatozoa by sex chromosome, a fluorescent *in situ* hybridization study. *Human Reprod.* **8**: 1740-43.

Zavos, M. and Dawson, K.A. 1991. Method for X and Y spermatozoa separation. U.S. patent no. 4, 999, 283.

The Formulative Theory of Determination of Sex

The deviations in sex ratios

Deviations in sex ratios are an established phenomenon in humans and animals as well. In U.K. the secondary sex ratio (S.S.R.) is 105.6: 100; males to females (Hunter, 1995). Weiner *et al.* (1968) reported S.S.R. of 106:100, 103:100, 101:100 and 113:100; males to females, in cases of white Americans, American Negroes, Cubans and in Greece, respectively. Van De Walle (1968) also quoted above secondary sex ratios for white Americans and American Negroes. Similar ratios for live births are archived in U.N. publications (statistical office, U.N. 1976) for a number of countries like France, Japan, Sweden, Bulgaria, former U.S.S.R. etc. Secondary sex ratios are usually male oriented as given above but there are examples of female oriented or near unity secondary sex ratios e.g. a secondary sex ratio of 97.8:100, males to females in Congo (Romaniuk, 1968), a near unity secondary sex ratio in Nigeria (Van De Walle, 1968), a female oriented secondary sex ratio in Angola (Statistical Office, U.N. 1976). In animals, Williamson and Payne (1967) reported secondary sex ratio of 51.5:50 and 52.3:50, males to females for cattle and pigs. Johnson (1932), in a study covering 124166 live births in Scandinavian herds born from natural service, reported a secondary sex ratio of 106.2:100, males to females and Gardener (1950) reported a secondary sex ratio of 105.9:100, males to females, in a study of 1783 births in Holstein cattle. There is no satisfactory explanation as to why this phenomenon exists and the various reasons for its existence are unknown.

Presence of sex ratios-More male predominant at primary sex ratio level

The sex of the offspring is confirmed at conception by the random combinations of X or Y chromosome bearing sperms with the ova, containing essentially an X chromosome according to chromosomal theory of determination of sex (Fig. 22 & 23); giving almost 50:50, males to females, sex ratio. Slight fluctuations from this point of equality may be explained as being purely accidental occurrences. But if these fluctuations are one sided *i.e.* on one side of sexual manifestation; either male or female, and lingers for sufficient longer time periods or in large enough population groups, it cannot be simply overlooked as being a matter of sheer chance. And this is what exactly the phenomenon of sex ratio deviations is all about. The phenomenon should be discussed at secondary and primary sex ratio levels differently. The phenomenon at secondary sex ratio level or at birth is self explanatory as it is supported by authentic data in various studies. But as far as primary sex ratios (P.S.R.'s) are concerned; there are no figures available. In order to explain this phenomenon at primary sex ratio level, some other factors have to be taken into account. The most prominent among them is the comparatively greater male mortality in all stages of life. The other side of the coin in this case is the greater survival capacity of the females in all stages of life. These figures are not available abundantly in animal species but in humans, there are dependable and authentic data available in the form of population records by the various governmental demographic records across the globe and the United Nations records of human populations at different and regular time periods throughout the world.

This is what is depicted in the Table 1 as the greater survival capacity of females as compared to males in humans indicating all along more male mortality in all stages of life starting from comparatively more male populations in lower age groups, resulting in the end in sufficient greater number of females as compared to males in senior age groups, especially in age groups of 80-84 and plus 85 in years.

There are more males as compared to females especially in lower age groups e.g. there are 1538000 males for every 1468000 females in less than an year, age group in USA but as we proceed to higher age groups, say plus 85 in years in the same country, then the number of females is 1166000 for every 577000 males. The corresponding data are similar for other countries like France, Bulgaria, UK, Japan etc. A point here to note is that these countries were randomly picked for this study from other countries of the world and the corresponding population statistics are similar for other countries of the world as well. This clearly shows that females have greater survival capacity as compared to males.

For obvious reasons, the mortality rates are inversely related to the survival capacities and this thing becomes very clear if we study the mortality data for populations in these countries selected above and depicted in table numbers 2 and 3. Here again, for other countries not covered in present study as depicted in these tables; the population trends as regards the mortality rates, are similar to the ones given in these tables for the selected countries of USA, France, Bulgaria, UK, Japan, etc. Such results

Table 1. The human populations in five randomly selected countries of the world showing real population figures; male and female in different age groups starting from less than an year to plus 85 years in age

Year of Census	Country	Gender	Age group in years								
			-1	5-9	15-19	20-24	25-29	40-44	70-74	80-84	85+
1974	USA	Males	1538	8964	10508	9196	7983	5545	2411	954	577
		Females	1468	8629	10252	9311	8147	5814	3291	1652	1166
1972	France	Males	445	1972	2147	2259	1913	1713	805	243	134
		Females	423	1823	2035	2134	1759	1661	1160	560	383
1972	Bulgaria	Males	322	337	348	333	321	123	29	17	
		Females		306	324	336	325	317	143	44	26
1975	UK	Males		2003	1825	1718	1871	1405	802	222	115
		Females		1900	1743	1659	1836	1375	1164	515	350
1975	Japan	Males	1054	4487	4108	4803	4982	4030	1119	286	115
		Females	992	4281	3966	4793	5065	4020	1384	472	253

(These data are taken from Demographic Year Book 1975 published by Statistical Office, Department of Economic and Social Affairs, United Nations, New York, 1976, pp. 188-223. The usual adjustments, errors, specifications etc that relate to the accuracy or authenticity of data given in above publication also conform to this study *e.g.* due to rounding the totals are not in all cases the sum of the parts in case of Japan and USA and figures given in the table are excluding diplomatic personnel outside the country in case of Japan and France etc.)

of study into the infant mortality and child mortality in lower age groups, show greater mortality of males in age groups ranging from first days of infancy to age groups 20-24 years in age. For example in Table 2; for USA, 12116 male infants die as compared to 9041 female infants with natural death for age group of less than a day and in age group of 28-364 days for the same country, 7975 male infants die as compared to 6063 female infants, of natural death. The data are similar for other countries depicted in the table and for other countries not covered in present study and not depicted in this table, are also similar ones.

In Table 3 the mortality rates are shown for age groups of less than a year to the age group of 20-24 years in age. For example in USA, 30305 male children die for every 22471 female children with natural death under the age of less than an year and for the age group of 20-24 years, 19500 male young people die for every 6321 female young people with natural causes of death.

Above data prove that males are comparatively more susceptible to mortality in all stages of life as shown in tables above. This comparative more mortality seems to continue even in embryonic states as Johnson (1932) reported a sex ratio of 132 males to 100 females, lost as abortions and still births in cattle. Chapman *et al.* (1938) also reported higher

Table 2. Real figures for infant mortality for less than one year in age in five randomly selected countries in the world

Year of Census	Country	Gender	Age group in days				
			-365	-1	1-6	7-27	28-364
1974	USA	Males	30305	12116	7346	2868	7975
		Females	22471	9041	5232	2135	6063
1972	France	Males	6726	576	2716	1031	2403
		Females	4943	378	1868	848	1849
1974	Bulgaria	Males	2162	241	645	357	919
		Females	1639	156	409	297	777
1973	UK	Males	6599	2161	1637	595	2206
		Females	4808	1571	1069	495	1673
1974	Japan	Males	12654	2524	4312	1768	4050
		Females	9234	1755	2803	1310	3366

(These data are taken from Demographic Year Book 1975 published by Statistical Office, Department of Economic and Social Affairs, United Nations, New York, 1976, pp. 289-302. The usual adjustments, errors, specifications etc that relate to the accuracy or authenticity of data given in above publication also conform to this study e.g. data in France excludes the deaths of infants dying before registration of birth and data for Japan relates to Japan nationals in Japan only)

proportion of males in young fetuses lost, in calves. Hence, it appears that comparatively more male mortality after birth is consistent with the greater embryonic death rates among males (Salisbury and Van De Mark, 1961) and at least in the later stages of pregnancy, male fetuses are more likely to be lost than female ones (Weiner *et al.*, 1968). It suggests that more males than females are conceived (Salisbury and Van De Mark, 1961) and primary sex ratio is in favor of even a greater proportion of males than the secondary one (Weiner *et al.*, 1966). The same generalization keeping in view the mortality sex ratio of 110 to 135 males for 100 females for children dead under less than one month of age and a ratio of 109:100, males to females for children who died in less than one year of age (Romaniuk, 1968) in Congo, indicating greater mortality of males in fetal stages also, suggests that the secondary sex ratio of 97.8:100, males to females (Romaniuk, 1968), also seems to shift comparatively in favor of males at the primary sex ratio level. A simple conclusion can be drawn from here that the deviation in the sex ratio does exist at the primary sex ratio level. May be it is on the side of male predominance or female predominance, is a secondary question but the very existence of it is proved by the above discussion. The same can be true for the sex ratios in other

Table 3. Exact figures for mortality in lower age groups starting from less than an year to 24 years in age in five randomly selected countries of the world

Year of Census	Country	Gender	Age group in years					
			-1	1-4	5-9	10-14	15-19	20-24
1974	USA	Males	30305	5587	4014	5150	16229	19500
		Females	22471	6244	2613	2859	5753	6321
1972	France	Males	6726	1471	897	895	2690	4063
		Females	4943	1050	648	571	1133	1408
1974	Bulgaria	Males	2162	290	185	156	311	457
		Females	1639	231	122	82	146	165
1973	UK	Males	6599	1203	817	673	1510	1836
		Females	4808	896	510	393	681	740
1974	Japan	Males	12654	4097	2023	1257	3666	5278
		Females	9234	3097	1208	782	1453	2792

(These data are taken from Demographic Year Book 1975 published by Statistical Office, Department of Economic and Social Affairs, United Nations, New York, 1976, pp. 322-343. The usual adjustments, errors, specifications etc that relate to the accuracy or authenticity of data given in above publication also conform to this study *e.g.* data in France includes deaths of unknown age and rates based on deaths registered within one year of occurrence and data in Japan based on reports for 12 months period preceding 1961 sample survey excluding slightly more than one percent of total population not covered by the survey. Data are for towns having a total population of 1155006)

countries like Nigeria, Angola, etc. Hence, the existence of sex ratio deviations at primary sex ratio level is confirmed.

Causes for sex ratio deviations

The exact reasons as to why this phenomenon exists, are not known. It seems that the natural selection is operating on the sex ratio, in a rather interesting way to produce an equality of sexes at maturity (Weiner *et al.*, 1968). It might be that nature in a process of balancing equally the biological expense incurred to produce male and female sex, is producing more males as male mortality is comparatively greater in all stages of life as compared to the female mortality. Hence, more males than females can be produced for the same biological expense (Fischer, 1968).

So, it can be inferred that nature is working in a way of its own to produce this phenomenon but even then the exact mechanism of its working is not clear and just how the ratio is disturbed from equality is not known (Weiner *et al.*, 1968).

However, the very fact of inequality of sex ratios, clearly proves that there must be some reasons and factors of any nature other than chromosomal ones that are responsible for the existence of this phenomenon. These factors may be working on or along with the chromosomal factors to cause deviations in sex ratios from equality. These factors may be of any nature namely, physical, chemical, biochemical, physiological, genetic, environmental etc. A brief discussion can be as follows:

1. **Physical factors:** These factors may include the physical and morphological features etc of the sex organs, gametes etc. The different size, mass, etc of gametes. The wetness or dryness, the thickness or depth of tracts of vaginal or genital organs etc, the amounts of liquid discharges of various glands of the male and female sex organs etc.
2. **Chemical factors:** Any special or occasional presence of a particular element or compound in the genital or body fluids of the sexual partners. Some special foods or drinks, taken at such occasions prior to or at the time of sexual activity may become a cause for this. The presence or utilization of various chemicals, fertility enhancers, coital facilitators etc for males or females. Some alkaline or acidic preparations used for above purposes may also have their effects.
3. **Biochemical factors:** Harmones, enzymes, fluids, secretions, blood fluids and their compositions, sex harmone and their concentration levels etc.
4. **Physiological factors:** Gamete motility or transportation activity, in normal or altered way. The state of genital tract activity in females and males. Normal or altered functions of body like respiration, digestion, perspiration, excretion, muscle tones etc in sexual partners.
5. **Genetic factors:** Some germplasm abnormality or function alteration. The expression or depression of certain genes or genetic factors. The genetic transfer during second anaphase during meiosis or deletion of a certain chromatin matter, some mutation etc. There is a gene in *Drosophila pseudo-obscura* which eliminates nearly all Y bearing sperms (Weiner *et al.*, 1968). May be that there are some unidentified such genes in mammals.
6. **Environmental factors:** The volcanic belts or valleys, the aquatic or arid zones, the humid, hot temperate zones or the dry cold Tundra regions, radiations in the atmosphere, sudden changes in the climates etc. may form some of such factors.

The presence or absence of above mentioned factors individually, collectively or in any combination that can have an effect on the mechanism of determination of sex or it may be that none of the above mentioned factors is effective and some other factors of any nature, possibly may play their roles in the determination of the sex of the offspring. So, it can be inferred that deviation of sex ratios from equality is due to the presence of certain factors in nature as discussed above.

Inability of chromosomal theory to explain the sex ratio deviations and need for a new theory to explain this phenomenon:

Chromosomal theory postulates two things. First, the presence of sex chromosomes in equal and opposite numbers and second, the random combinations of these equal and opposite sex chromosome containing sperms with the ovum, always with a particular sex chromosome. This random combination by its natural ramifications, yields the almost equal chances of having offsprings of any genetic configuration *i.e.* either male or female. The number of females or males generated this way as progenies of mammals should be equal always or there may be slight fluctuations if any and those also not in bigger sets of populations. Even if fluctuations occur this way resulting in the disturbance of sex ratios from unity; they should be on both sides *i.e.* if at a given time it is on the male side then on some other occasion, it should be on the female side. Such occasional male dominated sex ratio or on the other time, female dominated ratio, if cumulatively calculated, of prolonged time periods; say of 100 years or so, should yield exactly unity or near unity sex ratios.

But in practice it is not so. Sex ratios are not 100:100 or 50:50. On the contrary, they are always on one side of sexual preponderance; either male or female and this phenomenon has displayed the property of remaining so for sufficient longer time periods. And this is exactly what the phenomenon of sex ratios is.

The derivations on the basis of the presence of chromosomal factors and their random combinations, lay the foundation for expecting the sex ratio to be on a unitary or near unitary value. This is simply because of the fact that opposite sex chromosomes are in equal numbers and they in turn, should yield the opposite sexed offsprings in equal numbers also.

And this is exactly what does not happen.

And this is exactly the essence of the phenomenon of sex ratios.

And this is exactly what is the greatest drawback of the chromosomal theory to lead us to the conclusion that that this should not have happened and now because this has happened; so this turns out to be the failure of the chromosomal theory.

Chromosomal theory has failed because it had assumed that:

1. There is existence of sperms with different X or Y sex chromosomes; in equal numbers, generated by meiosis in germ cells.

2. These X or Y chromosome bearing sperms are released into the female genital tract to reach the ovum essentially with an X chromosome in a way that only one sperm will eventually reach the ovum and fertilize it resulting into an XX or XY zygote.
3. That these X or Y chromosome bearing sperms are produced in great numbers; usually in millions and they travel in almost unidirectional mass clusters or groups towards their destination *i.e.* the ovum. The rush of millions of entities towards a common goal, further creates conditions that the chances of either party *i.e.* X or Y chromosome bearing sperms are equal in making the combination with the ovum.
4. This results in the almost equal chances of formation of an XX or XY zygote.
5. If fluctuations from this equality condition do happen; they are definitely temporary and given a longer time period, they should tend to balance to a unitary sex ratio level.
6. This also concludes from the discussion above that if the so expected unitary sex ratio may turn to one side of sexual manifestation at one time; then on another interval of time, it should tilt to the other side of sexual manifestation.
7. These conditions essentially conclude that in larger enough populations in the process of progeny delivering in mammals, and in longer enough time periods, the ultimate sex ratios should balance at the unitary or near unitary values.

There are some other factors also to be considered. Recent studies have shown that there may be a difference in the size or motility of X or Y bearing sperms; resulting in the greater mobility of Y bearing sperms to combine with the ovum to give male predominant sex ratio as Y bearing sperm has been found to be more mobile and smaller in size than the X bearing sperm.

But these findings don't stand the test of time and facts. As pointed by Hafez (1985), that the findings quoted above cannot be generalized because they seem to be the results of the isolated experiments reported by few of the workers and do not form the outcome of the bulk of experiments in this case and they may be the random aberrations rather than being the general rule.

So, the sum up turns out to be as follows:

The chromosomal theory of determination of sex is unable to explain the phenomenon of sex ratios as results obtained on its basis of almost 50:50; male to female sex ratios do simply not exist and the deviations in sex ratios are not uncommon. This point of helplessness of chromosomal theory of determination of sex to explain this deviation is the greatest drawback of this theory and the very fact that this drawback does exist; it means that that some new explanation has to come into the play to explain this phenomenon.

New explanation

There has to be certain forces or factors present in nature that have a say in the process of determination of sex. This is the most suitable derivation to be derived from the above discussion. These factors may be of any nature namely physical, chemical, biochemical, physiological, genetic, environmental etc nature. If the existence of these added forces or factors is conceded; then the chromosomal theory has to be suitably fortified with these new found factors to make it able to explain the phenomenon of sex ratios. So, the chromosomal theory has to be relevantly modified, altered, adjusted, changed or fortified with the addition of these new forces or factors. The chromosomal theory with these new additions will be a new or modified theory. The outcome of the above discussion may be as follows:

1. Chromosomal theory of determination of sex, as such is unable to explain the phenomenon of sex ratio deviations as chromosomal factors alone cannot explain this phenomenon.
2. In order to explain this phenomenon; certain other factors of physical, chemical, biochemical, physiological, genetic, environmental etc nature should also be taken into consideration.
3. The mechanism of the determination of sex is determined by random chromosomal combinations of X or Y chromosome bearing sperms, with the ovum containing an X chromosome giving rise to either XX or XY zygote in equal chances to produce progenies of male or female sex in almost equal numbers. This forms the basis of expecting the offsprings in almost equal numbers and the usual sex ratios to be on a unitary or near unitary values.
4. If these chromosomal factors mentioned in 3. are working in ideal conditions; there are absolute chances of the sex ratio to be on a unitary or a near unitary value but the moment the ideal conditions are disturbed, then certain other factors come into play to exert their pressure to deviate the sex ratios from unity. These factors may be of physical, chemical, biochemical, physiological, genetic, environmental etc nature but in order for the sex ratios to be deviated from unity, these factors and their effects have to come into play to cause their influence for the sex ratios to be deviated from unity.
5. The sex of the offspring is determined by some sort of a balance between chromosomal and other factors as discussed above. These

factors are the forces of nature and generated and controlled by nature by processes like say natural selection or the survival of the fittest etc.

6. There can be a stage, when the effect of these factors, other than chromosomal ones, can be negligible and the sex of the offsprings is determined by the random combinations of X or Y chromosome bearing sperms with the X bearing ovum but on other times there can also be chances when these factors exert their influence on the process of determination of sex of the offsprings.
7. And this effect of these factors other than the chromosomal ones can range from small to considerable extent.
8. And there can be a stage when these factors can influence the process of determination of sex, even completely to a level of 100 percent altering of sex ratios.
9. This leads to the conclusion that if these factors are studied properly, then certain factors can be identified and isolated which can have maximum effect on the determination of sex and these factors when put to practical use in the evolving a viable methodology or treatment which if applied, can yield the expected and predicted shift in the sex ratio in the desired direction.
10. This viable methodology or treatment can even provide the basis for developing certain drug preparations or pharmaceutical formulations which when administered to mammals (even humans); could provide them with the desired sexed offspring.
11. These factors which can form the basis of developing certain medicinal formulations to willfully deviate the sex ratios as and how desired; can be called formulative factors and this theory which explains elaborately the failure of the chromosomal theory to explain the phenomenon of sex ratios and which further explains in detail the relevance of these factors to have the ability to influence the deviations in sex ratios and which in fact establishes the utility of these factors in addition to or over the chromosomal ones; can be called the formulative theory of determination of sex.

Study of these formulative factors

The various factors; which can influence the process of determination of sex, should be studied in detail. First of all, more data should be made available in areas where there are reports of sex ratios having got deviated to greater extents from unity. This is more likely to be achieved in case of human species. Human data are more accurately collected, analyzed and kept. They are collected in various demographic surveys in different countries of the world. Even the United Nations agencies and organizations keep records of such surveys. In today's date, there are elaborate demographic records of almost all the countries in the world in United Nations statistical demographic books.

Such countries should be selected for further exploration of facts which have different and opposite sex ratios. The male oriented and female oriented sex ratio countries should be earmarked and they should form the basis for the further study into the various conditions of life that exist in such countries and they have likely effects on the different sex ratios in these different population groups or countries. The various life styles and conditions of life in these countries or the population groups can yield the leads in the form of understanding these various factors which can have effects in the form of outcome of differently oriented sex ratios in different countries or population groups. This will help us to differentiate certain target areas with different and opposite sex ratios in further understanding these factors.

Target areas to study these formulative factors

The study of these factors can be made quite easy if some target areas are identified where the sex ratios exist in contrasting nature to each other. If a study is made into these contrasting sex ratios, for example the less than unity (female oriented) or equal to unity sex ratios of African climates like Congo (Romaniuk, 1968), Nigeria (Van De Walle, 1968) and Angola (Statistical office UN, 1976) and more than unity (male oriented) sex ratios in countries like U.K., Greece, white Americans of U.S.A. (Weiner *et al.*, 1968) etc and a thorough analysis of various forces or factors responsible for this and which in fact do exist in nature in above populations and are producing their effect in the form of these contrasting sex ratios, is made, such factors can be identified, studied and their mechanism of action understood.

This condition when achieved will definitely lead to a situation where it will be possible to form certain rules, theories, etc to explain their working and on the basis of this, one day certain formulations (medicinal preparations) or treatments will be devised which when applied will definitely manipulate the sex ratios as and when and how and to any extent desired.

Another example of such target areas is the existence of different sex ratios in white Americans and American Negroes in the same habitat of United States (Weiner *et al.*, 1968; Van De Walle, 1968). Such further target areas should be identified and studied in this way in case of humans and animals also so as to accelerate the research in the field of practical manipulation of sex ratios and predetermination of the sex of the mammalian offsprings.

The practical attempts to manipulate mammalian sex ratios

The idea to manipulate sex ratios in mammals has attracted the scientists and the reproductive physiologists since long. Earlier studies on the subject include the works of Lindahl (1956), Schroeder (1941), Gordon (1958) etc. But to date, no conclusive evidence of repeated success has come from these workers (Salisbury and Van De Mark, 1961). The recent attempts include the works of Ericsson *et al.* (1973), Gledhill (1983), Meistrich (1982), Bhattacharya *et al.* (1966), Corson *et al.* (1984) etc. These experiments were based on the technique to produce sexed semen. But experimental attempts have been hampered by the lack of laboratory tests to evaluate the degree of sperm separation (Hafez, 1987). So, it seems that until now the regulation (manipulation) of primary sex ratio remains a mirage on the horizon (Short, 1982) but still it can be hoped that in the future, it may be possible to specify the sex of the offspring (Hunter, 1982).

Conclusion

The process of determination of sex is controlled by chromosomal factors as long as certain ideal conditions exist and under these conditions the primary sex ratio should remain strictly near or at equality i.e. 50:50, males to females. But as soon as these ideal conditions are disturbed from normal, the sex ratio gets deviated from equality and this situation occurs quite frequently in nature as is evident from the phenomenon of sex ratio deviations in different human and animal populations and under these conditions certain factors other than chromosomal ones come into play and influence the process of determination of sex and this influence can be considerable as is evident from a secondary sex ratio of 113:100, males to females in Greece (Weiner *et al.*, 1968) indicating a further higher primary sex ratio, given the fact that male mortality is comparatively greater in all stages of life (Weiner *et al.*, 1968; Salisbury and Van De Mark, 1961). So, a study should be made to identify, analyze and understand the mechanism of these factors one by one and collectively. This can be made easier if the target areas of contrasting sex ratios are identified. Once, the mechanism of action of these factors and their collective/individual effects are known, it will be possible to devise certain formulations or medicinal treatments etc which when applied, can alter the sex ratios as desired and if this is achieved, this can have so many benefits like a big boost to milk, meat and wool etc production across the world and definitely having a big positive impact on the development of milk and meat industry (Seidel and Amann, 1982). Such developments of formulations to manipulate sex ratios in mammals can also be applied on human species and studied.

References

Bhattacharya *et al.*, 1966. An attempt to determine the sex ratio of calves by insemination with spermatozoa separated by sedimentation. *Nature* **211**: 863.

Chapman, A.B. *et al.*, 1938. Sex ratio of fetal calves. Am. Soc. Animal Prod. Proc. 31: 303-04.

Corson *et al.*, 1984. *Fertil. Steril.***42**: 756.

Ericsson, R.J. *et al.*, 1973. Isolation of fractions rich in human Y sperm. *Nature* **246**: 421.

Fischer, R.A. as quoted by Weiner *et al.*, 1968. In Weiner, J.S. *et al.*, 1968. *Human Biology*. Reprint edition. Oxford university press. London, pp. 117-19.

Gardener, K.E. 1950. The sex ratio in calves resulting from artificial insemination. *J. Dairy Sci.* **33**: 391.

Gledhill, B.L. 1983. The control of mammalian sex ratio by sexing sperm. *Fertil. Steril.* **40(5)**: 572-574.

Gordon, M.J. 1958. The control of sex. *Scientific Am.* **199**: 87-94.

Hafez, E.S.E. 1982. Reproduction in farm animals. Reprint fifth edition. Lea and Fabiger, p. 499.

Hunter, R.H.F. 1982. Reproduction of farm animals. Longman. London, pp. 138-39.

Johnson, I. 1932. The sex ratio and multiple births in cattle. Z. Tierzucht. *Zuchtungsbiol.* **B24**: 183-268.

Lindahl, P.E. 1956. Counter streaming centrifugation of bull spermatozoa. *Nature* **178**: 491-492.

Meistrich, M.L. 1982. Prspects of sexing mammalian sperm in R.P. Amann and G.E. Seidel, Jr. (*eds.*) Colorado University Associated Press.

Romaniuk, A. 1968. Case studies: Congo. *In* the Demography of tropical Africa. Princeton university press. Princeton., New Jersey, pp. 244-87.

Salisbury, G.W. and Van De Mark, N.Z. 1961. Physiology of reproduction and artificial insemination in cattle. WH Freeman & Co. San Francisco, p. 95.

Schroeder, V. 1941. Physico-chemical methods of sex regulation of progeny of mammals. Russian contributions. 1939 Genetics Congr. Am. Documentation Inst. Doc. 1565. Abstr. *In* J. Hered. 32: 248.

Seidel, G.E. and Amann, R.P. 1982. Prospects for sexing mammalian sperm. Colorado University Associated Press.

Short, R.V. 1983. As quoted in Reproduction of farm animals by RHF Hunter. Longman. London, pp. 138-39.

Statistical office. Department of economic and social affairs. United Nations. 1976. Demographic year book 1975. 27th issue. Publishing service, United Nations, pp. 432-61.

Van De Walle, E. 1968. Fertility in Nigeria. *In* the Demography of tropical Africa. Princeton University press. Princeton., New Jersey, pp. 516-19.

Weiner, J.S. *et al.*, 1968. *Human Biology*. Reprint edition. Oxford university press. London, pp. 117-19.

Williamson, G. and Payne, G.A. 1987. An introduction to animal husbandry in tropics. 3rd edn. Indian reprint. English Language Book Society. London, pp. 169-70.

Prior Art

Scientific history is replete with attempts, to manipulate mammalian sex ratios. A longstanding objective of the reproductive physiologists has been to separate X- and Y-bearing spermatozoa so that insemination of ovum with rich in either X- or Y-sperm fractions, would lead to off-springs of known sex. There have been many claims of success, and certainly the sex ratio has periodically been displaced from approximately 50:50 males to females. However no methodology has yet yielded a predictable, repeatable and significant shift, neither in the sex ratio, nor at present are there rational grounds for supposing that this can be achieved (Hunter, 1982).

In lower chordates and ratites of course, there has been proven history of possibility of altering the sex ratio (Bull *et al.*, 1980). There have been instances of changing the sex ratios in ratites and lower chordates with the help of certain chemicals such as estradiol, 17 beta and dihydrotestosterone etc. discussed and disclosed in American patent numbers 5, 20, 280 and 5, 377, 618. There have been instances of TDSD (Temperature Dependant Sexual Determination) in lizards, crocodiles and other related species (Bull *et al.*, 1980). This may be attributed to the lack of organized chromatin in these animals but in mammals which have a much organized chromatin, such instances are not reported.

There have been generally three types of methodologies and strategies applied to achieve the shift in sex ratio in mammals.

A. First were the attempts to produce sexed semen. Such attempts are as Gordon (1958), Lindahl (1956), Bhattacharya *et al.* (1966), Meistrich (1982), Ericsson (1973), Gledhil (1983), Corson *et al.* (1984).

There was a mention in the U.S. patent office in patent number 4999283 that H-Y antigen could be used to bind and hence separate Y bearing spermatozoa. But the technologies aimed at producing sexed semen have their own drawbacks. The main difficulty in above techniques is that experimental attempts have been hampered by the lack of laboratory tests to evaluate the degree of sperm separation (Hafez, 1982). Also, there has been the problem that spermatozoa are haploid cells possessing only half the usual number of chromosomes for the species, and there is no evidence that haploid-as distinct from diploid cells express their genetic constitution in surface characteristics. This being the case, it is difficult to imagine how chemical, electrical or bio-physical separation of X or Y spermatozoa might be obtained (Hunter, 1982). Other problems associated with these techniques were poor conception rate due to decrease in concentration and functional activity of spermatozoa, genetic tempering, complexity of these techniques, expensiveness, difficulty in handling, transport etc. of resultant sexed semen, if any? And embryos so obtained? Etc.

B. Secondly there were attempts to use certain compounds, anti-sera, globulins etc. to achieve a shift in the sex ratio. These attempts included Beernik *et al.* (1984), Barrat and Leger (1979), Sampson *et al.* (1983), Mitra and Chowdhury (1990), etc.

The sex ratio shift by these methods was not much and it continued to fluctuate slightly up or down the usual sex ratio. The resultant efficacy of these methods was poor and could not be replicated in further experimentations and it was never fit to put it to a practical use, on larger and practical scales.

C. Thirdly, there were attempts to alter the physiological and functional environment in the vagina and related parts of sexual tract in female animals. This approach had attracted many takers and right from the days of Schroeder, it has been a subject of attraction to the reproductive physiologists. Changes in the normal sex ratio of animals have been attributed to such conditions as season, diet, stress, –pH of the female tract, ageing or irradiation of gametes, and age of the parents. Although there is much disagreement on the extent of the effect of these factors, many attempts at controlling sex ratio have been based on them. A number of attempts made about 1940, (Schroder, 1941) were based on the theory that vaginal –pH controlled sex. An acid condition was supposed to affect the sperm such that a preponderance of females would result; an alkaline condition would theoretically produce males. Although some publicity gaining claims were made, conclusive evidence was not found in careful scientific investigations (Salisbury and Van de Mark, 1961). Hence, these techniques based on changing –pH in the vaginal tract were destined to rejection and desertion even before the end of the first half of the last century (Lawrence, 1941). A lot of water has flown down the river since then but the fact remains that these techniques could never be revived, proved or followed in careful, logical and systematic scientific experiments and evaluations after that (McPhee and Eaton, 1942). Diasio and Glass studied the effect of –pH on sperm motility by allowing sperm migration through capillaries containing media of differential –pH, but showed motility of Y bearing sperm were not affected by the -pH of the media. There was no scope left for such adventures and only some childish or foolish wild hints and imaginations remained to exist on the scientific horizon regarding production of sexed off-springs in mammals by manipulating –pH value in the vaginal tract of female animals. In 1960, Shettles first distinguished two morphological types of sperm, the large oval headed sperm correlating with the X chromosome, and the smaller rounded headed sperm correlating with the Y chromosome. He proposed that abstinence prior to ovulation with intercourse occurring at ovulation only, alkaline baking soda vaginal douching, female orgasm, deep penile penetration and preferably vaginal penetration from the rear, will assist the smaller faster Y sperm, while no prior abstinence, intercourse two to three days before ovulation, acid white vinegar vaginal douching, no female orgasm, shallow penetration in the missionary position will favor larger X bearing sperm. The problem with his approach was that two many confounding variables were used in his study, and further large scale testing could not produce the same results (To and Chan, 1994).

The prominent methodology involved was to change the –pH levels in the vaginal and related parts of reproductive tracts of female animals. Even, there has been a mention as a vague and wild hint in American patent number 4, 336, 801, that a change in –pH of vagina and related parts of genital tracts of certain mammals can effect the sex of the off-spring but nothing has been corroborated by the concrete practical and experimental data as tested on real animals in clinical and field trials in careful scientific studies. This has found a mention merely as a wild guess. There has been certain mentions of some methodologies regarding certain equipments, douche-nozzles, ovulation monitors etc. in patents, 4336801, 3749089, 4159718 and 3219525 which could be used to administer materials in mammalian vaginas with ease and convenience. But the fact remains that the scientific community, the world over is not a taker of this idea any more. Another fact that there can be at least a thousand or tens of thousands if not a million chemicals that can be employed to produce a change in the –pH of vagina or other parts of female genital tract of mammals. To come to a logical conclusion as to which is the most suitable compound that will surely produce a certain sexed off-spring by changing the –pH in vagina and related parts, will require the employment of at least tens of thousands of scientific experimental studies and that will require the entire life spans of many thousands of scientists, all brilliant in the field of mammalian reproductive physiology to arrive at a logical conclusion that such and such experimentation with a given chemical can yield a verifiable, predictable and repeatable methodology (Hunter, 1982).

Given all said and done, this is the most common knowledge on earth that manipulation of mammalian sex ratio is not a reality and still remains a mirage on the horizon (RV Short, 1982). The entire human history and thinking is definitely bound to have a sea change when there will be a single pin pointing of a particular compound that will produce a definite and predictable change in the sex of the mammalian off-spring, either by vaginal or otherwise administration and the present research is of this category. This is a different criteria and fact that this research in no way employs or intends to employ a route of administration other than the oral one. This is entirely different from the earlier vaginal routes that produced vague and ambiguous results and which now stand no where and could never be replicated in the subsequent scientific studies or at the most they were only imaginative hints, the types of which exist as a running mention in U.S. Patent number 4336801 discussed above. The present research is only a practical and field study based on a set of experiments that were conducted by the people in the Gregor Mendel Institute for Research in Genetics, Ludhiana, India and this is only the result of actual clinical trials and the outcome is reported as such and nothing has been concealed or exaggerated and the reference is drawn to a single compound, sodium ethanoate which was actually used to produce female off-springs in cows, then buffaloes and then as inspired by the success, further experiments were conducted on other mammals as reported in the present research.

References

Barrat and Leger. 1979. *J. Gynecol. Obstet. Biol. Reprod.* (Paris) **8**: 332.

Beernik *et al.*, 1984. Presented at the annual meeting of American Fertility Society.

Bhattacharya *et al.*, 1966. *Nature* **211**: 863.

Bull, J.J. *et al.*, 1980. *Qrtly. Rev Biol.* **53**: 3-21.

Corson *et al.*, 1984. *Fertil. Steril.* **42**: 756.

Diasio, R.B., Glass, R.H. 1971. Effects of –pH on the migration of X and Y sperm. *Fertil. Steril.* **22**: 303-07

Ericsson, R.J. 1973. *Nature* **246**: 421.

Gledhil, B.L.1983. *Fertil. Sterl.* **40**(5): 572-74.

Gordon, M.J. 1958. The control of sex. *Scientific Am.* **199**: 87-94.

Hafez, E.S.E. 1982. Reproducttion in farm animals. Reprint 5th edn. Lea and Fabiger, p. 499.

Hunter, R.H.F. 1982. Reproduction of farm anmals. Longman. London, pp. 138-39.

Lawrence, P.S. 1941. The sex ratio, fertility and ancestral longevity. *Quart.Rev. Biol.* **16**: 35-79.

Lindahl, P.E. 1956. *Nature* **178**: 91-492.

McPhee, H.C. and Eaton, O.N. 1942. Experimental attempts to modify the sex ratio. *J. Herred.* **33**: 429-33.

Meistrich, M.L. 1982. Prspects of sexing mammalian sperm in R.P. Amann and G.E. Seidel, Jr. (eds.) Colorado University Associated Press.

Mitra and Chowdhury. 1990. Abstr. 2nd edn. *Animal Breed. Abstrs.* **58**(4): 2354. etc.

Salisbury, G.W. and Van De Mark, N.Z. 1961. Physiology of reproduction and artificial insemination in cattle. WH Freeman & Co. San Franncisco, p. 95.

Sampson *et al.*, 1983. *Fertil. Steril.* **40**: 481.

Schroeder, V. 1941. Physico chemical methods of sex-regulation of progeny of mammals. Russian contributions. 1939 Genetics Congress. Am Documentation Inst. DOC.1565. Abstr. *In J. Hered.* **32**: 248.

Shettles, L.B. 1970. Factors influencing sex ratio. Int. *J. Gynecol Obstet* **8**: 643-48.

Short, R.V. 1982. Quoted by Hunter RHF in Reproduction of farm animals. Longman. London, pp. 138-39.

To, W.W.K. and Chan, Y.F. 1994. Gender preselection. *J. Hong Kong Med. Assoc.* **46**: 4.

U.S. patent no. **3**, 219, 525.

U.S. patent no. **3**, 749, 089.

U.S. patent no. **4**, 159, 718.

U.S. patent no. **4**, 336, 801.

U.S. patent no. **4**, 999, 283.

U.S. patent no. **5**, 20, 1280.

U.S. patent no. **5**, 377, 618.

PART - D

Reflections and Critique

- The Dawn of a New Era
- Need of the Hour
- Opening New Vistas of Research in Pure and Applied Sciences
- Aulprofem: A Strong Case for a Nobel Prize in Medicine for India
- The Missing Cow: A Key to Success

The Dawn of a New Era

This discovery of Aulprofem has in the real sense **dawned an era of sex fixing in the world.**

The future of Biology in general and Reproductive Physiology in particular is definitely the Sex Fixing.

The effect of this research is bound to be such that not only there will be changes in the structure and functioning of future society but it will have profound effects on the ways, the people think and even their behavior will be guided accordingly.

An existence of future society will even be difficult to be imagined without the presence of sex fixing phenomenon in the lives of the people. A day to day or the normal fabric of life will cease to carry a meaning without sex fixing.

The sex fixing will be there everywhere, knowingly or unknowingly, intentionally or unintentionally or in anyway in the lives, behavior, thinking, instincts or even in the ideologies of the people.

The new world definitely is going to prove to be an era of sex fixing and this phenomenon with its all prevalent presence and its ability to impress human thought and action, will definitely be one of the greatest forces that will constitute the strength of future world society.

These are some of the ways that the discussion on the impact of sex fixing on the future society can be summed up. Let us go a little deeper.

Today we live in an era of science. There is no refuting it. But so has man lived from the day he picked his first stone to throw at some predator or he used the first stick to hunt down a prey. The science has always lived with him right from that day. It has accompanied him in the shape of friends like a sharp bone, a sharpened stone or a stone hammer or spear fixed to one end of a stick, thus starting the race into the never ending world of technology. Then came fire and on one more fine day there came the wheel also. The union had started. The unity of man with science. They lived from now onwards, like good friends.

Totally inseparable.

Science always lived with man in the form of his formidable arsenal of stone weaponry, although crude one but it was certainly dependable. Then there was the stick, the fire, sharpened bones and other forms of this magnificent goddess, the science.

Yes, science was the first goddess on earth!

Man loved it, owned it, liked it and liked to possess it. The spirit of possession came with science. The sense of property has certainly a root in this.

The science had changed man but how much? The question remains and may be the subject of endless discussions but a derivation seems certain that the world of man had definitely changed and a new era had dawned on him and on the earth and the most interesting and important journey of the entire universe had begun.

The journey of the development of man.

The journey of the start of the dominance of this planet earth by this *Homo sapiens* who until now had been a pitiable creature and was always in awe from the most dangerous powers of the world like dagger toothed tiger, mammoths, snakes, pythons, crocodiles, bisons and all such deities of destruction.

Technology developed further. Man discovered clothing, housing and farming and societies got formed. Communes had developed. Then communes started fighting with one another. Tribes came into existence. Then there was the epic age and kings came on the scene. The struggle continued further till the rise of big empires. The science played a dominant part in these journeys of man by developing new applications of it in the progress of humanity and also by discovering newer and sophisticated weaponry arsenals. Every such major step was a new historical milestone and the announcement of man's entry in respective new eras. The story is a long one but one thing is very clear that an each new era dawned on earth with the adaptation by man of the each new big exploit of science.

The life stabilized a little after the advent of big empires. There came the relative peace and prosperity with it. Of course there were curses like subordination of entire races and nationalities and the resultant enslavement of man by man but the age of big empires in Europe, Africa and Asia is still considered as even the golden age by some of the historians the world over. The Greek and Roman empires in Europe and the Gupta Empire in India are examples of it. But one thing that followed the path of man in the age of epics, kings and the big empires was the discovery of written language and scriptures and books. This set the advent of an era of knowledge and learning. And not surprisingly, this gave rise to many interesting outcomes. There came thinking and philosophy and God also came with that as the mandatory by-product. There were city states, republics and monarchies. There were games of power, subordination, victories and enslavements. The tussle went on for thousands of years. Religions rose. The crusades were fought. Man was slaughtered for the sake of his own creation; the God, in countless numbers. The God had by now even subordinated thought and virtually captured every human body, heart, brain and thought. But suddenly the new discovery of God started displaying signs of a destructive power going out of the control and hands of man. Then it really went out of control and on the contrary started to control its generator, the man. Even it demonstrated its killing skills and appetite. It had by now gone berserk on a killing spree. The show had started. Infact, a deadly show. God annihilated man in a big way. But there was no stopping that because another era had dawned on earth. The era of God and its religions. A religion by its very nature, is at war with other religions and hence the *krieg* of religions started with one another and also with man. A lot of blood flowed with that as the necessary outcome.

Then came another development. The renaissance came. There was spinning jenny and printing press. There was steam engine and telescope. The industrial revolution had started and it was again the dawn of a new era. Then came the '*laissez faire*' and Hegel and of course there was over production of commodities and imperialism and it was again the dawn of a new era. The era of rise of imperialism and enslavements of almost entire of Asia, Africa and Latin Americas. Then came the culmination of this all. Almost eighty percent of world populations and land were under enslavement.

It was an era of imperialism. But it was an end in itself and very soon this new system earned disrepute and notoriety.

The crux, the climax had been reached.

There was the blowing off of this bubble of imperialism.

Then

There were the mega economic recessions of 1911 and 1929.

All life halted to a standstill.

There was suffocation of unemployment, resentment and chaos.

Everybody wanted a way out and a ray of hope was soon visible on the horizon. It was the philosophy of blind nationalism. A respectful name was coined for it as 'fascism'.

The fascism appeared to be the immediate salvation and it gripped the entire nations and then engulfed them. Engulfed them into the darkness of dictatorships, agony, torture and subordination. The Italy fell the first prey and then Germany, Spain and Japan went the same way. The philosophy travelled further. It was very soon reincarnated into a new and the worst face of it, the Nazism. The Nazism perfected the art of political hate generation and the worst grouping of communities on partisan, divisive lines. The racial discrimination became a political weapon. The blind nationalism of old fashioned fascism had suddenly turned into a vicious tree of communal hatred with this metamorphosis of it into Nazism and very soon this tree grew. It grew in size and developed branches called Gestapo and SA and SS. It was not long that it had overgrown and overshadowed everything.

Everything and everywhere.

Almost the entire world.

There were Mussolini and Hitler, the new prophets.

The Fuehrer sneered at everything of lower status.

Of these, were the men without blue eyes and red hair.

Those were the fearful days. Fear lurked in the air. The darkness seemed to have grown eyes and every wall had invented mysterious ears. Something devastating and ominous seemed to happen. The destiny of the world appeared to be the destruction. All roads seemed to merge on one point called annihilation.

Annihilation and destruction.

And it was expected. One fateful day it started.

The world war had started. It was named World War 2 and was comparable only to World War 1 and to none other as concerned the devastation and damage to men and materials in the entire history of mankind. In fact it was at least dozens of times more in magnitude in destruction to the World War 1.

The war was the imminent outcome and the entire mankind was bathed with blood.

Blood of humans.

Blood of six million Jews. Blood of sixteen hundred thousand innocent children of all ages. Some big and some smaller ones. Some toddlers and some infants. Some few days older. All were snatched from the arms of their wailing and helpless mothers and thrown to incinerators of methyl-iso-cyanide bathed with super hot steam heated to more than two hundred degree Celsius to bake them like roasted chickens. This was named the cleaning exercise nicknamed 'ethnic cleansing', with the best of intentions that only the commanding core of Nazi generals and ideologues could understand.

All done in good faith, to please the Fuehrer.

After all, a price has to be paid for the development.

The world at this time seemed to echo like the howling parade of wolves of Dracula.

During all this exercise, only one thing smiled.

The dreadful moustaches of the Fuehrer.

They no longer looked funny. Charlie Chaplin (see Appendix I) had earlier made a useless effort to make them look funny but now on the contrary they inspired fear and awe and sent scintillating chills across the spines of people of all categories and seemed to numb their feelings and thoughts with immediate effect. They did other things as well.

Sometimes they seemed to snare. They were fitted with ancillary nostrils also.

The entire world shivered whenever they snared. It appeared as if they had grown eyes on their tips. In fact really they had grown eyes on tops of every hair that constituted them and these eyes came to be known as Gestapo.

But things changed.

The war was over and the outcome was freedom of almost all the slave nations and this was the start of another era. These are the tales from the gory pages of history.

The pace of development of mankind is dependent mainly on three things. First is the development in science. Second is the development in philosophy and third is the application of these new principles in science and philosophy in practical revolutionary emancipation of mankind *i.e.* the political movements. This is the reason that most of the great men fall in these three categories. They are either scientists or philosophers or they are revolutionaries. The three all-time great personalities who had the most profound impact on the lives of the men in all times and societies; were Charles Darwin, a scientist, Sigmund Freud, a philosopher and Karl Marx, a revolutionary ideologue (see Appendix I).

The modern age that starts with the advent of nineteenth century AD till today is characterized by the fastest growth in the field of science and this growth has rightly been transformed with equivalent proportions into development in philosophical advancements and the rise and development of political thoughts and systems. The world society has changed to a greater extent. Today in 2008-9, a news of a bank bankrupted in USA is flashed in minutes throughout the world and so does the one of fall in share markets in New York or Hong Kong or in Tokyo. A religious *pogrom* in Orissa or Karnataka (see Appendix II, states of India) is broadcast within hours to all corners of the world. The world has changed. Science has made tremendous strides in development of mankind.

There was a time when people died of diseases from microbes and one fine day a person called Alexander Fleming (see Appendix I) discovered Penicillin and it dawned a new era in the history of medical sciences. So did the discoveries like automobiles, aero planes, radars, green revolution, radio, television, flush sewerage, mobile telephony, internet etc in their respective times and turns. They all heralded new eras in the development of societies, mankind and above all history of this planet earth. The world changed with them and so did the lives of the people and the societies developed with them as they could not have done without them and the world community as a whole grew with fortified vigor with the prophylactic dose of these discoveries.

But these tales have many narrative angles. Like poly faceted gods. You get a different picture when you view from different sides of the same coin. There is another angle. Let's discuss this also.

Then came the end of twentieth century. The world populations had grown to plus six billion people. There were only one billion people right from the start of the mankind to 1879 A.D. across the globe but things had changed suddenly in the last one hundred years or so.

There were food shortages all over. President Bush (see Appendix I) blamed people of countries like India and China who had suddenly started consuming more. The irony remaining that nearly 300 million people lived in India in semi or total starvation below poverty line even into the march of civilization in twenty first century..

The ill fed people.
The ill fed nation!
There were many.
India, Somalia, Rwanda, Ethiopia, Sudan, Philippines and many more.
There is one more angle to the story.
Another twist.
The discovery of Aulprofem takes place in India.
An error, a correction or a blurr or boon. Who knows?
The world is starving.
The populations are growing.
Lots of points......dash......dash ......can be quoted here. We catch the continuity again.

The arable land is limited. In fact it is diminishing. India in 2000 had 170 million hectare of arable land with 1000 million people. Another India in 2008 has 147 million hectares of land with 1150 million people. And still another India in 2020 will have 100 million hectares of arable land with 1400 million people to feed. The projections even paint a still worsened picture. The agricultural green revolution has failed and the second green revolution is no where in view. A horrible sight!

The second green revolution is missing. In fact it has sent some ominous signals. The genetic manipulation in second green revolution has already proved to be a villain. In fact it has even turned the second green revolution a bigger villain. Worse than the Dracula. And without this genetic manipulation, this revolution is a balloon without air.

In fact it is not even a balloon. And there is no air in it also.

The balloon is deflated.

In reality it has burst already. A complete hollow exists now in and around it.

Who will stitch it?

Many governments ask.

In fact many have asked already. President Abdul Kalam (see Appendix I) had asked many times. But unfortunately no tailor came. They had died already. They had committed suicide with the Bt cotton farmers supplied by gods of genetic engineering.

Some were alive but they could not come. They were terminally ill with diseases such as neuro and renal and cardio failures born from the residues of insecticides, herbicides, pesticides etc sprayed in the first green revolution and then the gene manipulations in the second green revolution had plagued them with genetic disorders and all had developed cancer and LGBS (Laudre Gabriel Body syndrome) and who knows what ?

And then they died. They died all. An end of a tale.

President Kalam quit office in pain and despair, waiting for them.

But they never came. Now there is a rumor that they will never come. Even they will not be born again. Their spirits also were incinerated by the merciful God in Heavens with the thought that they should not infect the other spirits of holy people with the diseases mentioned above; who have come to celebrate their inter life vacations with Him in the holy kingdom.

A sorry derivation!

There will never be a genetically modified green revolution.

This is the end of the story.

And moral?

This is better not told.

The Indian government raised area under wheat cultivation in 2007-2008 and the prices of pulses soared. The oilseeds, the vegetables and the fruits and onions, potatoes and salads.

All are flying.

Prime Minister Manmohan Singh and finance minister P. Chidambaram (see Appendix I) are both clinging to the blanket of soaring prices of Aladdin. But it is not coming down.

A beautiful and well orchestrated effort in vain!

And they say, he is a world class economist.

So, he has accordingly gifted India with a world class inflation rate.

Where will India go?

A wrong question. In fact 'where should the ordinary citizens of India go?' It should have been like this.

The food, the clothing and shelter is needed.

But not to worry. Another era has dawned. Aulprofem is discovered.

There will be plentiful-ness of everything.

"The aulprofem comes.

Lots of milk and lots of wool and lots of meat.

Food and clothing, all in one.

Fixing the sex. Playing God.

Calling gods and not the tailors.

Tailors of illusion, myths all. Prophets of blasphemy and peril.

No more Surinder Kohlis and Maninder Pandhers".

A nice poem. Kohli and Pandher (see Appendix I) spoil the taste. But even then, let's continue.

"No more pollution, no more insecticides.

No more cancers, no more kidney and liver and neuro and optic and cardio fiascos.

No more deaths from these terminal diseases.

A new era dawns !"

"Milk for all. Health for all.

Happiness and prosperity every where.

Cheerful children, carefree elders. Contented aged.

A *bahist* of happiness. Paradise regained".

"The advent of Aulprofem.

The dawn of a new era !

No more explaining. Everything is clear.

Full of light. Full of joy.

A new era dawns!"

Could'nt have happened better than this!!

Need of the Hour

What does a blind man need?

'He needs the spects', seems the immediate answer.

But no, he needs the eyes. The real sound eyes with eyesight in them.

This is the answer that should have been.

Who is bigger, Freud or Marx?

The eternal question for intellectuals.

Another way to put this question, 'which is dearer, love or bread?'

A man makes love when he is not hungry.

Or

A hungry man seldom makes love.

There may be thousand ways to put this. One of them being as follows:

A starving man loves nothing but the bread.

A proverb in Punjabi runs like this:

All thoughts are hollow when belly is empty.

A man is man because he is civilized and he is civilized because he has founded civilization based on the three formidable pillars of food, clothing and shelter.

Given that three is a triangle.

And no triangular structure can be still if one of the pillars is shaky.

The pillar of food is withering and shaky.

So, civilization is in danger.

The danger of annihilation.

Because man is hungry.

Hungry and starving.

The greatest reality of today's world is the hovering monster of hunger, looming large over the heads of billions of poor masses in the third world in Asia, Africa and Latin America. Food security is a serious question even in rich and developed nations these days.

Hunger.

A dreaded word.

Does Aulprofem offer an answer?

A tiniest discussion can find an answer to this.

The healthiest food of mankind is still the milk.

And Aulprofem can really produce unlimited quantities of it.

Sufficient to feed the entire mankind.

Feed and overfeed.

It has the quality to do it.

Quality and credentials.

And overall, the promise to achieve this formidable feat.

So, Aulprofem can save the civilization from dying.

Dying from starvation.

A discussion on bigger questions can be started like this but if the discussion is really big; it should take some other turns and twists as well. Let us see whether this discussion on the better applications of Aulprofem can have some pleasant, surprising and plentiful twists.

Another angle.

We can go discussing like this.

What about pollution and pals?

Pals like chemicals, poisons and poison residues.

Thinks, we are missing their talk. Talk and company.

Let's join them.

Sex fixing is the need of the hour.

Starting question.

Green revolution has failed.

Another point.

Instead it has laden the precious soil on mother earth with poisons of insecticides, pesticides and chemical fertilizers.

Now, it seems the discussion started rolling. Let's continue.

The extent of toxicity has risen to such alarming levels that the entire food grains we consume, the each grain of pulses that forms a part of our diet and every constituent of vegetables, fruits etc that we ingest, contains the deadly chemicals and residues, to the horrifying levels.

Good discussion. Bravo!

Keep speed.

The milk we drink contains urea, chemical sulphates and related chemicals. Even the groundwater has become a disposal destination for harmful poisons. The alarming levels of pesticidal and insecticidal poisoning witnessed recently in the Pepsi and Coke waters in India (see Appendix III), is mainly the gift of green revolution. The food we eat, the water we drink and even the air we breathe in; is infested with chemical toxicities.

Who is to be blamed for this?

Some aliens?

Some supernatural entity?

Some external forces? Some imperialism? Somebody enemy of progress or the progressives themselves?

Who will answer this?

I think nobody has the answer.

There are 33 agriculture universities and almost double the number, agricultural research institutes in India. All are involved in the precious and prestigious job of propagating further and further the most pious agenda of green revolution and in this way spreading the reign of toxicity and poisoning to the every possible tract and trace of life that is still left untouched from the angelical touch of chemical death that green revolution has come to stand for today, in the true sense of the word.

Dr. Norman Borlaug
(Glimpses of greatness from annals of history)

Dr. Borlaug ((see Appendix I)), could have never imagined in the farthest of his thoughts that the hybrid dwarf varieties that he and his clout were advocating and propagating could become a bane of humanity. Instead, he was a man of best traits of humanitarian thought and practice and he stood for liberating humanity from the pangs of hunger and starvation. And he succeeded in that. That is why the world still acknowledges him as the father of green revolutions across the globe. He was awarded a Nobel Prize and he deserved it. But the years passed by. And one fine day, it became known in the world that exactly the opposite of what was earlier expected; happened. It turned out that the dwarf hybrids that had become the flag bearers of green revolutions were demanding more and more of urea and other chemical fertilizers. They were genetically not the nature's 'natural choice'. They were 'man made' and to top of that they were developed for an immediate need; a need to rid the starving humanity of the shortage in food availability. They were designed and developed for an immediate and short term goal of increasing the nitrogen based protein yields and equipped accordingly. Or in the true sense, equipped poorly enough. So they lacked the prolonged evolutionary development and capability of living species that is generally the rule in this universe and that takes millions of years for a single species to develop to a perfect adaptation to the environment and practical realities of existence. The natural selection and rejection procedures and postulates as told in the Charles Darwin's famous 'Natural Selection Theory' (see Appendix III) could not be applied in their case in the true sense of the word. So, they lacked the natural strength to adapt and to survive in the practical conditions of the living systems in this real world.

The result?

It was exactly as destined and nothing as expected. These species could not stand nature's natural forces of endurance, adaptation and survival. They were a total misfit as regards the postulates of the great theory of Darwin. The truth was that they needed some additional protection from the ruthless onslaught of the forces of nature other than the one existing in a natural system. Such a protection simply can not be provided in a natural system particularly for prolonged periods of time. And these species flopped. They were always in need of protection and aid. In fact, crutches. Their position was very similar to the economy of developing country like India that is always in need of protection from the forces of liberal economic competition. And that never can come out of the status of a 'help needy state's economy'. So, these species were provided help and protection. This protection in this case was the excessive provision of nitrogen and it came in the form of urea. They needed more minerals and radicals and it came in the form of zinc, magnesium, calcium and their phosphates, sulphates, murites etc. These were the chemical fertilizers. They could not fight the natural weeds and herbs and they were fortified with weedicides and herbicides. They could not cope with the challenges of natural insects and pests and they were armed with the insecticides and pesticides. And the outcome of the exercise was that the chemicals and poisons became the shield of these varieties. This need increased with the passage of time and in the end they turned out to be totally dependant on these shields. Crops that were loudly being touted as the victors for humanity in the battle against hunger, were miserable losers themselves in the fight for survival in the natural environments. The more and more use of chemical fertilizers and insecticides etc and their resultant accumulation as chemical and poison residues in the living systems, soils and environment; turned the whole practice of green revolution into poison revolution. This was nothing of the sort that those of most pious souls like Borlaug and company could have imagined.

The outcome of all this was horrible.

Poisons and their residues in the living systems became culture mediums and the breeding grounds for countless diseases that seemed to have been waiting for an opportunity to demonstrate their skills to traumatize and havoc the humanity. They in turn put up a splendid show.

Everybody was spellbound. Spellbound to see the outcome.

Outcome and the result.

The result were the diseases, humanity had never witnessed or even imagined. The AIDS, the hepatitis, the dengue, the bird flu, the mad cow, the LGBS and who knows what, what.....? The failure of kidneys has become an epidemic in India and the world. The occurrence of diabetes, the heart diseases, eye disorders, blood pressures, epilepsy, leucoderma, asthmas of innumerable varieties, migraines and new diseases of harmonal disturbances especially in the neo borns and children. Children grow very fast into adolescents and youths age very fast to death. All these things are witnessed today. Few months back there was a story on Zee news (see Appendix III) in October or November, 2006 from Gujarat that a four year old had matured into a youthful person. And who knows that in the coming four years he will age into an old man in the twilight of his life? Has the life calendar been changed to the worst meaning in his living system or what? Such stories will be too a many in the very near future to come.

Another green revolution?

Do we need it?

And the slogan is that India needs it, the world needs it and the every developing country needs it. Prime minister needs it. The policy makers in the Krishi Bhavan (see Appendix III) need it. They need it in such futile gossip houses in almost all the nations. The technocrats need it in ICAR (see Appendix III) and the needy nation needs it. The populations are growing. We in India were 1000 millions in 2000 and we will be more than 1450 millions in 2020.

We need food.

Really, a lot of food.

A painful derivation?

Alas! There could have been another alternative.

And the agricultural land is shrinking. From170 million hectares of arable land in 2000, it will shrink to nearly 100 million hectares in 2020.

A drastic change, it seems.

But not the one on a positive side.

This is the clear indication that things are going to deteriorate in the worst direction in the times to come.

The world should be saved from a situation like this.

And, the savior?

The answer is very clear.

There should not be a guessing about this. All hopes get pinned on another green revolution. The like of which, the world has already seen, or a different one this time? There have come so many horrors of the first green revolution to light that nobody wants the repetition of the like of the first one. So, this time again it will be a green revolution but not the old one. It will be new. It will be the second green revolution.

Just like Maoists. They add the word neo with everything like neo-imperialism, neo-colonism, neo-democracy and all neos like this.

We have added the word second instead of neo and the outcome is the word 'second green revolution' which undoubtedly is a new word. A nice christening. Isn't it?

A name change, so no questioning.

The savior has arrived and he is with a different name.

Who has the clue that he is not new?

The no nonsense answer and an end to speculation of all types.

We have reached at the universal answer to all our problems. Ask the PMO. Ask the Krishi Bhavan (see Appendix III). Ask the technocrats. But do not ask the nation and the people. They are not the persons from whom the answers are sought but on the contrary they are showered with the answers.

They have no option except that.

They are not the policy planners. At least, not in the near future. Even the lecture writers in President's secretariat are told this story of the birth of the 'Savior'. And they start writing the same things in turn in the official lectures of the honorable president and he goes on rattling them out, word by word in the press and in front of the electronic media. The missile man fires green missiles (Monsanto Imagine, Appendix III). It is a different story that these missiles have turned brown and pale due to high proportion of genetic fuel laden in them.The nation listens and everybody agrees to the holy gospel. After all, wise men are to be believed, rather followed.

So, the choice is undisputable. It is green revolution. And this time it is with a change, at least in the christening. This time it is green revolution number two. Another name for it is the G.M. green revolution. By G.M it means not the General Motors but the genetically modified. The stress this time is on the genetic modification of crops. Has any body forgotten the experience of Bt cotton in Andhra Pradesh and Maharashtra? Thousands of farmers had to opt for suicides due to this praiseworthy misadventure of G.M. crop developed by Monsanto (see Appendix III). Few thousand human lives. This does not seem to be a bigger sacrifice at least in a country like India where able competitors like Narendra Modi and Bal Thackeray (see Appendix I) have already made big names and displayed shining examples of mass genocides in Bombay and Gujarat. The first encounter has been sour but still we should carry on. After all

Charles Darvin
(Glimpses of greatness from annals of history)

for what circumstances is the proverb christened 'try try again'. So, we should have more G.M. crops. This time it will be done on broader dimensions in a vide variety of crops, even the eatable ones. Yes, the new thrust will be on crops that fall in the category of food grains and eatables, not only by humans but animals as well.

The problem?
There will be many.

There is no logical conclusion to believe that these new genetically modified crops will not face the opposition from the forces of nature that the dwarf hybrids of first green revolution faced. And it is a universal truth that it is the nature that triumphs in the end and no one else. The point where crops of first green revolution failed will have to be encountered by the crops generated by second G.M. green revolution. And they will also need crutches. They will demand shields from the onslaught of forces of nature. They will have to face natural selection. The power of this indomitable principle discovered by the greatest of greats' scientist Charges Darwin has remained and will remain for ever as long as there is life on earth and the nature is the master of all forces in the universe.

So, the shields?

Again there will be urea fertilizers. There will be chemical fertilizers. There will be insecticides and there will be pesticides, weedicides and the lot full family. The soil will be rained with poisons and it will be toxicity all around.

And under such conditions, where will the human race go?

Nowhere, is the answer.

The most dreadful conclusion.

Genetically modified foods and fodders will not become a basis for the generation of genetic modifications in humans and animals that feed on such crops; there is no logic to believe it or even the otherwise. There may be a possibility that this will not happen. But the more chances will be there that genetic modifications in the first phase *i.e.* in the foods of humans or animals will have a tendency to induce genetic mutations in the humans and animals that feed on such crops. A recent study in Russia (see Appendix III, Global Research) has revealed that consequences of G.M. foods on pregnancies and on the off-springs in mice have not been favorable. So, a day is not far off when the foods generated by the so well touted but truly controversial second green revolution, will herald an era of health hazards for humanity but also it has the potential even to put the entire foundations of humanity on threat. Who knows that one day the kinds of genetically mutated Maninder Pandhers or Surinder Kohlis will abound the earth when humanity starts feeding on genetically modified foods. A person who relishes on human flesh may be a genetic mutant or a product of modern civilization? The answer is not a vague one when the entire modern civilization feeds on poisons generated by green revolutions. And how many more Maninder Pandhers and Surinder Kohlis (see Appendix I) are there in the world; there are no statistics available. So, the solution to future food security planning and management, should in no way delve on the foundations of genetic modification, poisoning and toxification.

So, the solution?

Where does it lie?

Nature provides a solution, as it always has.

A new concept of sex fixing has found its germination and is likely to acquire the shape of a big movement and take the entire world in its stride. South African patent office at Pretoria has very recently cleared a drug patent of the world's first mammalian sex fixer drug, Aulprofem. Australian patent office has also accepted it and very soon the patent offices in U.S. and Canada are going to follow this.

Mammals are an important family and any success in altering the sex ratio in them is bound to have profound social and economic value. This drug is a liquid and it produces female offsprings in mammals with a near 100% success. For the first time there has appeared a sex fixer that followed a route, other than the genital one. Probably this is the secret of its tremendous success. Many scientists in the past had tried to manipulate sex ratios in mammals but they could not succeed.

The success of Aulprofem gains prominence in the background of world already on the verge of a major food crisis. This drug, is basically designed to produce female offsprings in cattle and buffaloes; the two major milk yielders in the world. On an estimate if even a meager 10% of the total 300 million cattle and buffaloes of India, are taken for the application of Aulprofem at a cost of rupees 295 per dose, it will boost milk production in this country as follows.

India has an animal wealth of nearly 300 million dairy animals; cows and buffaloes (nearly 210 million cows and 90 million buffaloes). If a meager 1% of them are taken for the application of Aulprofem every year at a cost of Rs. 295 per dose, totaling to 88.5 crores of rupees. This will give rise to 9.0 lakh more female calves per year, keeping in view the nearly 35-40% national conception failure rate. Provided this phenomenon is continued for a minimum of 10 years, and keeping in mind 20% calf morality rate and 30% more general mortality; this will mean 50.40 lakh more female animals annually at the end of such an exercise. So, keeping in mind a conservative 8 liters of milk per day per animal on an average, this means 403.20 lakh liters of more milk per day. Calculated at a rate of rupees 20 per liter, this amounts to rupees 8064 lakhs (or 80.64 crores daily) or 29433.6 crore rupees annually, more in the GDP of India. This is an increase of 33258 %. By no means, this is a smaller achievement! If this application is increased to say 10 percent, then the corresponding increases in GDP of India can be understood accordingly. On a simplistic estimate this means that we will be getting back Rs. 2, 94,336 crores back from Aulprofem in the form of increase in annual GDP of India by spending only a small amount of Rs. 885 crores per annum.

This is not short of revolutionizing the entire agricultural economy of India!!

The conclusion is a mammoth success story. The percentage of application of Aulprofem can be increased form 10 to 20 or 30 percent.

And the relative increase in milk production can easily be double or triple times.

There is another angle to it.

Suppose a stage arrives when the availability of milk production per person in India becomes more than 500 grams or more. With increased nutrition due to this a person has to skip at least one meal a day. This means roughly, one third of food grains saved for three times a day eating Indians. If this happens, we will be saving more than 70-80 million tonnes of food grains per annum. Apart from this we will be saving nearly 10 million tonnes of pulses, 100 million tonnes of vegetables, more than 05 million liters of cooking oil, a bulk of spices, condiments and other related materials. There will also be saving of cooking gas, coke, wood and electricity to the corresponding amounts. The end result will be a giant saving bonanza for the nation. The same story for other nations can be expected as well.

These days, there exists another sad story related to the health of the nation. That is the tremendous use of metabolic harmone oxytocin. This uterine contractor drug was designed to help in delivering babies, a use that has come under a veil of controversies these days due to the many side effects of the drug. But the dairy owners in India, has found a novel use of this material in milking the cows. They are very economy minded people. They calculate that the precious milk consumed by the calf, can be saved if the calf dies and the cow is milked with the help of an oxytocin injection that costs only a few paisa. They have some affection for the female calf which one day will mature into a full grown cow but they are total love lost for the male calf which according to them is a futile biological exercise. They kill it and they specialize in it. They have developed novel methodologies for this. A sum up brief detail is as follows.

Nearly 10 to 15 million male calves are done to death in India every year because they are born in this world as males instead of being born as females and male calves are no longer needed in this era of farm mechanization. They are subjected to unbearable, torturous deaths, worse than the crucification, in the following ways:

1. Luckiest amongst the killed calves are those who are suffocated to death just after birth to fetch more money for their nascent and un-infected skin to be used as comparatively costlier leather.
2. In the hot scorching winds of May-June, when the whole world rests in the cool environs of air-conditioners, coolers and fans, those new-born calves are pegged outsides in the burning sun, so that they are sun-stroked to death.

3. In the chilling cold of December-January, these calves are pegged out in the open throughout the long snowy nights so that they are frozen to death before dawn.
4. Rest of the calves are subjected to the most heinous and un-tolerable death ever reported in history – the total starvation to death. Most of the calves are destined this tortuous end which sometimes prolongs to several weeks.
5. The torture of the male calves for veal industry is well known and there is a lot of hue and cry about this the world over and the Indian dairy also has discovered the benefits of this beastly practice. This is another story that this entire practice is a big blot on the face of the entire mankind.

A society or a nation where such a cruel holocaust goes unabated, no longer deserves to be called a civilized society or a nation. The humans living in such a society, definitely lose their precious right to be called humans.

So, it is one of the prime objects for us all, involved in this magnificent project of Aulprofem to wage a relentless war against this un-natural, man's greed generated holocaust. A point here to note is that the metabolic harmone of oxytocin facilitates this holocaust as it becomes possible to milk the cow by injecting this biological poison to the animal at the time of milking. This hormone's metabolites even pass through the circulatory system of the animal to its milk and same milk is drunk by the humans, including children and adults, thereby poisoning their body systems to varying levels. Being a metabolic harmone in nature, it enhances the metabolism or in other words the pace of metabolic pathways and even the cycle of life gets accelerated. A child matures into a youth fast and the youth passes into the old age even faster and the ultimate march to death of man is hastened. The kidneys are the first victims of such a biological poisoning as they are the organs trusted with the prime task of throwing unwanted material in blood and body fluids, out. They are overworked and exhausted. The other organs which are the victims are the liver, the cardiovascular and the optic systems. This oxytocin injection was banned by the Indian government in the year 2002 but due to the all prevalent corruption in the drug control department and the drug inspectors occupied in the sole objective of minting money by corrupt means, this nefarious injection is available in almost all drug and grocery stores throughout India. Hence, we all should strive to put this evil practice to end. This no doubt, will be a most pious cause if achieved.

The drug lobbies and the governments working to the tunes of them, feel the heat of the time and studies after studies in the physiology and biochemistry laboratories and departments in universities and research institutes, churn out data and findings that this harmone is not at all harmful either for the animal or for the humans that feed on the milk delivered by it. They forget that the ingredient that act as a uterine movement enhancer and can numb and still the animal for a minimum of ten to fifteen minutes period, is definitely bound to cause some affect or reaction in the bodies of the animal in which it is injected and definitely it will also be producing many bio products that will definitely live in the system even after its first effect in the body of the animal has ended and definitely also some of them may find a route through the milk into the bodies of humans feeding on the milk delivered by this animal. But who bothers?

Aulprofem can eradicate this evil by producing those unlucky male calves as the female ones and thereby saving their lives. The use of oxytocin will automatically be stopped as nobody will need the harmone to milk the cow. And the calf will do the natural job of letting down of milk. The female calf will flourish and will give the dairy owner a repayment of his money spent on the feeding of the calf in the form of a full fledged cow in due course of time. This in turn will boost the milk productions in the country and across the globe, to many times the present value, thereby increasing the diet and nutritional potential of the mankind and putting the menace of hunger and starvation to a permanent end.

So, the need of the hour is that the sex fixing has to be adopted on a massive scale to feed the hundreds of crores of growing populations and to provide the most nutritious food to them, the milk, the nectar of life and the guarantor and savior of mankind; today, tomorrow and in all times to come.

Exploring New Vistas of Research in Pure and Applied Sciences

A thirsty crow and pebbles story.

The moral?

Necessity is the mother of invention.

But necessity is unlimited. So, are the desires?

Lord Buddha preached to keep the desires in check.

A wise derivation.

Rightly so, he is considered one of the great men ever born.

But necessities are unlimited. So are the desires.

Man desired to fly.

And he did.

He did a lot of things.

He wanted to free the womenfolk.

So, he discovered contraceptives.

He or she; may be a question to debate but some *Homo sapiens* did it.

Osho: (see Appendix I) told once, "The best freedom for the fairer sex was won with the discovery of the oral estrogen pill".

There are other examples as well.

Ships, automobiles, air conditioners, gramophones, mobile phones and internet.

All desires converted to practical achievements. The list is endless.

Man is a formidable achiever. It can be rightly concluded. But there are certain limits to it. Death and disease haunt him still, in the face. There are many faces to it.

Cancer, epilepsy, AIDS, hepatitis LGBS, asthma, diabetes and virus families.

He needs to face them. Necessity has arisen.

Invention will follow, or not.

Let's think and discuss.

There are only two ways a research can be done. The theory anticipates the possibility of a new field of exploration or the spontaneous discovery is made or more accurately detected by a person whose intended concentration was focused on something else.

The entire glorious history of scientific developments is classified thus.

But now we have a different story.

The crux of the matter is that it is difficult to base the discovery of Aulprofem in any of these two categories. First, theoretically, more scientists worked on the possibilities to develop the technologies for mammalian sex fixing, both on the drug as well as non drug side. So, it is difficult to conclude that no scientific foundation existed to develop a mammalian sex fixer. There were attempts to manipulate mammalian sex ratios, *in vivo* and *in vitro* and frequency of scientific experimentation was such that it appeared that every big scientist was on the verge of declaring the discovery of world's first mammalian sex manipulation success. The ultimate achievement always appeared just one step ahead and man could be God! Or at least; the son of the Almighty, because until now the domain was kept strictly by Him under his portfolio of executive manifestation. This exclusive right of God to provide a baby of choice; male or female in man or other mammals, will be claimed by man was the imminent expectation of innumerable number of scientific experimental schedules and every big scientist wanted to become the master of the world. There existed no scientific logic to guide them so; it is impossibly difficult to believe. So it can be inferred that theoretically there existed no foundation, on which the discovery of a sex fixer could be based; is not true. The picture on the other side was also very complicated. The greatest chromosomal theory of determination of sex clearly lays certain guidelines and some of them are very clear and binding. So, chromosomal part is a reality and has to be a thumb rule. The million dollar question was that how it was possible to have a baby of choice when the segregation of chromosomes with X- or Y-types was not possible. As great international authority on the subject E.S.E., Hafez (1985) has put it 'the greatest difficulty to manipulate the mammalian sex ratio is the absence of laboratory tests to evaluate the degree of sperm separation'. Earlier a lot of attempts were made in the forties and fifties based on the concept that vaginal – pH controlled sex ratio but as Salisbury and Van De Mark (1961) has put it "nothing conclusive was found in the subsequent careful scientific investigations". Despite many publicity gaining claims, including the attempts of heavyweights like Ericsson *et al.* (1973), Lindahl (1956), Gordon (1958), Meistrich (1982), Corson *et al.* (1984), Barrat and Leger (1979), Sampson *et al.* (1983) etc.; the outcome was negligible as Prof. R.F.

Hunter (1982) has quoted R.V. Short as saying that "manipulation of mammalian sex ratio still remains a mirage on the scientific horizon". So, it is not an easy job to classify the discovery of Aulprofem as an achievement in pure or applied sciences. The important point is that it is an important discovery and is a much awaited one but keeping in mind the total jeopardy and uncertain haphazardness of scientific environment and experimental history at the time of success of Aulprofem to manipulate mammalian sex ratios and produce female off-springs in mammals; it is better a mum than a hundred wise comments on the subject.

One more aspect, the very field of drug development is a much complicated one and still the research is in the domain of a sort of 'hit and miss'. There can be a drug for a disease, rather for every disease. But the question is that which drug for which disease? This is the eternal question before mankind and it has been and perhaps will remain so, the biggest question haunting man. The cancer experience is an example. We simply cannot go on screening endless number of chemicals, natural or unnatural to find their utility in healing an ailment or to find the treatment for a disease. The number is endless. There are hundreds of thousands if not in millions of compounds that has been discovered and identified or more accurately that fall under the categories of bio-chemicals and pharmaceutical compounds and the number is increasing every day. So, a compound which is most suitable to cure a particular ailment is very difficult to diagnose. Even a category of compounds effective against a particular disease or a group of diseases does not make it mandatory for all the members of that category to be effective against a particular disease or the category of diseases. There are sulphonamides that do not have antibiotic properties and there are barbiturates that are not hypnotic. So, the big conclusion is that the ways of drug research are unpredictable and difficult to tread and this ending is not a happy one.

A sad story. Some more discussion?

There are diseases like sleeping sickness, AIDS, dengue, asthma, leucoderma, epilepsy, MDRTB, MDR malaria, post disease tetanus, hepatitis, polio, pox and a host of viral diseases, the inherited diseases like thalassemia, color blindness, hemophilia etc. and a brand new series of lethal bird flues, LGBS and mad cows. Developing drug treatments for such diseases has always been and still remains the most daunting task before mankind. The vaccine development has rather been an easier job and many scientists worked successfully to develop vaccines for many diseases but of late there also have been certain difficulties. The AIDS virus mutates very fast and there is no guarantee that a particular vaccine effective against one trait will be effective against all strains. Similar is the case for other virus diseases. We have not succeeded against dengue which these days has acquired the proportions of an annual epidemic in countries like India and even against petty diseases like common cold, there is not a vaccine.

The discovery of Aulprofem faced such challenges. The possibility or no possibility of the development of a drug chemical, the selectivity of a drug action that could do the formidable job of sex fixing, were the most painstaking questions that demanded answers and it was by grace of God or by a particular will of Mor Mukatdhari Murali Manohar (see Appendix I) that such a discovery could be made. The million dollar question here is that should a scientist working in the field of drug research take a flightist's attitude that he should graze with the flock? After all these are the numbers that guarantee security or should he agree with Tolstoy that numbers invite annihilations. The scenario today in the vistas of drug research is the former one and this has become the biggest enemy of drug research it self. A scientist should not be a coward. He should be a man of guts just like the para gliders of Second World War. He should always be ready to tread on the most difficult paths that nobody opts or even more, he should create his own ways before he takes a decision to walk. He should take this bearing in mind that there exists no laboratory on earth where a perfect scientist has to work. He should conceive and make his own laboratory where he will make the world's biggest discovery. This is the universal truth in science and is repeatedly proven time and again when each and every time the greatest discoveries are reported. Be it Louis Pasteur, be it Alexander Fleming, be it Albert Einstein or a C.V.Raman or a Gregor Mendel or Charles Darwin (see Appendix I) or whosoever? The myth of solitary inventor has lived, is very much alive and will continue to live forever till the last moment of the age of the universe. The world lives or not but the greatness of a virtuous person will always live!

But things have changed.

People have become more intelligent.

They do not believe in the lofty ideas. They have become practical.

Everybody wants to be successful.

Just give a slightest consideration to the fact that innumerable number of diseases are there, a small number is enlisted in the previous paragraph; which have no treatments. It is not the question that nobody bothers. People bother but they want the guarantee to success, a thing that never existed as far as great achievements are concerned. Nobody is to be blamed. Only the system. A system has emerged over the years where everybody wants to be a great scientist. He should be equal to an Alexander Fleming or a Frederick Banting (see Appendix I). If Louis Pasteur (see Appendix I) could discover a vaccine, then why couldn't he? The entire climate has become deteriorated. Something has collapsed in the system. The humanity is faced with big challenges of death and disease and the bigger ones are yet to come. Will there be the types of more mad cows or more dengue or AIDSs? The answer is a big yes. There will be more diseases, the types of which are already there in the form of hepatitis's and bird flues and the only difference is that they will be more lethal this time and still more dangerous in future, every time they make their entry, as such or with a new entrant.

The scientists in the field are busy and they will remain so in their important jobs of drug research as far as they offer a hope of finding a therapy to cure a disease with shield of working in safer pastures of vaccine developing with very well knowing that such exercises have proved futile and will yield no outcomes in future also. At least the position remains so in the field of diseases due to fast mutating virus.

Sir C.V. Raman

(Glimpses of greatness from annals of history)

A new companion to this futile drug development is the emergence of **genetic sensationalism**. A similar phenomenon existed few decades ago in the form of **physiological** research which could not yield drug discoveries for most of the diseases that existed nearly a hundred years ago and still are in the class of un-curable. But these things combined, have claimed the most number of Nobel Prizes in medicine. Every year a Nobel Prize in medicine is awarded, it is mentioned in the citation that such and such research will help in discovering a treatment for a given intended ailment. The genes right from the days of Watson and Crick and Khorana and Nirenbrg are promising cure for thalassemia, color blindness and hemophilia but till date no treatment has appeared on the scene but dozens of Nobel Prizes in medicine have been awarded for research on genes and they continue to be given for purposes that have no relevance with the discovery of drug treatments for these killer diseases of mankind.

The lesser the luggage, the more pleasant is the journey.

This is the slogan of the Indian railways.

Rightly said, by all parameters.

The more complicated is the medicinal research, the lesser the chances are of finding the paths to drug treatments for diseases which are incurable. This seems to be the answer if a logical conclusion is to be drawn from the discussion.

Why to pomp and show when simplicity is more attractive and yielding. But this is the most difficult truth that can be sent right down the throats of most of the people in the world.

Har Govind Khorana

(Glimpses of greatness from annals of history)

Cardiovascular diseases kill nearly 15 million people worldwide and a bigger chunk of these deaths, nearly 25% is due to diabetes. Diabetes on alone is a bigger killer. Then there is epilepsy which claims nearly 6 hundred thousand lives a year. Then there is MDR malaria which claims nearly 7 hundred thousand lives and MDRTB claims nearly one million lives and the number is increasing every year. Then there is sleeping sickness that kills nearly three hundred thousand people with the most horrifying death on earth. There are many more on the deadly list which appears to be endless at least for the time being.

But the prizes were awarded for physiological research before, then for neurological and other anatomical etc research and now for the genes' sensationalism, they continue to confer honors on persons connected with these magnificent arenas of scientific exploration but the truth remains that these personalities and their achievements have stood no way and they have negligible contributions to work out treatments for the ailments plaguing mankind. The list of diseases incurable as given above stands such and tall and mocks at the very face of Nobel Authority whenever it announces a new Nobel Prize in medicine. The list was such nearly hundred years back, so was it some fifty years back, same was the position twenty years back and still it is so and if the current trend continues it will be the same list some fifty years to come; of course with more entrants, the likes of which already exist as mad cows and bird flues and LGBS and AIDS etc.

Then the salvation?

We should be believers and not non believers. The Mother Nature is watching. It has been and will always be the greatest force on earth and it has been kind to man. The genuine answer to the modern day drug research is that we should revert to the basics of fundamental and original drugs research. That may not be that much decorated, celebrated, ornamental, attractive and above all acknowledged as the fascinating gene research, physiological research or some most appealing vaccine development promises but it will definitely yield treatments for some of the most horrifying ailments plaguing mankind and alleviate many of the sufferings of the human race. The original drug research should focus on the working on chemicals, compounds; both in the fields of inorganic and organic domains and to center studies on them with a view to find certain hidden virtues and properties linked with them that can be utilized to find cures for the diseases that has posed before us the damnest and distressfully challenging death and disease phenomenon.

The discovery of Aulprofem can lead the way in this regard. After all this is the success story of working on simple chemicals to find greater solutions in saving mankind from the dreadening claws of death and despair that haunt every human today in every corner of the world. Until a few years back, who could have imagined that world's first mammalian sex fixer would be discovered with the focusing of medicinal research on simple chemicals. All big names in the field of reproductive physiology and genetics were glued to the magic of sophisticated IVF techniques. But the success came from a very simple and down to earth original technique. The working on a very simple compound, the sodium ethanoate. The curse on humanity of diseases has been more because it has abandoned the natural and original ways of finding cures but the truth remains that the greatest blessings and happiness' of Mother Nature bestowed on man on earth, lie in its simplicity. The moral of the story lies in the truth that nature likes the simpler and original ways of life. The drugs research and scientists across the globe has to acknowledge this humble fact. This only will open before mankind the countless vistas of fruitful medical research and this will be the greatest blessing by the forces of nature upon science, scientists, mankind and of course the ultimate creature, the man.

References

Barrat, J. and Leger, D. 1979. Aveir des grossesses obstenues après stimulation de l' ovulation. A propos de 519 grossesses. *J. Gynecol. Obstet. Biol. Reprod. (Paris)* **8**: 332.

Corson, S.L. *et al.*, 1984. Sex selection by sperm separation and insemination. *Fertil. Steril.* **42**: 756.

Ericsson, R.J. *et al.*, 1973. Isolation of fractions rich in human Y sperm. *Nature* **246**: 421.

Gordon, M.J. 1958. The control of sex. *Scientific Am.* **199**: 87.

Hafez, E.S.E. 1982. Reproduction in farm animals. Reprint 5th edn. Lea and Fabiger, p. 499.

Hunter, R.H.F. 1982. Reproduction of farm animals. Longman. London, pp. 138-39.

Meistrich, M.L. 1982. Prspects of sexing mammalian sperm *in* R.P. Amann and G.E. Seidel, Jr. (*eds.*) Colorado University Associated Press.

Sampson, J.H. *et al.*, 1983. Gender after induction of ovulation and artificial insemination. *Fertil. Steril.* **40**: 481.

Aulprofem: A strong Case for a Nobel Prize in Medicine for India

"....The whole of my remaining realizable estate shall be dealt with in the following way. The capital shall be invested by my executors in safe securities and shall constitute a fund, the interest on which shall be annually distributed in the form of prizes to those who, during the preceding year, shall have conferred the greatest benefit on mankind...; one part to the person who shall have made the most important discovery within the domain of physiology or medicine;...The prizes for.....shall be awarded ....that for physiology or medicine by the Carolinska Institute in Stockholm;..."

This is exactly the wording of the translation into English of Alfred Nobel's will which was signed on November 27, 1895 in Paris and officially approved by the Swedish govt. on June 29, 1900. Soon after the will became known publicly, it was criticized for legal and other reasons. One reason was that it lacked in precision. An example of it is the use of words, physiology or medicine. The term physiology in those years was used to describe what today are a number of biological disciplines. Interpreting the term, has left the Nobel Authority with a wide range of options. The prize in 1973 to Karl von Frisch, Konrad Lorenz and Nikolass Timbergen can best be understood a prize in behavioral sciences. The prize in 1979 to Cormack Hounsfield for developing computer assisted tomography fits well in the ambit of a Nobel Prize in physics. And the prize in 1983 to Barbarra Mc Clintock for discovery of mobile genetic elements can be thought of as the prize in plant genetics.

However the emphasis is on two words; discovery and greatest benefit on mankind. The discoveries are easier to define in basic sciences than in clinical practice whereas the greatest benefit on mankind is more obvious in clinical medicine rather than in basic sciences. Charles Nicolle received the prize in in1928 for identifying the vectorness of body louse in typhus and it were the simple delousing procedures that saved several hundreds of thousands of human lives during the period of World War One. Paul

Hermann Muller received he prize in 1948 for discovery of dichloro, diphenyl, trichloro, methyl, methane (DDT) as an efficient insecticide. It made man master to control epidemics like typhus, malaria and others. The WHO has estimated hat nearly 25 million human lives were saved by this chemical alone before its use was banned due to its bioaccumulation tendency in living tissues and thereby causing environmental damage. The greatest benefit on mankind factor is quite obvious, in these examples.

Candidature for a Nobel Prize

Really a big question and the greatest honor of the world!

A Nobel Prize in medicine is the ultimate achievement that man can aspire of, on earth. A Nobel Prize is given in six different categories; one in each, every year. These categories are Physics, Chemistry, Literature, Economics, Peace and Medicine. Although it is difficult to judge that which category is the best and all prizes are considered to be of equal importance but still the Nobel Prize in medicine can be considered the best of all in the sense that it is immediately directed towards the alleviation of disease and suffering from humanity. It is the humanitarian angle that puts the prize in medicine category on edge to the other ones. So, it may seem aiming too higher for anybody or anything to claim or even to hope for the greatest achievement on earth. The world is simply not small as a consideration is made on the world level especially in the field of medicine research. There are thousands of medical research institutes across the globe and hundreds of thousands of expert people working in them and many of them are very talented and qualified. When the larger numbers like them are working in the field, the competition becomes toughest and it becomes really hard to zero in on the most suitable contender, when there are too many genuine candidates on the scene. That is the reason that it has become very difficult for the Nobel Authority to decide on the individual greatness of discoveries in the medical field and mostly the prizes have been awarded to more than one achievers on a sharing basis. This was not so in the first 50 years of the last century but it has become more of a routine in the last 50 years and in the initial years of the present century. The Nobel Prize in medicine has usually been a shared one during these years.

Another important point that the medical science is an important discipline of science and a simple success in it, goes a long way in the service of mankind. So, a lot of attraction is naturally directed towards this field. Still death and disease are the biggest tormentors of man. Whereas it is the noblest aim of medical science to rid the humanity of all diseases and related sufferings and there has also been a significant progress but still the destination of achieving a disease free society, appears to be a far fetched one. It was the great Lord Lister who stressed on the need of sterilization and due to his efforts, countless infections mainly in the surgical procedures came under control. Ronald Ross, an officer with the royal army made pioneering studies on malaria and got the prize in 1902. Vaccination therapies were developed due to the efforts of all time greats like Louis Pasteur, Adolf von Behring and others. Adolf von Behring got the Nobel Prize in 1901. In 1903, Niels Finsen got the prize for discovering light treatment for lupus vulgaris. Friedrick Banting discovered insulin and was given Nobel Prize in 1923. George Richards Minot and William Parry Murphy shared the prize in 1934 for discovering liver therapy for anemia. Gerhard Domagk discovered the world's first sulphonamide, Prontosil and got the prize in 1939. The discovery of Penicillin was made by Alexander Fleming and it became possible to cure numerous infections by Gram positive and Gram negative bacteria. He was awarded Nobel Prize in 1945, nearly 14 years after he had made one of the most important discoveries ever reported in the entire history of mankind. Paul Hermann Muller was awarded the prize in 1948 for the discovery of insecticide, DDT. Although not a drug, it proved very effective in preventing spread of typhus, malaria and other insect vector diseases. All other Nobel Prizes given in the first half of last century were for physiological, histological, neurological and related research except the one in 1933 to Thomas Hunt Morgan for his discovery of the first gene mutation in fruit fly, *Drosophila*. The greatness of this discovery was in proving the effectiveness of Hugo de Vries' theory of biological mutations.

Then with the start of the second half of twentieth century the first prize in 1951 went to Max Theiler for developing a vaccine for yellow fever and the very next year Selman Abraham Waksman got the prize for the discovery of streptomycin, the first antibiotic effective against tuberculosis. In 1957 Daniel Bovet was awarded the Nobel Prize for discovering synthetic drugs such as antihistamines that block the action of biological amines. In 1966 Peyton Rous and Charles Huggins shared the prize for discovering viruses that induce tumors and for working out a treatment for the cure of prostate cancer with harmones. In 1998 Robert Furchgott, L.J. Ignarro and Ferid Murad got the prize for discovering signaling properties of nitrous oxide. All other Nobel Prizes went to discoverers connected with physiological and genetic research with the exceptions of a Nobel Prize in 2003 to Paul Lautebar and Sir Peter Mansfield for discoveries concerning magnetic resonance imaging and the prize in 1973 to Karl von Frisch, Konrad Lorenz and Nikolass Timbergen for the study of social animal behavior especially the dance language of bees and how young birds become fixated to their mothers.

Of the important prizes given in the field of physiological research include the one to Ivan Pavlov in 1904 for his work on the physiology of digestive system, one to Krogh in 1920 for discovering gas exchange in lungs, to Heymans in 1938 for showing how blood pressure and oxygen content are measured and transmitted to brain etc. Of the important prizes in genetics, include the one in 1962 to Watson and Crick, one in 1968 to Holley, Khorana and Nirenberg, one in1993 to Roberts and Sharp for split genes. Many other scientists got this distinction of getting or sharing a Nobel Prize and it will be very difficult to discuss here about each and every discovery and the discoverer involved with this prestigious prize.

Ronald Ross

(Glimpses of greatness from annals of history)

This is a brief description of the history of Nobel Prizes awarding. Because it is not possible to discuss the entire history of Nobel Prizes awarding in the present essay, so only few examples are mentioned.

Aulprofem: Contender for the prize?

A difficult question? and still the more difficult an answer.

The wish of Alfred Nobel. And, the writing of the wish in the will. The prime most conditions laid by the founder himself to be applied to judge the suitability of a research to be able to win the honor of getting a Nobel Prize are described in the words; the greatest benefit on mankind and the most important discovery. By far these are the most binding guidelines. The other binding 'during the preceding year' was interpreted by the Nobel Authority as to mean that the benefit of the research became obvious to the prize awarding authority in the preceding year. By and far, these three rules have formed the basis for the award of this most sought after distinction of the world. There have been certain swings and toes and froes from this rule but the general line of addiction to the guidelines have stuck to this stand. For example the discovery of insecticidal property of DDT was not such a bigger scientific achievement but on the other hand it saved tens of millions of precious human lives, thereby justifying its candidature. Another example is the discovery of genetic code and the other gene related discoveries which are really great discoveries but the most important promise they made to the mankind that they would form a basis for working out treatments for the congenital and hereditary diseases has not come true and still they have not saved even a single human life. But it may be contented that in some far fetched or near future, they will be able to achieve that feat. So, it sounds appropriate to award such discoveries. The moral of the story is that there should be a proper balance between the terms; the most important discovery and the greatest benefit on mankind. The individual or combined weightage of one or both these terms in the final outcome of result seems to form the basis for the awarding of the prize.

James D. Watson
(Glimpses of greatness from annals of history)

The Aulprofem as a contender of the prize? First we have to ascertain weather it is a fundamental discovery in basic sciences or a clinical invention or both, or neither of the two? The problem with Aulprofem is that it is as much a discovery in basic sciences as it is an invention in the applied ones. The sex fixers; the fore runner of which and the first successful proud member that Aulprofem is, and it has come as a jolt to the formidable chromosomal theory of determination of sex. The sex of the off-spring is decided by the random combination of X or Y chromosome bearing sperms with the X chromosome bearing oocyte, is the most common knowledge on earth today. Almost every literate person who can spell the name of his or her father or of his or her motherland, knows about the combinations of these chromosomes and this knowledge prevails across the globe like a myth. And this myth will break with the advent of an era of sex fixing; the concrete foundation of which has already been laid by the arrival of the Aulprofem on the scene. Very soon, it will be the most common truth known to all that the sex of the off-springs is decided by certain factors and sex fixers are those forces that lay and decide the guidelines of working of these factors which play the crucial role in deciding the sex of the off-springs. And this has happened with the mammals for the first time in the entire history of mankind. So, sex fixers are the forces in deciding. And, be it Aulprofem and it be the female sexed off-spring in mammals. This is the message, very loud and clear of the new era cast in full with the dawn of Aulprofem on the horizon of scientific achievements. Where are the chromosomes? And what is their role in the classical decision making in genetics, deciding the sex of the progenies, will be asked or even on a fine day, cease to be asked? This is the bitter truth of tomorrow. May the role of chromosomes be fully curtained with the advent of sex fixers in times to come? And, this thing happening in mammals! Which have a much organized chromatin, not like lower chordates, the ratites and reptiles which show the phenomenon of TSD (temperature dependent sexual determination), is a bigger bombshell that can be dropped from the mighty cannon of scientific explorations. The truth, even if we try to conceive today, sends shivers down the spine of scientific realization. The sex fixers will become not only the subject of talk of everyone but the talk itself. The sex is the most talked about subject on earth and that forms the basis of sex determination to be a matter of common and general knowledge for everyone and now this duty will be done by the sex fixers and they will be the most common and general knowledge on earth. This is the power of Aulprofem and this will start manifesting itself in giving more and more power to this discovery with the passing of each individual day and very soon it will be the most powerful truth and belief on earth and almost every citizen of the world with even a little knowledge of the state of affairs of the world will be very well aware of the truth and power of Aulprofem in the lives of human race in this magnificent world.

So, the position of Aulprofem is decided in the basic sciences. The knocking down of the mighty chromosomal theory of determination of sex, at least qualifies the Aulprofem for this. Now we will discuss some other aspects as well. The Aulprofem is the first sex fixer of the world. So, a new field of exploration of the newer sex fixers, opens in the scientific disciplines. A research which has the power to open a new field of scientific exploration can not be underestimated and it is definitely an advancement in the basic knowledge of the scientific community. So another qualification for being the original and mega research in basic sciences.

Aulprofem produces female sexed off-springs in mammals and this has a lot of social and economical considerations. So, it is almost superfluous to discuss the applied angle which is very well clear with the known knowledge that a lot of benefits to the society can be incurred due to this property of Aulprofem. From the commercial and clinical applications of this research it is very well a conclusion that Aulprofem is an invention in the applied sciences as well. So, the dual nature? This is what the Aulprofem is all about and this is by any standards, a great achievement of a scientific research. A scientific discovery in pure sciences is an invention in the applied sciences and *vice versa*. These added credentials of Aulprofem make this discovery simply unique. And this is also adding value to the discovery.

Let us have a small discussion on the scope of Aulprofem as a benefactor of mankind. The world population is increasing at an alarming rate. In 1947 the population of combined India was 370 millions and three countries were carved out of it, Bangladesh, Pakistan and India. Now in 2007 *i.e.* merely 60 years after that, the population of India alone is 1120 millions. The population of Bangladesh is nearly 180 millions and that of Pakistan is nearly 160 millions. The combined population of all these countries is something around 1450 millions *i.e.* an increase of nearly 400 percent. Such is the story of many a developing countries. The dismal picture on food grains front in 1950's or 1960's, got improved with the start of green revolutions in 1970's in these countries, which continued increasing food productions well into 1980's but due to decrease in the hybrid vigor and limiting of farm lands in these countries because most of the utilizable surplus lands were converted into arable ones in 1980's. Even the precious forest lands, so necessary for maintenance of ecological balance were converted into farm lands, thereby endangering the natural and ecological balances of *fauna* and *flora* in these countries. Now a stage has come when these countries are already on the verge of an ecological disaster and no more surplus or forest lands convertible into arable ones exist without causing an ecological disaster of the biggest proportions, perhaps even endangering the very existence of human life in these geographical habitats. One more thing, that due to the use of excessive urea and otherwise fertilizers and insecticides, pesticides, weedicides etc. in these countries, the soils have become dangerously contaminated with toxic poisons of these fertilizers, weedicides etc. Not only the soils but the underground water beds and even the air in these countries is drastically contaminated and not fit for a healthy human life. But the populations are still growing and the governments or the political establishments in these countries, do not just have the determination and the political will to manage this menace of growing populations. The food grains production has already started to decline and the trend seems to be irreversible, at least for the time being. The food shortages have already started to be felt by the people in these and similar nations and the dreadening times of 1950's, 1960's and in certain areas of the world the 1990's have already started to knock at the doors of these nations. The mammoth wheat imports of 5 million tonnes by India in year 2006 are just an indication and the picture of the times to come. So, the coming times are definitely going to be the tough ones. There will be hunger, starvation and death by starvation in abundance every where except, the developed nations. Then morally and ethically speaking, the death by starving is the death of the worst kind. Perhaps this is the reason that Almighty God has arranged for the appearance of Aulprofem on such a crucial juncture of time that the world really needs security and salvation from the impending food crisis that has already started to knock at the doors of the world inhabitants belonging to the *Homo sapiens* race.

The dawn of Aulprofem has occurred on such a time that the world really needs it. The female populations of the better breeds of milch cattle can be increased to unimaginable levels with the application of Aulprofem and milk production increased to feed the growing numbers of populace in these countries and the similar ones. It is a same sad story of food shortage and crisis in almost all the developing countries of the world. Nobody can forget the famines in Somalia, Namibia, Sudan and other countries in the 1990's.

On a simplistic estimate if we calculate the application of Aulprofem on a 100 million dairy cattle of good yielding varieties like HF, Jersey and others, the corresponding increase in the milk production can be easily understood. With 35-40% conception failure rate the worldwide, this means the added 30 million more female calves of these varieties with the application of Aulprofem. If 20% of them succumb to calf mortality; and the 30% of remaining 24 millions die before reaching maturity into full mothers, thus leaving us with 16.8 million milch cattle. If on an average they yield 20 liters of milk per day per animal; the increase in milk production will be 336 million liters per day. If roughly, half liter of milk is provided to a starving member on the verge of death by hunger, of the third world developing or underdeveloped countries; because developing countries has been and forever will be the developing ones (and the developed countries will be developed ones for ever). This automatically means the healthy nutritional diet to 672 million people. Bringing 672 million people out of the garb of acute hunger, starvation and poverty can not be considered a smaller achievement! And the scientific discovery responsible for this, can claim to be one of the biggest benefactors of mankind. If this application is spread to say 200, 300 or 500 million animals per annum, then the corresponding numbers of people coming out of the clutches of starvation and death by hunger, will be 1344, 2016 or 3360 millions.

These figures are nothing but sufficient to put the Aulprofem on the colossal stand of the biggest ever benefactor of mankind.

This makes the point of 'greatest benefit to mankind' very, very and really, very absolutely clear!!

The Missing Cow: A Key to Success

A beautiful dream can run like this.

India has a dairy animal population of 100 million average category milch cattle and buffaloes, each yielding about 3500 to 4000 liters of milk per lactation. Suppose every animal delivers on an average inter calving gap of 12 to 15 months, then we can have an output of about 1000 million liters of milk per day. Enough to feed a nation of 1500 million people, nearly 600 to 700 grams of milk per person per day in year 2025 A.D.

And if the grade of cattle and buffaloes is of exotic degree, each animal yielding plus 5000 liters of milk per lactation, then the outcomes need not be imagined.

They can in fact be felt!

Many day dreamers usually hallucinate into illusions.

Many tout the American models for development and declare that India will be a super power by the year 2013.

But when America falls, they denounce American styles but still believe that by aping *laisez faire* India will fly over the Alps.

Some say that by doing free market *aerobics* and *yoga*, India has a shining future.

But they do not know that no *yoga* can rectify the myth of the fluctuating voltage of the left eye.

All seem to be flights into fantasy.

Alice in the wonderlands.

But not the first one.

The story of 100 million cows and buffs.

If only a nation can have it!

A nation of 1500 million people by the year 2025.

Day dreamers sometimes create history.

Abraham Lincoln did it once and so did Garibaldi, Galileo, Newton, Pasteur, Wright brothers, Curies, Mandela (see Appendix I) and the list is quite long......

Let's dream a little.

Dream and believe in the story of 100 million dairy animals.

Let's think and plan to make our national dairy a successful working proposition. Working to achieve the prosperity of a nation.

A nation already on the verge of starvation.

God be with us, let's hope.

God has to be with us. There is no escape.

Let's work and think like a nation and not like a crowd.

We have to build a nation.

And make a White House on top of the Himalayas.

And a niche for every Indian, food to feed and milk to drink.

And plenty of hope and happiness.

We can lay the foundations............of a great nation.

The God is definitely with us and Mother Nature is smiling.

One day She was in very good mood and She nodded and gifted the great *Bharatmata* with the power of Aulprofem.

Now there will be floods of milk, She whispered.....and smiled again.

There is a story in *Mahabharata* (see Appendix III). One night lord Krishna and Bhima, the powerful He man were passing through a jungle. As the night fell, they decided that they change duties on guard alternatively. First Bhima decided to stand guard and lord Krishna went to sleep. As soon as the lord was deep into sleep, a tiny monster looking creature appeared. He approached Bhima and asked him to fight a brawl. As Bhima looked at him, he smiled and told him to go away as he was too small to fight such a powerful man like Bhima. But the monster insisted upon fighting. Bhima requested him to go away but he kept on insisting and challenging him to fight. Soon Bhima starting losing cool and became angry and called him for a fight. But he was surprised to find that the tiny monster had grown to double his former size. The monster again challenged him and once again Bhima flew into anger. The monster again grew double in size. Now he was enough in size that he could at least engage Bhima. The fight started. The monster was very smart. When Bhima struck, he would slip away and when he struck, it would give Bhima lot of pain. Bhima again flew into anger and the monster once again grew in size. Now he was almost equal in size to Bhima. Bhima was puzzled. He had found the monster in a very small stature but now he was in quite equal and powerful built. Now the signs of bewilderment started to turn into fatigue and desperation and Bhima once again became angrier. The monster once again grew in size and now he was double in size to Bhima. This process was replicated once more and now great Bhima was a tiny character in front of that monster. But suddenly the lord woke up and the monster slid away. Now it was the turn of the Bhima to sleep and the lord to keep vigil. But before sleeping he narrated the story of the monster to the lord. Lord listened and smiled and told Bhima not to worry. Bhima went to sleep. As soon as Bhima went to sleep, the giant monster appeared again and asked the lord to fight with him. The lord looked at him and smiled and strange enough, the monster shrank to half in size. He once again challenged the lord to fight and the lord smiled again. The monster shrank again to one half. The process was repeated couple of times and the monster shrank in size such that it could easily be controlled and the lord caught it and tied it in the corner of the cloth he was wearing to cover the lower portion of his body. The lord rested in peace after that. When, in the morning Bhima woke up and he recollected about the fight he had with the monster, he asked the lord about the monster. The lord smiled and opened the knot and displayed the monster and asked Bhima to identify him. Bhima was surprised and requested the lord to tell him the story. The lord replied that the monster was nothing except one's own thought and reflection of his inner being. As Bhima was a very powerful man, he was naturally proud of being a powerful man. This feeling had crept in his mind as an essential part of his inner being and personality. So, this feeling of being proud got converted into a feeling of false anger. As he found no competitor to his power and status as a strong man, he naturally was carried away by the feeling of anger that was nothing but an extension of the feeling of being proud. So the anger manifested itself as the monster and fanned his inner hollowness of false pride and whole night he was fighting with his own boastful pride and anger. Bhima got to understand the crux of the matter and bowed to the lord and thanked him for giving him such a valuable narration and advice.

Can we convert our national assets into the anger monster of the story so that they grow in size and strength to double the existing levels with the each endeavor of zeal and dutiful enterprise that we undertake, aimed at the national development of our dear country?

With lord Krishna already happy and smiling with us as he has already destined the gifting of Aulprofem to our nation and this whole mankind, this is not going to be much difficult a task to be achieved.

The national dairy project based on the basic infrastructure and foundation of 100 million milch cattle will employ about 20 to 25 million acres of land for fodder *i.e.* less than or about 10 million hectares. India has about 145 million hectares of arable land in 2008. It had about 170 million hectares in year 2000 and by 2020 there will be about 100 million hectares of arable land left for cultivation. So, less than 10 million hectares of land if allotted for dairy purposes, is not a major problem if it has the capability or viability to feed every Indian citizen 600 to 700 grams of milk per day.

In fact it looks bribingly enticing.

The other alternatives can be worked out. For example the foliage of the forest trees can be studied for its nutritional values and suitable trees can be planted in the reserve forests and lot of national waste space alongside the national highways and link roads, the rivers, canals and their smaller subsidiaries and distributaries, village panchayat lands and municipal lands and other places such as parks, ponds, lakes and public offices etc. If the government is determined to do something, then nothing can play stopping it from doing so.

The other necessities like infrastructure and sheds etc construction, setting up of standardized feed and pharma sectors and units and dairy related accessories like milking machines, fencing for the dairy farms, temperature conditioning for dairy farms and localities, pest and disease and hygiene controlling etc should also be undertaken on macro scales and scientific levels.

The exotic breeding and production of better lactating progenies should become a national priority. The offsprings with more yields, may be of foreign origin, mixed with native traits to help them adapt to the local environments, should be the aim of the crossbreeding programs. In addition to this, it should be taken care that the varieties of cattle that are very low yielders should be scheduled for sterilization procedures and end of their dynasty lines should be the aim of such maneuvers. It should be borne in mind that in the broader context, such bulls and even cows should not be able to produce further progenies. For this purpose, they can be sterilized, tubectomised or segregated from the animals of opposite sex. The cattle or buffaloes which are low yielders with less than average milk production, should be treated as a burden on national economy and the aim of the nation should be to get rid of them in the longer perspective. It may not be the killing of such animals but they can be refrained from further reproducing their progenies. Whatever may be the thought and action behind such action, it should be aimed at the elimination of them from the national dairy scene as early as possible.

The cross breeding programs produce progenies as males and females; in one to one ratio. Half of the progenies are males and half females. In fact the proportion of males to females is always more, usually in the ratio of 106 males to 100 females born. This is the most common sex ratio in dairy animals as well as humans. The reasons behind the occurrence of sex ratio may be many and discussed in a different place and chapter of this book but it is a fact that they do exist and the sex ratio is more male dominated. So, the cross breeding programs produce more males, which are more or less futile in the propagation of further progenies and in the yield of milk and lesser females, which have productive values. We ought to reverse this trend and go in for techniques that produce more females, should be the concern in these exercises. Otherwise this makes these programs more programmed for failure than for success. At least, more than half of them are destined to be wasteful and surplus and less than half are fruitful for us humans. This is a great drawback and handicap of usual cross breeding programs. These programs result into the production of less than half the number of females that it could have produced if all the progenies would have been produced as females with some female sex fixing technology. Nations as well as the people need the females as these are the females which yield milk that is so important for the human race to feed itself upon. This is the milk that has been popularly touted as being the perfect food for man from centuries. Recent studies have shown that it is not a perfect food but lacks few of the necessary ingredients. But even then, these ingredients are in such negligible quantities that their absence or presence can be overlooked. These necessary ingredients are not in such a big measure and quality that they still decrease the status of the milk as being the best food for man. The truth remains that even then, no food on earth matches the nutritional qualities inherent in this precious gift of Mother Nature to man; the milk.

The productions of progenies of better and desired traits as wholesome females will definitely increase the efficiency of such programs to more than double its existing value. This can be a true working reality with the Aulprofem and its magic in hand. Now we will be able to double the size and strength of anger monster to not only double but more than double its existing value with each and every stroke of the exotic cross breeding exercise. The results and benefits of such outcomes will be clear to all at the first instance and there will be no explanation needed to be sought and delivered from the kindest lord Krishna. Then the nation, the mighty Bhima will be blessed with the sermon of prosperity and understanding. There will be all around plentiful materials and happiness.

The mystery of the missing cow is nothing but the materialization of this otherwise lost benefits of exotic breeding programs in an ordinary run of the routine working of these programs into the more than double increase in efficiencies of these exercises with the application of Aulprofem in these programs to produce all prevalent female progenies. Then four such exercises will produce four female offsprings instead of two, the resultant eight will produce eight female offsprings instead of three female offsprings being produced by six females in the first instance (the first four having produced two females in the first case). In the third step the sixteen Aulprofem applied females will produce sixteen female offsprings and the non Aulprofem nine females will deliver four and half female offsprings. In the next generation the Aulprofem application will yield thirty two female offsprings instead of seven female offsprings (produced by 9+5=14 females) in the non Aulprofem group. The difference of 32 to 7 is well understood in only four generations of breeding. This is the power of Aulprofem and this, well explains the mystery of the missing cow. The cow that is otherwise missed or not produced (in a case of non application of Aulprofem). This missed or missing cow, creates a lot of gap in the generation of milch females and the difference becomes many folds, only in four generations of breeding as we have seen in the above example producing 32 cows instead of 7 being produced otherwise. The missing cow is the cow not produced in ordinary breeding programs and the

Table 1. Table showing the female cattle progenies per a particular year with the application of Aulprofem and the female progenies produced when no Aulprofem is administered

No. of years	No. of female cattle produced during a particular year without aulprofem usage	No. of female cattle produced during a particular year with aulprofem usage
1.	0	1
2.	1	1
3.	0	1
4.	1	1
5.	1	2
6.	1	3
7.	1	4
8.	2	5
9.	1	8
10.	4	12
11.	1	7
12.	7	24
13.	0	35
14.	11	51
15.	0	73
16.	16	105
17.	11	152
18.	13	220
19.	11	317
20.	35	457
21.	11	660
22.	51	953
23.	11	1375
24.	97	1984
25.	0	2864

Table 2. The table showing the total number of female cow progenies produced as a result of the application of Aulprofem and the total female progenies produced when no Aulprofem is given

No. of years	No. of total female cattle produced till a particular year without aulprofem usage	No. of total female cattle produced till a particular year with aulprofem usage
1.	0	1
2.	1	2
3.	1	3
4.	2	4
5.	3	6
6.	4	9
7.	5	13
8.	7	18
9.	8	26
10.	12	38
11.	13	55
12.	20	79
13.	20	114
14.	31	166
15	31	239
16.	47	344
17.	58	496
18.	71	716
19.	82	1033
20.	117	1490
21.	128	2150
22.	179	3130
23.	190	4478
24.	287	6462
25.	287	9326

application of Aulprofem tends to materialize the birth of this female offspring which otherwise is to be produced as a male calf. This is the mystery of the missing cow. Now we will try to illustrate this with one more example, given below.

The Table 1 above depicts the comparative analysis in the number of female progenies produced during a particular year when the Aulprofem is administered to a single female cow at her first mating till her total life period of seven lactations and the resultant female progenies further produced to its progenies when they grow up to age of three years and deliver; to the number of female progenies produced to it or its further female progenies when the Aulprofem is not administered. The Table 2 gives the total number of female progenies produced till a particular year under same conditions as told above in Table 1. It is calculated with the legend of 7 lactations per a single cattle in mind and the starting of the lactating life for the one animal with the first delivery of the male calf and for the second animal to start its lactating life with a female calf in case of non application of Aulprofem. It is seen that at the end of such an exercise at the culmination of year 25th, we have 9326 total female animals produced in comparison to the 287 total female animals produced when no Aulprofem administration is undertaken. This is a difference of 32.5 times. So, the administration or non administration of Aulprofem, on longer periods of strategy and policy planning make a lot of difference. The male calves produced are just a burden on the economy of a nation and the wisest alternative open for a person to opt on the national or larger levels, is to go for the administration of Aulprofem.

The dairy business for a small and marginal farmer is the need of the hour, the lands are diminishing. In fact they have already diminished to such levels; going still down from them will be suicidal for the farming community. Already the suicides of the farmers in India is a national issue with more than 70,000 of them resorting to this wonderful *modus operandi* per year to get immediate relief from all worldly worries and sufferings. There are two creatures in India or any poor agro based economy; which present the most merciful of an expression, are number one, the farmer and number two, his bull. The bull in fact is the top scorer when it comes to the performance in this regard because he number one, leads the life of a bonded slave of Roman era when it comes to the job of pulling ploughs, carts and getting whipped and secondly, he plays an instrument of soothing and entertaining for the farmer when

Table 3. Table showing the total number of female cattle produced with and without the application of Aulprofem

Sr. No.	Pattern	Total number of female cattle after				
		5 years	**10 years**	**15 years**	**20 years**	**25 years**
1	A	2	10	23	25	153
2	B	3	12	31	117	287
3	C	4	19	72	265	945
4	D	6	38	239	1490	9326

{Legend to the table: Pattern D shows the total number of female cattle produced with the application of Aulprofem. Patterns A to C show the number of female cattle produced after years 5, 10 15, 20 and 25 with pattern A showing the cattle produce with each first progeny of every new cow starting its productive life cycle with the fist calf as being the male one and the pattern C with first calf as female one and the pattern B with alternate first calving as male and female for each alternate new cow at the start of the productive cycle} the later thinks that there is a creature created in this wonderful creation of *omnipresent* Almighty that plays to dance to his whims and fancies and obediently like the one two three army soldier, obeys his commands to heart. The technology came as a relief for the bull but it presented another side of the coin also. The bull became irrelevant in this total story of mechanized development. It became a burden on the economy. The more are the numbers of the bulls, the bigger is the problem. In India where people have religious sentiments that they even can not allow their cow progenies to be slaughtered; the problem of taking decision of either keeping the bull and running into loss or to send it for killing to a slaughter house and win the wrath of God, religion and the despise of the society; is a most difficult one. The farmers usually opt to keep the bull and be on the side of God. But this also involves ill feeding this creature right from the birth and quite often subjecting it to conditions of ill-treatment and a sort of 'forced natural death' so that the farmer gets rid of it, as soon as possible.

The dairying for such farmers becomes an against the odds concern when the cows deliver male calves and if the practice lingers on for even small prolonged periods, the situation becomes difficult to sustain (see Table 3). The business runs into a loss. But the conditions can be much favorable when the cows deliver female calves. The more are the number of the female calves, the more are the assets in the form of cows when these calves grow into adults. The farmer under such conditions, is relatively much well off than the farmer who on other hand gets the male calves as the delivered progenies. The missing cow that as a result of such a male delivery, is the missing link to prosperity can be a decisive player if it is really delivered as the female one.

There is another angle to it. There is the general trend to replace the low yielder cattle with the high yielder ones as the desired target in an exotic cross breeding program. But as soon as the progeny delivered is the male, the entire exercise gets fruitless. Under such conditions, if there is a guarantee of some sort that the to be delivered progeny will be of the female sex with a hundred percent certainness, then the farmer will be encouraged to undertake such programs of breed improvement. The end result will be the more production of animals with high yielding quotient. So, under such conditions the emphasis will be on the production of animals of high yielding traits. Let's consider that under a given situation, 287 such cows yielding a daily output of 10 liters will be replaced into 287 cows of 15 liters daily output, with double the speed and efficiency. The lower speed and outcome in such exercises usually kills, the enthusiasm and enterprising to undertake such exercises. With Aulprofem in hand and the power of it at our disposal, we will certainly have this weapon and edge on our side to do the wonders for us. Good yielders and bad yielders are almost equal consumers of resources and care but they certainly yield to varying levels.

So, with the Aulprofem in hand, we will be able to get the maximum out of the same amount of expenditure of resources and efforts that we put into the enterprise. With more number of females, the same resources which are otherwise spent on the upbringing and maintenance of male calves and bulls, can be rightly spent on the female calves and cows to give us the maximum milk in return. This is the usual aim in the dairy profession as well in the efforts to go for breed improvements. The dairy profession which can run otherwise into a loss can thus be converted into a profitable working economic concern. This can be the gift of Aulprofem or the blessing of God. Whatever we take it, it will definitely be another milestone in the journey of progress of mankind to perfection and away from the hunger, starvation, poverty and miseries.

PART - E

Scope and Applications: A Bonanza of Benefits to Mankind

- Saving Lives–Fighting Death
- Alleviating Poverty
- A Biggest Environment Friendly
- A Case for Rural and National Development
- Saving the Female Gender: A Great Social Boon
- Economic Recession and the Way Out
- Fighting Terrorism-A National Salvation
- A Strong Case Against Narcotics and Addiction
- Alleviating Poverty: A Humanitarian Perspective
- Delivering Designer Babies

Saving Lives–Fighting Death

Is the capital punishment justified?

Another way this question can be presented.

Should somebody be allowed to kill a human being?

Good arguments.

But these are some of the very crude and uncultured types of debating.

A genuine way to speak may run like this.

Does human life carry a value?

Or,

Do we believe that all human lives are precious?

Yes, indeed; is the most appropriate answer.

A story told by grandmas of all planets and continents is put as follows.

Mother Nature once produced the most beautiful princess. She was not born to a king or his minister or the chieftain. On the contrary she was born to a poor fisherman family. The fisherman was very happy when he found that he was blessed with the most beautiful child on earth. So, was his wife also. They thanked God and pledged that they would never let a grief befall on their lovely child. The child grew as the years passed and when the seventh year was over, she was a seven years old girl.

She started going to school.

The most beautiful girl, everybody told.

The most beautiful child, everybody thought.

The days passed by.

The fisherman went into the sea everyday and came back with plenty of fish. The fish he sold in the market and fetched lot of good money. The money was used to feed the family and all were happy.

The fisherman, his wife and the most beautiful princess.

All three of them.

All were very happy.

The years passed on and the tenth year came. And when the tenth year was over, she was a ten years old girl.

She was in class four.

The fisherman daily went into the sea and brought plenty of fish back home. The fish he sold in the market and fetched lot of good money. The money was used to feed the family and all were happy.

The fisherman, his wife and the most beautiful princess.

All three of them.

All were very happy.

The years passed on and the twelfth year came. And when the twelfth year was over, she was a twelve years old girl.

She was in class six.

The most beautiful princess.

The fisherman still went into the sea everyday and brought plenty of fish back home. The fish he sold in the market and fetched lot of good money. The money was used to feed the family and all were happy.

All three of them.

One day he went out fishing and was in the sea when a big hurricane came. The sea was furious and the waves rose high and the fisherman was still in the sea. In fact he was deep in the sea. The fisherman was very bold and skilled. He made his way out of the sea. But his boat never came with him. It was destroyed in the sea. That day he returned home without any fish.

So, he could not go in the market and fetched no money.

There was no food in the house and they cooked nothing and nobody ate anything. That day no body was happy in the house.

He, his wife and the most beautiful princess.

All were hungry.

The next day came and the fisherman could no longer go out fishing. So, he went to the money lender for some money so that he could buy a new boat.

There was a hungry wolf living in that village and it was the money lender.

The money lender gave him money on high interest and made him write the deed with his blood and promise. The conditions were very tough but he borrowed money because he had a wife and the most beautiful princess to feed.

The next day the fisherman went into the sea and came back with plenty of fish and everybody was happy. He started paying back the installments of the borrowed loan also.

The years passed on and the fourteenth year came. And when the fourteenth year was over, she was a fourteen years old girl.

The most beautiful princess.

She was in class eighth.

One night when all were asleep, the thieves came and stole the boat. The next morning when they woke up, the boat was no more. So, the fisherman could no longer go into the sea for fishing. Everybody thought and whispered in the village that the wolf was behind the theft but nobody dared say anything because everybody was afraid of the wolf. The fisherman once again had to go to the wolf and borrow more money to buy a new boat and this time the wolf made him sign an even more difficult deed. He signed with his blood and promise and bought a new boat.

The next day onwards he was once again in the sea, fishing and caught plenty of fish that fetched him lot of money but all that money went to the wolf because he had to pay back the loan he had taken on the promise of his blood. Very few money remained in the house and they cooked very less and ate and slept half hungry and no body was happy.

He, his wife and the most beautiful princess.

All three of them.

This very soon became a norm of life. A big chunk of money went daily to the wolf and the fisherman and his family started to starve. The things could not go on like this for a long time and everything changed out very fast.

The fisherman and his wife were made of some hard stuff and they had seen the hardships earlier in life but the princess.

She was not made for a tougher life.

She started to wither. Her parents even made her to jump school. She was now in class nine and fifteen years in age but she no longer went to school. Soon her name was cut off from the school records.

Twenty five days passed from this day and she fell ill.

The reason? Nobody knew.

The fisherman knew and wept. His wife knew and she wept also for that they could do no other thing than to weep. They had no money to get her treated. They didn't even have the money to feed her. The merciful neighbourers and wise men and women came and saw her. They concluded that she was highly undernourished. She was pale and sick from being semi starved. But nothing could be done. All the money her father was earning was going to the coffers of the wolf. All people knew it but they also could not do anything except to be sad. They went back to their homes.

A merciful doctor knew of this and he came to see her. He gave some medicines and instructed the parents to feed her properly and maintained that she be sent back to school also.

But how could that happen?

Her parents had no money. They could not feed her properly. Then how could they afford her going to school to study?

The days passed by but her condition did not improve. Everybody was sad and unhappy but nothing could be done. Her parents had no money.

Money is very necessary for feeding the children.

Money is very necessary to send the children to school and...

The money is very necessary to treat them when they are sick.

The money never came.

It was hijacked by the wolf. In fact, the wolf daily hijacked it from her father. Whatever he earned, went to the wolf for the installments of the loan he had borrowed with a promise of his blood.

The wolf sucked blood, said the people all around but no body dared to speak it in the open as everybody was aware of the blood thirst of the wolf. So, everybody kept quiet.

But the health of the princess did not keep quiet.

It deteriorated gradually. Deteriorated to the worse.

More days passed and it was the twenty fifth day after she had fallen ill. She was very sick on that day.

And on the twenty sixth day, she was no more.

The princess had died.

Everybody was sad. Her father wept. Her mother wept. Everybody wept. But the princess was no more.

So, she could not weep.

She was lying motionless and totally peaceful.

That day she was buried deep in the sand of the seashore.

And it was again quite and peaceful everywhere as if nothing had happened.

Few days afterwards, her mother died. May be of hunger or grief, nobody knew. Few more days afterwards, her father died also. May be of hunger or grief, no body knew.

But the legend lives.

Till now.

The story was hijacked by all the grandmas of the world.

It is still being told to the little princesses who are born to the mothers on every delivery of a girl child that human race records.

Every delivery gifts a little princess to the world if the child is a female one.

More princesses are born but a question remains.

It remains unanswered.

It has remained unanswered for centuries.

From ages in fact, it seems.

Who killed the princess?

The most beautiful princess.

The disease? The grief or ......

Her father? Her mother? The society? Or the doctor or the wolf?

.....are the questions......but one answer on which all seem to agree is that nobody killed her, even the disease and grief.

But poverty and hunger made her dead.

They killed her when both.

They always kill when they are together. They are the best of the pals since the days immemorial and the presence of one is the mandatory condition for the other to be there also. In fact if one of them is generated, it generates the favorable conditions for the creation of the other also at that particular time and place and together they are a deadly combination. They kill the humans in wholesale quantities.

The people agreed and all the wise men and women agreed also. Soon afterwards when contacted, the doctor agreed also. In fact he had opined and diagnosed such before even.

But of what use?

The days had passed and the sand had cooled where she was buried and lay motionless and quite and made no noise and movement and what remained was a story and the fact that the princess was dead ....the most beautiful princess.

Diabetes is a silent killer, say the medical sciences. It does not usually kill directly but creates circumstances such that the only escape from the situation is death and only death is the gate that opens to liberate such people from disease and life. On alone also it kills nearly one million people per annum. Cardiovascular complications account for plus 15 million deaths per annum in the world and a big chunk of these deaths (generally one fourth to one third) are due to diabetes only. According to medical literature the chances of a diabetic to become a cardio patient are as many as more than sixty percent. In fact diabetes and heart disease when run together, form a very deadly combination. They usually kill or in the most of cases they do. They are formidable allies in demolition of life. Their track record is very impressive. They have little competition in their line. They present a shining example of a death execution duo in action.

Just like poverty and hunger.

In fact they have a competition.

The hunger and poverty score over them, in skill as well as in numbers.

They always score much better and higher too.

The average life in India is about 60 years for men and about 65 years for women. The developed countries have it nearly 80 years for men and near 85 years for women. The same trend continues in most of the countries including USA, Japan, China, Australia, Russia, New Zealand and European countries. The United Nations publications of data regarding human demography and population give a clear and vivid view of this. This data presents some very interesting derivations when analyzed critically and logically. Let's have an analysis.

The life of almost every Indian is nearly 20 years shorter than other people of the world as listed above. This simply means that every Indian citizen lives 20 years lesser than the other people in the world, especially the developed parts of the world.

But he is also a human like other people of the world.

So, he should have also lived 20 years more like the other people in the world.

Where does he lose his life?

And why?

Or in other words, to whom does he donate this?

The answer is quite obvious.

The living conditions in India are such that are not worthy of sustaining a longer life. In fact they tend to terminate a human life much earlier.

Exactly 20 years before the actual death.

The answer is not as simple as it looks out to be.

Actually..........

Every Indian citizen is forced to die 20 years before his actual death on an average.

The conditions of life in India, murder every Indian citizen exactly 20 years before his normal biological death which otherwise is programmed by nature to live at least 20 years more on an average in the case of every Indian national like the human inhabitants of other countries of the world.

The circumstances of life in India are the killers of almost every Indian citizen simply 20 years before his real and actually destined death.

A simple question arises here.

Who controls such conditions?

Another way to ask.

In fact who creates such conditions? Who holds the lever?

The government or..... The ruling junta; is the immediate answer.

If not they, then who else?

No body.

There is a shameless proclaiming by the Indian government in press and electronic media that 70 percent of Indian population is from the age group of 18 to 35 years. With this they declare that India is a country of the young people.

Some sort of eternal youth prevails here.

Everybody who listens to their propaganda, draws this obvious conclusion.

But they never take it the other way that they here in India, simply do not allow people to age and live longer. The truth even if examined superficially, can reveal that if aged people are non existent in India then where at all do they go after all?

Where have all the aged people in India gone?

To death; is the answer?

There is no other answer. A person may be very intelligent or even if all the people of the world start thinking day in and day out and they go on with this exercise for one hundred and thousand long years; they will draw the irrefutable inference that all the people who suddenly disappear from the stage of life when they reach a little mature age of sixty or sixty five, go and simply find a shelter and solace in the dearest lap of death.

Can there be another explaining?

No, never. There can't be.

The answer is very simple that they are subjected to such third class living conditions that they do not survive longer and die very soon and do not reach mature age and this is a fact that should bring only disrepute and blurred image but the ruling clan and policy makers seem to have crossed all the limits beyond the working spectra of reasoning and intelligent mind set.

But still they are going on with this all. We have no other option except to leave this discussing here only and proceed further in our derivation.

So, the Indian government or the ruling elite simply murders every Indian national exactly some twenty years earlier than his otherwise real death.

Can somebody do something to stop this?

Should we pray and plead upon God to send a rescue to put an end to this? Many people have already done this but He has not been moved, even a little. He has not done anything. Even God is not doing anything.

Surprised?

God and destiny? Where are they?

They are functionless as they have already destined a worthless and incapable government for the people of India. In fact they have gifted a faulty system of governance to rule this once great nation.

Let's come back once again to the pivotal question.

A nation of 120 crore people.

An average life span of 60/65 years for men/women.

This simply means that 60 to 65 crore people are killed, each one 20 years before his/her actual death.

Such 60 to 65 crore people would have lived to 80 to 85 years each in other countries. This means that there is a loss of 20 crore human lives in a span of about 60 to 65 years on an average.

In a country of about 120 crore people, this loss becomes of about 40 crore human lives in 60 to 65 years.

This means that nearly 6 to 7 million human lives are lost every year in India due to poor living conditions.

And these poor conditions are the abject poverty and hunger.

And...... nothing else.

Poverty and hunger are the grips that hold these poor living conditions like the strong foundation platform that holds the great and magnificent structure of the Empire Estate Building in New York.

But? Another point.

Aulprofem is a known poverty fighter.

Fighter and eliminator.

So, when Aulprofem will be in action, the poverty of India will fume to disappearance and along with this, all the suffering and deaths due to poverty and hunger, will also disappear.

There are many countries where the living conditions that prevail in reality are akin to the conditions in India. In these countries, the application of Aulprofem can be undertaken on macro levels to eliminate poverty and hunger and improve the life spans of the people there.

There are vast majority of the mankind living in third world countries which account for a near seventy percent of world population and life conditions in these countries are almost similar to the ones existing in India. Depending upon the mortality equation that we have discovered for India and expanding it to the third world, we can expect to save nearly 15 to 20 million humans that are subjected to termination of their life spans every year due to abject poverty and hunger conditions prevailing in these countries. The subject does not end here and it tends to expand further and take many things into its fold. Just we step to look a little further.

A clause in the will of Alfred Nobel lays down two conditions for the award of the prestigious Nobel Prize in Medicine. A prominent condition says that award will go to the person who has conferred 'greatest benefit on mankind' with his discovery. Based on this Charles Nicole received the prize in 1928 for his work on typhus and establishing a role for the simple body louse in the spread of this disease. Thanks to his discovery and the simple delousing procedures that during the First World War, hundreds of thousands of human lives were saved. Again in 1948, the prize went to Hermann Muller for the discovery of dichloro-diphenyl-trichloromethylmethane, commonly known as DDT. This chemical was instrumental in saving human lives from insect transmitted diseases such as malaria. The World Health Organization (WHO) estimated that during the period of its use, approximately 25 million lives were saved.

Today, DDT is a banned chemical due to its bio-accumulation property in the living tissues in plants and animals but at the time of the award of the prize, the point of 'greatest benefit on mankind' was quite obvious.

The discovery of Aulprofem, being a formidable poverty fighter and hence playing a pivotal role in stopping deaths due to poverty and hunger and starvation; will be a strong contender for the title of the 'greatest benefit on mankind'. There lies the most powerful explanation for claiming such honor for a discovery that will be saving some 15 to 20 million precious human lives per year and every.....every year.

Now we have cornered upon a discovery that can easily save equivalent numbers of humans from clutches of death every year that this magnificent discovery of DDT did in nearly 25 years *i.e.* nearly one million lives per year. So, the discovery of Aulprofem easily proves to be a considerably bigger achievement if viewed from the angle of saving human lives from hellish clutches of death.

There are many witnesses to the testimony of the great killing powers of poverty and hunger. Maxim Gorky (see Appendix I) is one among them. This annihilation duo can do wonders in the killing prowlness, was elaborated in his novels and other books like 'Mother'. Tolstoy also testified this in his greatest creation 'Resurrection'. Charles Dickens is another such witness and Munshi Prem Chand (see Appendix I) is one more. In fact, all wise and intelligent men bear a testimony to these magnificent qualities of these allies in death. May be he is Rousseau, may be he is Voltaire, Hegel, Fast, Lincoln, Neruda, Russell or Laski (see Appendix I). We need not explain much on this topic. The world literature is full of such stories and descriptions; all told in much greater clarity, length and details.

To exercise a rein on the wild stampede of these biggest enemies of mankind; the poverty and hunger, we have to do nothing but to promote and propagate the applications of the wonderful discovery of Aulprofem on the dairy and related animal wealth of the world. This will open before mankind the unlimited bounties of prosperity and grace in all possible ways of the utmost smiling and blessings of the Mother Nature, showered upon mankind. It will be an Eden rediscovered and the most beautiful dream witnessed with the open and wide awake eyes on the arenas of the practical striving and zealousness, to achieve and attain to excel.

We can make it a living reality with the power of Aulprofem on our side.

Let's pledge and commit ourselves to this mission of human salvation.

Salvation from hunger and torment and lack and suffering.

For ever and ever........and forever.

Alleviating Poverty

Some wise men argue like this:

When money is gone, nothing is gone.

When health is gone, something is gone and

When character is gone, everything is gone.

This is a universal truth and there is no debating over it.

So, intelligent people in high places, especially in the governments in countries like India, have taken it as a principle and made their people really moneyless and destitute.

This is exactly how wise people should behave.

It can be argued like this.

People do not have housing, proper clothing and definitely are underfed most of the time and even starved sometimes and starved regularly when the great forces of nature like super-monsoons, cyclones, draughts, become strategic allies of their governments. A draught is officially declared in some of the most prosperous of states like Gujarat in 2000 and Punjab, Haryana, Uttar Pradesh (see Appendix II, state of India) etc in 2002 and a famine is declared in Rajasthan in 2007 in India.

The world famous famines of Ethiopia, Somalia, Sudan, Rwanda and other such countries are part of magnificent development and management achievements of the governments in these countries.

Good governments, bad governments.

Developing nations and developed nations; all are terms.

Good terms, then what is poverty?

There are chawls in India.

There are ghettos in the U.S.

Destitution and lack, are the two monsters that seem to have the entire world in their deadly claws.

Poverty is defined as the condition of living in humans when material provisions for normal life; like food, clothing, shelter etc. cease to exist at levels, required for sustenance of healthy and proper life conditions and going still down beyond them, means difficulty in continuing of human life itself. These days non availability of medical help in case of need for it, has also become a criterion for poverty. The criteria of poverty are different in different countries and concern is also shown for living standards or lack of them. The presence or absence of certain pivotal criteria like availability of clean drinking water, sanitation conditions, means of transportation, social security etc, also form a basis on which poverty of a given country, area or people is judged. But basically in most of the cases the fundamental condition of unavailability of one or all; of food, shelter and clothing, is still considered the basis, on which poverty is judged.

How to fight poverty is a great question.

Can Aulprofem do something about it?

A discussion can follow like this.

The Aulprofem surely guarantees the availability of the first condition for the non existence of poverty *i.e.* abundance of food and that also of the healthiest type for mankind. The excess availability of healthy food and nutrition can act as a permanent barrier for the poverty to set in. Aulprofem also guarantees the availability of other criteria for non existence of poverty *i.e.* clothing by ensuring excess production' of wool across the globe. The wool is the best clothing in cold climates and regions. A large number of humans live in such conditions world wide. The third parameter for the state of poverty to be present is the non availability of shelter and from ages the humanity has learned well how to tackle the problem of the provision of shelter. The rural areas in India has seen the masses living in houses made from straw, thatch, mud and they still live in houses made from these filthy compositions. Nearly plus fifty percent of populations in India live in these wonderful inhabitations without electricity, potable water, proper sanitation and other civilized criteria for the classification of modern *Homo sapiens septens* as the most evolved creature on earth. The Aulprofem directly does not offer a solution for provision of shelter to mankind but indirectly it offers a big solution for it, by solving the first two of the humanity's most basic needs *i.e.* food and clothing and thereby making the minds of humans' free of the worries associated in procuring the sufficient supplies of these two and so, they can now concentrate their attention on getting the third one i.e. shelter. So, it can be concluded that with the power of Aulprofem in hand, the mankind will always be free of the worries related with the availability of the three most basic necessities of life, *i.e.* food, clothing and shelter and the most fundamental truth of the world with Aulprofem will be the non existence of poverty for ever, ever and ever.

Exaggeration?

It may seem. But the discussion tends to continue.

One more point in this discussion is regarding the provision of the type of nutrition with the help of Aulprofem to mankind. It is a universal knowledge that milk is one of the most nutritious foods that mankind has ever consumed, only comparable with meat. And milk and meat are the two foods whose production will be increased with the application of Aulprofem. A nutritional diet is the best defense of human body against disease. It builds immunity within the living systems. With milk and meat in abundant use, the humanity will be much armed against the various diseases with the result that people will fall ill very less and they will need medical help to the much lesser levels. The good prophylaxis is always preferred to medical treatments and provision of healthy and nutritious diet is the best prophylaxis that the humanity has ever explored. So, Aulprofem means the provision of medical care by ensuring the proper conditions for health prophylaxis. The conclusion of the story is that Aulprofem secures the mankind on the front of medical health also.

The question of poverty has some other angles also.

Can poverty live in collaboration with prosperity? The poverty often is not visible. In fact it is ignored. Ignored by the civilized world, ignored by media, ignored by the rich who see everywhere the wealth, luxuries and great privileges of life that can only be offered by this wonderful phenomenon of poverty. Cheap and abundant servantry, the needy people who afford to go to any length of submissiveness and surrender which can also be of moral type most of the times; is the greatest boon offered by the poverty to the rich *persona populace*. The glitter of share markets, the fascination of shopping malls, the glamour of tinsel towns of filmdom, the multi star hotels, pubs, discotheques, the call girls, the underworld and business escorts so glorified to unlimited proportions by movies like '*laga chunri mein daag*' etc; all are hall marks of development that can not be built without the concrete mixing of poverty in the foundations of the modern civilization. Judging from above details, the poverty seems to be the ultimate bliss that could be bestowed upon man by the **Almighty Himself**.

A country where poverty and prosperity coexist, nobody wants to confess poverty. Even the poor do not want to claim that they are poor. A spirit of *poor-o-phobia* exists in the society. When some self possessed and egoistic rich hold public demonstrations of charitable acts; only then we come to know that there are long queues of people standing for hours together in burning sun of May and June to just claim a loaf or two of stiff and stale bread which usually is not of hygienically potable grades or some old ladies waiting for day longs to accept in honorable charities, a few kilograms of dry rations or vegetables in front of the famed flashlights of media cameras and enthused press reporters. The beggars in front of temples, railway stations, over bridges, market squares, are the only visible certificates of poverty that can be judged without a prejudice or mental bias.

But Aulprofem certainly is the enemy of all this. Whereas it is quite obvious that it can put an end to almost all the glorious virtues of poverty; it is also very evident that it can lay the formidable force of poverty in the universe to a permanent rest of eternal death for ever and ever.

So, the Aulprofem secures the mankind on vital fronts of provision of the most basic necessities of life *i.e.* food, clothing, shelter and medical health. The world with sufficient supplies of these amenities, will be a world free from poverty.

Free from worries.

Free from mental, physical and social trauma that starts with the birth of a human to some poor parents and extends full course throughout his life until he reaches his final destination of the cool and peaceful environments in his grave.

Aulprofem guarantees the abolition of all the vices and evil manifestations associated with the gigantic phenomenon of poverty. So, the existence of Aulprofem is the biggest safety condition for the non existence of poverty in this world. This makes the discovery of Aulprofem, a great savior force and true guardian of mankind in the war against poverty.

The Biggest Environment Friendly

There is folklore in North India saying:

'Which is the best way to face the God's wrath?'

The answer to it says:

'Change the bad God with the good one'.

Back in late seventies or early eighties when India was still a very backward, semi-feudal agrarian economy and it is still more or so; we in villages, where there was no proper electrical appliances; fans or room coolers etc in rural homes, used to go roof tops; on houses, made partially of mud and kiln bricks cohered with mud instead of cement and with non concrete roofs, to have sleep during hot and humid summer nights. Fighting all nights with swarms of mosquitoes, maneuvering like jet fighters planes and virtually displaying all tactics of splendid air combats, to attack us continuously throughout the nights; we were nothing but helpless, defenseless Austrians facing the wrath of powerful *Nazi Blitzkrieg*. The only demarcation was that in our case there was no lull in the war and every night turned out to be a sleepless, aching torture. However, during certain nights when there was some wind blowing, we could get temporary relief from mosquitoes and heat. If such happened, people used to say from all rooftops around,

'Today someone who has daughters is the chief of gods' cabinet'.

On other days, they used to say,

'Today someone with daughters-in-law is heading the cabinet of gods'.

Listening that everybody would smile as the difference between daughters and daughters-in-law was clear-cut in that semi-feudal society.

Aulprofem does not have a direct effect on the environment.

But on the other hand, it has the power and potential to change the entire bad gods' cabinet or the bad God with the good one.

How?

The answer is not complicated at all.

In fact, there are many answers to it. Some of them are as below:

India is an agrarian country.

Humanity needs food.

The world needs agriculture to feed plus seven billion people.

The agriculture needs efficient cultivation.

The efficient cultivation needs genetic dwarfs.

The genetic dwarfs need fertilizers.

They also need pesticides, insecticides and herbicides etc.

Fertilizers and insecticides etc cause chemical poisoning and toxic residuism of soils.

The entire soil strata and underground water beds get polluted in the process.

The genetic dwarfs cause genetic poisoning in the species that eat them or their parts including animals and humans alike.

The more bitter truth is that we in our desperation to feed more and more growing populations, have started to develop and rely upon genetically modified crops.

These genetically modified crops are more likely to cause genetic poisoning in humans and animals that feed upon them or their parts. There may be a spurge in the evolving of many a mutant species in *flora* and *fauna* due to this. Many of them being extremely harmful and deforming.

Many of them being new species of disease causing bacteria and virus.

Nobody ever heard of AIDS, hepatitis or LGBS some forty years before.

And many of such deadly gifts may soon be in store for the descendants of Adam.

Still we have genetic dwarfs.

But we are helpless dependants on them.

And we have placed manufacturing orders for the genetically modified crops with the Devil's factory of scientific developments.

And there is no way out!

We could have gone to the good needs store of the Mother Nature.

But............? Not....the excuses factory is never less performing.

And....

The situation is a really horrible one.

The populations are still growing. In fact they have started to explode.

A deluge of people!

Who will feed them, is the billion dollar question?

Another question still looms large.

Who will save the environment?

A patillion dollar question. Isn't it?

Who will answer these questions? Probably the nature has an answer.

And the nature has already answered it.

It has developed Aulprofem.

O' ho! A stupid derivation?

Isn't it?

Let's not run to the conclusion. The discussion needs some intelligent and elaborative bickering and a dive into the facts.

Facts that are really live and burning.

Burning with truth.

Another point. We have to start the discussion from, say a point.

And the point may be.

'Aulprofem is the savior'.

But how?

Let's discuss.

Grains (rice, wheat, coarse ones etc), pulses, vegetables and fruits, together form the main food of mankind. In addition to this, humans resort to two other foods; milk from dairy animals and meat from terrestrial or aquatic animals.

It is grains plus pulses, vegetables and fruits that form the basic diet of humans. Such a diet can be splendid but it cannot always fulfill the required quotients of nitrogen and sulphur or fats and proteins etc. Meat is a good supplement and it is used by humans all over the world in various forms like beef, mutton, chicken, fish, sea food and a whole variety of things. It is certainly superior and has a role when it comes to meet the dietary requirements of proteins, fats and vitamins etc, but it has an efficient and perhaps a better rival.

The milk.

The milk; a diet with which every human starts his life can be the better option every time we think of supplementing basic diet with anything. The argument has many takers and is believed and in fact practiced by many people across the globe. It has the certification of nature as the first ever food to be fed to every mammal when it comes as a neo-natal in this big bad world. The first and the best. Many people in the world believe in this. And the conclusion has; in a way, logic in it.

Meat is very rich in proteins, fats and other nutritional ingredients, necessary for the proper development and maintenance of physical health and mental well being. The quality and qualification of meat to act as human diet is unquestionable but it can not be taken continuously as humans fundamentally are not carnivores. Eating meat regularly for many days, causes certain problems to health in humans because man is basically not designed to live on flesh. His teeth are not like carnivores or felines. His digestive tract is not so suitable. The intestines are too longer and not suited to digest meats. In fact they are not suited to digest cellulose also. The digestive system of humans is perfectly designed to eat mixed diets containing vegetative and non-vegetative foods. The balance of human diet is such that it is basically vegetative and should be occasionally supplemented with non vegetative foods. All primates are made so and man being one among them, is also the same. So, the provision of meat as a basic food for humans is simply ruled out. It is at best a supplement and that also an occasional one.

But milk is different. Probably it is so because it is the first food that humans consume when they open their eyes in this splendid world, full of surprises, good or bad for all the inhabitants and newcomers to it. The first few minutes of tryst with the destinies of all human infants and they encounter their first experiences with milk. The milk is hence for with engrained with the rest of their lives and probably it is the best food that their intestines are designed for. After a few months or an year or so, the human kids start to take other diets such as soups or puddings, meshings, pulps, juices, food grains, cereals, pulses, vegetables, fruits etc. In fact they are supplemented with milk still as the basic diet. As the time passes on, they are pressed into service more and more to replace milk slowly out of the diet; sometimes completely and sometimes partially. It has to depend upon many things. One among them is the social status of the parents. The poor certainly can not afford to provide milk to their children as it is relatively costly and out of their reach but the wealthy and well to do certainly opt to go on with a diet with milk as an essential ingredient for their children for considerably longer periods. The more health conscious and rich people prefer to continue with milk as a necessary ingredient in their diets for life longs.

The milk does not have side effects as that of a meat diet as human digestive system is well suited to a diet, containing milk. It can be had daily without the usual side effects of a meat diet. In fact many people, who can afford this, continue with milk as a daily essential part of their diets, usually in bigger amounts. Probably many people skip meals at least once a day to opt for milk *in lieu* of meals. The usual three meals a day routine can very easily be substituted with two meals and milk one time a day and many people in fact do this. This does not give any bad side effects but on the contrary it adds to the good health development of the people who go for it. Milk is considered as one of the best foods ever discovered for man and is well and rightly touted and publicized as the balanced diet for humans. In fact it probably is a balanced diet, perhaps the only individual balanced diet ever discovered by man for him. The beauty of this situation is that humans can continue on this diet regularly and continuously for many days, weeks or years and still remain healthy and strong. This is such a privilege that can not be enjoyed with any diet, other than milk. Even if we continue with other diets and just supplement our eating three meals a day routine with two meals a day and one time milk diet, we can continue to remain comparatively healthier and stronger. In any case however, the daily intake of 400 milliliters of normal milk with 8.5% S.N.F. (solid material and natural fat) and 4.5% fat is sufficient to keep a man healthier and viable.

Milk contains proteins, fats, carbohydrates, vitamins and what not? It is one of the best diet supplements. In fact it is one of the best diets ever reported or even it may be claimed that it is the best of the bests. A daily diet without milk or meat may have certain handicaps. It may not contain proper proteins. It may not contain fats. It may not contain vitamins and other nutritional ingredients. In that case, we have to supplement the normal diet with milk or meat etc. The meat as discussed has certain digestive drawbacks and can not be used on a daily basis but the milk does not have these handicaps and can be continued with daily usage. So, the normal diet with food grains, vegetables, fruits etc can suitably be supplemented with milk on a daily basis without any side effects and loaded with benefits only. To sum up the above discussion, it can be said that milk is the best diet of the human race on earth and this point has been proved without doubt by the various nutritionists, scientists and wise men. From centuries, it has claimed and defended its status as the best food for humans and it deserves this standing rightly and logically.

The impending food crisis on humanity can be salvaged with the abundant production and supply of milk throughout the world. This is a point which is not so hard to press and make people understand. But the main question remains as how to produce enough milk on such bigger levels that it can be provided to one and all human inhabitants in the world? The answer is very difficult to find but the Mother Nature has answered it with the single word and that word is:

Aulprofem.

The Aulprofem is the solution and the salvager.

But how?

Let us discuss.

Aulprofem produces more female off-springs in dairy animals. In fact it produces female off-springs with a near 100% success rate. And it is a fact that the female calves grow to mature into female animals. The female animals then in turn get fertilized and deliver and yield milk to be consumed by their calves and humans also. This is the milk that is the best balanced diet for the humans. This is the milk that is wanted by humans more than anything as is concerned to meet their daily dietary needs. The equation becomes very simple. Aulprofem means more female calves. More female calves mean more cows and more cows in turn mean more milk. So, the Aulprofem means more milk for humans. This is the milk that can be produced and supplied to already exploded populations on the verge of starvation. The food crisis can be averted and in fact salvaged.

The environmental question comes into consideration at this point. The more milk means less dependence on food grains. Less dependence on food grains means less focus on intensive farming. Less focus on intensive farming means less dependence on hybrid dwarf varieties or the genetically mutant varieties. Lesser the use of these species, lesser is the use of fertilizers and pesticides, herbicides etc. Lesser the use of these chemicals and poisons, lesser is the damage to the soils and underground water beds.

In order to free the under ground water tables from receding and poisoning and to rid the fertile soils of the harmful chemicals and pesticides' etc residues that have been showered by the so called green revolutions over them, it is very important to tilt of the usual farming in favor of the dairy farming. In dairy farming, only fodders are grown which do not need any insecticides, herbicides and even the fertilizers are used to very low levels in their case. The lesser the use of these poisoning agents, the lesser will be the environmental damage.

The main danger to the world environment has come from the agricultural chemicals, fertilizers, pesticides, herbicides etc. With the weapon of Aulprofem in hand, we can undo this damage to the environment by totally shifting the focus of agriculture productions in favor of fodder cultivation from the present grains production. More fodders will be needed for the more female cows generated by the Aulprofem and more milk will be produced by these cows. More the milk, lesser will be the demand for food grains. Lesser will be the demand for the food grains, lesser will be the dependence on hybrid dwarf varieties. Lesser the dependence on these hybrid dwarfs, lesser will be the use of fertilizers and pesticides, herbicides etc. The end result will be the less damage to the environment. In the end, a situation can be achieved where there will be much decreased dependence on intensive agricultural performances. Under these situations, it will be possible to undo with the chemicals and hybrid dwarf breeds based agriculture and revert back to the old age organic farming where there was no damage to the environment. The world at that time will need less food grain production. There will not be much pressure on the farming sector. There will be no need for the genetically mutant or the hybrid dwarf varieties. There will not be the use of fertilizers and pesticides, herbicides etc. There will not be any chemical poisoning. The soils will be free of the poisons and the toxins. The under ground water beds will be free from the chemical poisons and toxins from insecticides, herbicides etc. The world will be a better one without poisons and toxins, without which we can very well do and live our routine works and duties in better way in climates and environments much improved.

Thanks to the discovery of Aulprofem, that Mother Nature has blessed the mankind with a salvager to save the environment.

The humans have only one Earth, the mother earth. The meaning is very well clear for one and all. In the blind race for progress and profits, the man has proved to be a bad son of the mother earth. There were major food crises' in the sixties and seventies in the twentieth century. The agriculturists and farming scientists all over the globe worked day in and day out to save the humanity from the dangers of starvation and hunger that plagued mankind and they were successful too in that. There were great scientific geniuses and visionaries like Lord Borlaug and others who developed models for the humanity to tide over those crises of food scarcity. They were successful achievers. They achieved their targets. The food productions across the globe increased. In fact they increased to much greater levels. At that time, these were the greatest conquests on the food front. It seemed that humanity had won the final war on hunger. Everybody was happy. Almost every citizen of the world rejoiced in those days. You could smell the sense of pride on the face of an Indian or a Pakistani or a Bangladeshi, Sri-Lankan, Nigerian, Libyan, Kenyan, Korean or a Filipino. People boasted of their respective governments' achievements to tide over the food crises. People had added one more word to their vocabularies, the 'Buffer Stocks'.

Everybody knew what was the meaning of the word 'Buffer Stocks'.

Buffer stocks meant something very familiar to everybody. It meant something like the boy next door. It seemed like a friend. It seemed like a very near and dear relative. It seemed like a friend in need. It seemed in deed the greatest shelter that you could fall back upon at the time of a crisis. Every poor could tell his family that he knew of a treasure unearthed and kept well preserved and in good shape for him and his family by the government. That they could stay unworried as buffer stocks was the biggest insurance policy that could be ever conceived of for their families. The best social security promise could never match in the contents, to the buffer stocks.

But the time trickled on and on and one fine day it was as clear as the broad daylight that the growing populations had consumed all the buffer stocks or had rendered them meaningless as compared to grown up population magnitudes. The faith in buffer stocks had evaporated like the heated camphor or the ammonium chloride. The buffer stocks were no where. The people no more talked about them. There was not only the air of disbelief but even a state of panic prevailed everywhere. A new vocabulary had become very popular and it prevailed every where.

Ominous words. People talked about them like the stories of some deadly earth-quakes. Like some train accidents of head on collisions. Like the Nazi gas chambers. Like the Taliban's bombing of World Trade Center.

New vocabularies had started to command every brain. In fact every human wanted to demonstrate his skill and mastery of the knowledge of new discoveries of destruction that the blind race of development of modern agriculture had produced as the illegitimate, by product progenies of thoughtless genetic tempering with crops and plantations.

Every human was bubbling with knowledge.

Knowledge of various words.

Insecticides, pesticides, herbicides, fertilizers, chemical poisoning, environment damage, pollution, contamination of water beds underground and all horrifying words of all types and sorts.

Strange words.

Words that generated fear.

Words that fired like German *MG-42* machine guns mounted on *tripod* gears in second world war or bombed like *Messers* fighter planes of *Luftwaffe* or like *saber jets* in the Korean war.

People were really frightened. The words fell like *carpet bombings* from their brains into the barren Afghanistan lands of their listeners' thoughts.

The environment was damaged. The pollution was the talk on everybody's lips. The green revolutions had gifted the mankind with increased food grains and buffer stocks but they had also rained the earth with pollutants of all types. The mother earth was no more beautiful as before. It was miserable, polluted and pitiable as a deserted whore with open syphilis sores on her body and mind. There were countless gifts of development. The kidney failures, the liver malfunctions, the cardiovascular diseases, the AIDS, the hepatitis, retinopathies, diabetic epidemics and what not? The new diseases of many types and dreadening body syndromes had abounded the world and had surprised and shocked the mankind. There was a sudden surge in these and other killer diseases. And it was not long before the reason for this was also known.

The reason was the pollution and the environmental damage.

Everybody was scared but concerned. The curse had befallen on the *Adam's breed*. The apple had turned sour and ominous. It was nobody's fault. Nobody had intended the outcome. Man had simply played the innocent monkey playing with the matchbox and the burnt city was in no way a fault of his doing. Who could have thought that the fruits of development would be so repulsive and unpalatable and vicious also?

Humanity now wanted a genuine solution. Many alternatives were suggested. Most prominent among them was to revert back to the old organic farming. That seemed promising but the only drawback was that the food productions could fall to considerably low levels that way. This alternative was much discussed but never followed. At best there were some isolated and localized adventures into this thought and practice but the concept could never be followed in its totality. The reason as told above is quite obvious. The organic farming was the best bet of humanity against chemical poisoning and poison residues generated by the most magnificent green revolutions. This was the most valuable intellectual asset touted by the so called naturalists, environmentalists and social reformers alike. But this soon turned out to be a hollow hallucination. It was not long that many people became aware of the futility of this isolated fascination of the intelligentia. This was dreadful and ominous fallout. The messiah was long awaited, advertised and well advocated for and the most prominent achievement it could gain was its drastic failure. This was a miserable crash of all humanity's hopes. Now all optimism seemed to had transpired into desperation.

Only a miracle could save the mankind but miracles seldom happen or even if they do so, they do it very rarely.

But the Mother Nature is very kind and benevolent.

It really performs miracles sometimes. And the discovery of Aulprofem is such a miracle. This is such a miracle that has come to the ultimate salvage of mankind from these twin problems of food scarcity and environment pollution. This is such a discovery that promises salvation from both these problems.

The Aulprofem will increase the milk productions to greater heights and can revert the intensive agriculture based economy into the dairy based one with more emphasis on animal husbandry. There will be more number of female cows. There will be more milk. There will be more fodder crops. There will be lesser intensive farming. There will be lesser use of fertilizers and insecticides, pesticides, herbicides etc and there will be less chemical poisoning and environmental damage.

Without Aulprofem, this achievement can at the most be a day dream and it can never be converted into a concrete reality. The Aulprofem ensures proper and nutritional food *i.e.* milk for the humanity and it can be thought to increase the milk productions to greater heights and so the agricultural grains productions can be brought to relatively low levels and the intensive, dwarf hybrid based and fertilizer and pesticide, herbicide etc dependent farming can be done away with. This will pave the way for real and sufficient natural farming. This natural farming will definitely be different from the much touted artificial organic farming of today. This will show the path away from hybrid dwarfs and fertilizers, pesticides, herbicides etc and at the same time it will ensure the abundant and healthier nutritional supplies for the mankind.

This will rid the humanity from chemical poisons, toxins and poison residues of all types.

And this will be a great achievement.

The salvation from chemical poisons, toxic residues from the crops, soils, water beds etc.

This will clear the way for cleaner environments.

This will be the success of Aulprofem and the biggest gift, a scientific discovery can provide to the mankind.

The gift to breathe free.

The gift to fill your lungs with fresh air.

The gift to feel free.

To feel free of the dangers of pollution in the air.

Pollution in the air of the flying poisons.

Flying poisons of sprays of insecticides, weedicides etc.

Poisons and poison residues.

The soils will be free.

The waters will be free.

The minds will be free.

The poets will sing.

Like Rabindranath Tagore:

'Into this ever widening and awakening of thought and mind,
O' my Father,
Let my country awake.

....
....

Where the mind is without fear'

And this will be the greatest blessing of Mother Nature to the human race on earth.

A Case for Rural and National Development

Introduction

Aulprofem is a mammoth project. It is a project that can cause an increase in the GDP of India to the tune of five lakh crores in rupees (or U.S. dollars 100 billions). This being the first sex fixer of the world, is all set to open a new era of biological research and exploration. The importance of this valuable development in medical research is tremendous and it has given an altogether different view and look to the concept of sex assignment and determination. The heavyweight scientists like Ericsson *et al.* (1973), Meistrich (1982), Corson (1984), Gordon (1958), Lindahl (1956), Beernik *et al.* (1984), Barrat and Leger (1979) etc were all trying to give a new meaning to the concept of alteration of sex ratios in mammals but all failed miserably as the repeatability, verifiability and consistency criteria of the results of their scientific achievements were concerned. This discovery based on its 100% success ratio will definitely give a new meaning to the most basic and important biological phenomenon of sex determination which is virtually the foundation of all living creations in the universe. The chromosomal theory of determination of sex will have to be changed accordingly in the light of the successful derivations drawn from this magnificent scientific breakthrough.

The world is just on the verge of a food crisis of colossal dimensions. The hunger, starvation and famines are going to be the order of the day if the current trends of falling food grain production and the failure of green revolutions due to decrease in hybrid vigor, are any indication of. The much touted second green revolution is nowhere in sight and considering its total dependence on genetic modifications, a concept that is never approved of by nature; there is certainly an inherent failure tag fixed to it even before its becoming a working reality. The nature has developed a genetic balance of its choice in the course of long evolutionary period spanning hundreds of millions of years and it has always been a perfect one for any specific period of biological existence depending upon the ways, nature has selected and evolved.

The world populations are growing. There is an ever increasing demand for food. Aulprofem can increase milk production across the globe many folds to the present levels by increasing the numbers of high yielding milch varieties of dairy animals. Similarly meat and wool productions can also be increased by increasing the numbers of animals like goats, pigs, horses and sheep etc. with the application of Aulprofem on these species. The green revolutions has already gifted the soils of progressive states like Punjab, Haryana, Andhra, Maharashtra, Western U.P. (see Appendix II, States of India) etc with heavy doses of insecticide and pesticide poisoning.. The foods, underground drinking water and even the atmospheric air has become loaded with toxicities of innumerable types. This has led to disease and suffering in animals and humans, of proportions and types unknown in history.

The animal wealth of India is 300 million dairy bovines. If only a 10% of them are taken for the application of Aulprofem for a short period of 10 years; then the annual increase in milk production will be nearly 160 million tonnes per annum, giving about 400 grams of milk for every Indian per day in addition to saving nearly 70-80 million tonnes of food grains, 10 million tonnes of pulses and nearly 100 million tonnes of vegetables per annum.

So, the Aulprofem is a milestone in biological research as well as a savior of mankind in feeding the ever increasing human populations, the world over and saving the world from toxicities of poisons of development like insecticides, pesticides etc and creating a clean, pollution free, hygienic and livable world.

Aims and Objectives

The aims and objectives of this project are as under:

1. **To increase milk production across India:** The main function of the project Aulprofem is to increase the milk production in India manifolds the present level. According to an estimate, if even the 10% of present animal wealth of India is brought under this project, then after about 10 years of its results transforming into practical reality the annual increase in milk production in India will be about 16 crore tonnes. That simply means an approximately additional half liter of milk for every Indian.
2. **To generate employment in rural India:** Aulprofem will be a mammoth employment generator in rural India. The generation of

employment with Aulprofem will be nearly six to seven thousand people fully employed and six to seven lakh people partially employed across India.

3. **To increase meat production across India:** With Aulprofem, we can produce more females of the meat producing species such as pigs, goats, yaks etc. and hence increase their numbers and this way there can be more meat production.

4. **To increase wool production:** With the application of Aulprofem on sheep, there can be produced more female sheep, thereby increasing their populations and the result will be that there will be more wool production.

5. **To produce healthy India:** With more production of nutritional foods like milk, meat etc., there will be enhanced health status of Indian citizens. This research will prove a milestone in creating a future healthy and strong India.

6. **To decrease poison levels of insecticides, herbicides, fertilizers etc:** These days there is too much of the insecticides', pesticides', herbicides' etc. use in agriculture. Under the slogan of green revolution, our agricultural universities and research institutes have developed, varieties that require the use of urea fertilizers and insecticides, herbicides etc in greater volumes. The result of this has been the excessively increased levels of these poisons and their residues in the soil. Most affected are the relatively progressive farming states like Punjab, Haryana, Andhra Pradesh, Maharashtra etc. With more application of Aulprofem, there will be more emphasis on fodder crops which usually do not need above mentioned poisonous materials. Hence our land will become free of poisoning by these harmful chemical agents.

7. **To clean the eco-system of India:** By laying stress on more dairy based farming and with the resultant need of fodder crops, there will be increasingly less use of substances, hazardous to health and eco-system such as insecticides, herbicides etc. This will go a long way in cleaning our polluted eco-system and putting the natural environment on the corrected and right track.

8. **To rid the farmers of the rural debt:** With national revamping of rural Indian economy and being an instrument of development and prosperity, Aulprofem will do a lot to strengthen the otherwise weakened financial position of the entire farming community of India.

9. **To develop dairy industry:** With more production of milk, there will be more development of dairy and dairy based industry.

10. **To develop meat processing industry:** With more production of meat producing animals, the result will be the more development of meat processing industry.

11. **To develop hosiery industry:** With more production of natural wool for increased numbers of sheep, there will be a boost to wool industry.

12. **To earn foreign exchange:** With more development of milk, meat and wool etc. based industries, there will be more production of the materials related with these industries, which can be exported to earn precious foreign exchange.

13. **To save foreign exchange by decreasing imports:** By producing more of the dairy, meat and wool etc. industry products, there will be no need to import these substances and hence a lot of foreign exchange will be saved.

14. **To stop farmers' suicides:** With the strengthening of financial status of the farmers, there will be more prosperity in rural India and the current trend of farmers getting trapped in debt burden, can be reversed and no more farmers will be forced to commit suicides.

15. **To enhance social status of the farmers and rural people:** By strengthening the financial standing of the farmers and the rural people, Aulprofem will enhance their social standing as respected and honorable citizens of India.

16. **To fight social menaces of terrorism and rural unrest:** By bringing wealth and prosperity to rural India, Aulprofem will certainly help to fight the growing resentment and de-alignment of rural young Indians from the main line society and hence it will prove an effective tool in fighting evils like terrorism and violent antinational movements.

17. **To fight growing crime in rural society:** With obvious economic benefits of Aulprofem to rural society, it will be an effective tool to stop the growing tendency of the rural youth getting attracted towards the crime and organized mafia etc.

18. **To create a sense of victory of femaleness:** Aulprofem is the world's first mammalian female sex fixer. So, it will definitely carve a feeling of the victory of femaleness in the minds of general people.

19. **To create a sense of national prestige:** Aulprofem is a mega discovery on Indian soil. So, it will produce in the minds of Indian people, a sense of national prestige and patriotism.

20. **To stop the import of costlier foreign techniques:** Just recently, Haryana government has imported a very costlier imported technology from the United States for the production of sexed semen. With

Aulprofem in use, there will be no need to import such costlier technologies.

Need for rural development

India lives in villages. This is the most fundamental truth about India. In 1901, 213 million people lived in villages as compared to 26 million in urban areas and in 1951, 299 people dwelled in villages as compared to 62 million people in urban areas and in 2001, 743 million people lived in rural areas as compared to 286 million in urban areas. The percentage of rural to total population in these corresponding years of 1901, 1951 and 2001 was 89.2, 82.7 and 72.2 respectively. This means that the great majority of Indian people has been and is still living in rural areas. So, the real development and upliftment of Indian people means the betterment in life and living conditions of rural people.

Status of rural and urban India

Despite the vast majority of Indian people living in rural areas, it is not a bright picture as concerned of the general life and progress in India. The position on ground level is particularly not good. Nearly 25 to 30 crore people live under the poverty line. There are total 36.9 crore workers in the unorganized sector in India and 23.7 crore workers are alone in unorganized sector in rural India. There are nearly 40 crore people nationwide who do not have access to clean drinking water. There are nearly 16 crore people who cook their daily meals on the fire of cow-dung and there are nearly eight crore people who have cooking gas in their houses in total India. Nearly 30 crore people live in kucha houses and there are only 17 crore people who have all the three facilities of electricity, toilet and safe drinking water in their homes across India.

Rural development with Aulprofem

Aulprofem can be instrumental in the development and betterment of living conditions of the rural people in following ways:

1. **Economic welfare of rural people:** The economic benefits of Aulprofem are very obvious. There will be more production of female cattle and buffaloes in the villages and the milk production will increase. This will give added incomes to the farmers, thus improving their financial status and conditions of living. The economic welfare is usually the lever of development. So, the benefits to rural people from Aulprofem are beyond questioning.
2. **Employment generation for rural people:** There will be direct employment of full time nature of 06 to 07 thousand people and to add to it there will be part time employment to nearly six to eight lakh people. The employment generation on such a large scale, will

add to the economic benefits of rural people and benefit them with more progress and development.

3. **Health related benefits for rural people:** With the increase in protein content in the daily diets of the rural people their health status and condition, will improve. The present intake of protein is 10 grams per day in the diet of an ordinary Indian whereas the corresponding figure is 25 grams for the world average. So, the protein intake is very less for the rural and general Indian citizens. With increased milk and meat productions, this daily intake of protein for the rural and general Indian citizens will increase. This will definitely improve their health conditions and status and even they will be more competent to fight the onslaught of diseases due to increased body immunities with increased protein contents in their daily diets.
4. **Freedom from poisoning from insecticides, pesticides etc:** With the more production of female cattle and buffaloes etc., there will be more need for the fodder crops which generally do not require the use of insecticides, pesticides, weedicides etc. Also the use of more fertilizers is not required in case of fodder crops. So, the result will be lesser use of these harmful poisons on the crops and lesser accumulation of their residues in the living tissues of plants and animals. With the result, there will be very less harmful poisons and toxins in the living systems of various *fauna* and *flora* species.
5. **Social benefits:** The social benefits of such a project of mammoth scale, are obvious. We will like to discuss a few of them:
   - **Freedom from cancer and other diseases:** The insecticidal, pesticidal, weedicidal residues along with chemical fertilizer residues in the food grains, fodder and other crops is a major cause of many diseases like cancer, liver problems, renal malfunctions, harm to small babies and pregnant women etc. With the adoption and use of Aulprofem on such a large scale, there will not be any need to use the various insecticides, pesticides, chemical fertilizers etc. on such a bigger scale. In fact, the use of such products will be greatly reduced with the increased emphasis on fodder crops which normally don't need the use of such harmful chemicals. The result will be the freedom from such toxic, hazardous by products of progressive farming so well touted by our various agricultural universities and related research institutes. The mankind will be free from the scourge of such lethal and deadly diseases mentioned a little above.
   - **Clean environments:** With so less insecticidal, pesticidal and weedicidal etc. poisoning, there will be very less chemical pollution

in the environment. The clean environment will be a great achievement of the project Aulprofem.

- **Freedom from rural debt:** The rural inhabitant and especially the farmer will be a much comfortable citizen with the result of obvious economic benefits of Aulprofem. This will prove very helpful in the production of a new culture of rural debtless-ness.
- **Freedom from worries:** With so many economical, environmental and health benefits of freedom from diseases, the village folks will be much relieved creatures, with so less to bother due to less preponderance of diseases and rural debtness.

Flow of funds to rural areas

The Aulprofem is wholly a story of rural development, fast materializing. With Aulprofem, there will be improvement in the lives of small and marginalized farmers who are the worst sufferers of present day agro economic conditions. These small and marginalized farmers are the dairy animal keepers and it is an important source of income for them to keep the vehicle of their lives rolling. It is extremely difficult for them to survive on the strength of agriculture alone which is based on the availability of just one or two acres of farm land. So, dairying keeps them alive and going. Thousands of them have already committed suicides and thousands are to follow. With the consolidation of dairy which in India, is a rural concern; there will be more and more of money flowing in rural areas from the supply of milk into cities. Even if a mere 10% of the total 30 crore dairy animal wealth of India is selected for the application of Aulprofem, then after a period of just 10 years from the start of such application; an additional **3 lakh crores of rupees** calculated @ rupees 20 per liter of milk, will be flowing from the urban areas into rural areas. This in itself is enough to revolutionize the entire agro based, rural economy of India.

Employment generation with Aulprofem

India is a big country. Nearly 72% of the population in this country lives in about six lakh villages across the vast stretches of land, spread across this large nation from Kashmir to Kanya Kumari, and from Gujarat to Kohima (see Appendix II). The focus of a project like Aulprofem should basically be centered on villages or the semi-urban and peri-urban areas. These are the places where the biggest portions of the animal wealth associated with dairy production, are found. We can devise the working units regarding Aulprofem in round about clusters; say each one of nearly 100 villages. We can have a full fledged whole time worker as in charge of these 100 villages. He can develop a network of his own in this designated territory. He can select one person each from a village who can represent the organization Aulprofem. In this way we expect the organization Aulprofem to have about 6,000 full time workers across India in rural areas only. Considering the semi urban and peri urban areas, we can have more 5 hundred to one thousand such people in the organization. To sum up, we will end up having nearly 6 to 7 thousand people working as whole timers in the project. There should be at least one person in each village as a local representative or roughly 100 such persons under a one full timer. Such people should be permanently stationed in the villages or the semi or peri urban areas, where they are employed. Even they may be semi employed prior to their enrollment in the project and they can carry on with their earlier status and routine, right after their entry into in the project. Considering such semi employed people that can be enrolled on partial working arrangement; we hope to join further 6,00,000 to 7,00,000 more people with the project Aulprofem. Now the organization really looks to have grown. The only crucial thing is to have this happen. In the following paragraphs, we will try to envisage the arrangement that will result as a practical manifestation of this concept of Aulprofem taking the real working shape in its true form.

With about 6 to7 thousand people engaged as full time workers, we need to pay each a minimum of 4 to 5 thousands of rupees per month. Adding to it will be the additional about 2 thousands of rupees as conveyance and connectivity charges. So, each person will end up costing the project to the tune of nearly six to seven thousand of rupees per month. Calculated at a rate of 7 thousand per month for 7,000 people it comes out to be nearly 5 crores per month or nearly 60 crores per annum.

The other 6 to 7 lakh people connected with the project should also earn a living, may be on a partial basis. Let us suppose that each one sells 5 to 10 bottles per month. Then supposing the kind of marketing we are in, by eliminating all middlemen and agents, we can easily give him a margin of rupees 100 per bottle. This means a profit of rupees 500 to 1000 per month or six to 12 thousand per annum. Considering the semi employed type of people engaged with us, this seems to be a good remuneration. Considering and comparing with the Indian govt's much touted employment guarantee plan which hopes to give every extremely poor Indian an annual employment of two months and an income of 6,000 per annum; this is no less an achievement.

Apart from this the more production of female cattle and buffaloes means more milch animals, more dairy production and more people engaged in the profession of dairying. This will add more value to the already increased employment levels with application of Aulprofem; in the form of more people involved in the dairy and dairy related activities, in villages and semi urban areas.

References

Barrat, J. and Leger, D. 1979. Aveir des grossesses obstenues après stimulation de l' ovulation. A propos de 519 grossesses. *J. Gynecol. Obstet. Biol. Reprod. (Paris)* **8**: 332.

Beernik, F.J. and Ericsson, R.J. 1982. Male sex preselection through sperm isolation. *Fertil. Steril.* **38**: 493.

Corson, S.L. *et al.*, 1984. Sex selection by sperm separation and insemination. *Fertil. Steril.* **42**: 756.

Ericsson, R.J. *et al.*, 1973. Isolation of fractions rich in human Y sperm. *Nature* **246**: 421.

Gordon, M.J. 1958. The control of sex. *Scientific Am.* **199**: 87.

Lindahl, P.E. 1956. Counterstreaming centrifufation of bull spermatozoa. *Nature* **178**: 491.

Meistrich, M.L. 1982. Prspects of sexing mammalian sperm *in* R.P. Amann and G.E. Seidel, Jr. (*eds.*) Colorado University Associated Press.

Saving the Female Gender – A Great Social Boon

The sex ratio is a measure to know the number of the females per a unit value say a thousand males in the populations of men and animals. In humans in India, it has gone considerably down. The sex ratio in 1901 was 972 females per 1000 males. In 1941 it was 945 females, in 1961 it was 941 females and in 2001 it is 933 females per 1000 males. In Haryana it is 861 and in Punjab it is 874 females per 1000 men in 2001 (courtesy India 2007 published by Ministry of Information & Broadcasting, Govt. of India, page 6-22).

The fall of sex ratio against the fair gender, to alarming levels is a matter to worry for the policy planners and thinkers. Many reasons are attributed to such a fall in sex ratio. The first and prime most is the societal mindset and working. The Indian society is men dominated and womenfolk are a subordinated lot. Same is the condition in many other developing countries of the world. A boy is preferred to a girl child and the birth of a girl is not welcome but usually is considered a sign of some ill luck befalling upon the family. Being the doing of nature or something out of the control of man, she is simply taken in, in the family. The long histories of many developing countries are full of lengthy details of innumerable stories to ill treatment to the girl child and its non acceptance by the society. Even the girl children of people holding very high positions in society were not given the type of respect and recognition that they deserved by the other people. The famous queen Razia Sultana of Delhi Sultanate (see Appendix I) had only one handicap, that of being a woman. Otherwise she was a competent and able ruler. This only handicap became the cause for her disliking by the elite courtiers and ministers as no one wanted to serve under a woman and ultimately it led to her downfall. There have been written records of the events when the queens of the dead kings had to undergo self immolation called *sati,* in India. The entire medieval times have witnessed the human female folks being treated as material commodities and even being sold like slaves. More than written history, the very existence of societies in many of the developing countries, is a living testimony to the practice of women subordination and subjugation to inhuman suppression.

Times have changed a lot but the basic things still remain the same. There has been spread of education and general awareness but what can one expect of the country where semi feudal system of existence of societies exist. Countries like India, Pakistan and Bangladesh are clear examples of societies like this. The gang rape of Mukhtaran Mai (see Appendix I) in Pakistan and the parading naked of many women in public in different events in Rajasthan and Madhya Pradesh etc in India, are self explanatory examples of working of the pro-male mindset and thought. The Taslima Nasreen (see Appendix I) episode in Bangladesh, is an example of how the society in general thinks of even of its educated and learned women. The banning of girls and women from getting education and doing jobs in Afghanistan by the now banished ruffian Talibans (see Appendix III), is another example of such a mindset in working.

Aulprofem which is virtually the victory of female sex, can come to the rescue of female world in this regard. Being the world's first mammalian sex fixer and a female one, Aulprofem is synonymous with the feeling of feminism strengthened and triumphed. Enough are the days when womenfolk were suppressed. There should be an end to them. That is the meaning perhaps that nature wants to convey to the mankind with the timing of the discovery of Aulprofem. That the nature is the ultimate master of the universe is beyond the scope of any discussion and is the most universal truth of all times. The all time great brains; may be it is Charles Darwin, may be it is Leo Tolstoy, may be it is Karl Marx or may be it is Sigmund Freud (see Appendix I); all are one on the point of all times supremacy of nature. The supremacy of nature is not only manifested in its being the greatest decision maker of the world but even the timing of everything is decided by nature. The global and universal happenings, all are directed and actuated by nature. Even the sequence in which they appear is decided by it. The setting in and out of ice ages, the formations of oceans and lands, the creation of biological periods of existence, the global warmings and coolings, the earthquakes and tsunamis, the formation of mountains and deserts; almost every act is directed and acted by nature. Even the great discoveries and researches are decided by nature and their timings and sequences too. May be it is the discovery of laws of motion by Newton. The discovery of theory of relativity by Albert Einstein, the discovery of genetic code by Watson and Crick (see Appendix I), the discovery of radioactivity, the discovery of vaccination theory or the discovery of antibiotics. A great discovery generally is not only a solution to many problems of the world but it also becomes a factor to facilitate the occurring of many new discoveries in future. That everything is decided and timed by nature, so the shaping of future societies and the world in general is dictated and directed by nature too. The discovery of Aulprofem which undoubtedly is a great discovery, is also perfected and timed by nature. There should absolutely be no questioning to it and there is no logic to think on the otherwise. This seems to be the dictation of nature to man not for the time being but for all times to come.

Victory of femaleness-decided by nature

So, nature decides the discovery of the victory of femaleness. The first female sex fixer of the world has a message for humanity. The weaker sex is weaker no more. There are the times coming, of the consolidation of the fair sex and the ultimate victory of being feminine. The mammals are the most important family in kingdom *animalia*. The discovery of sex fixing in them and that of the female sex fixing has a message for everyone. The females are going to be triumphant in not only the social consolidation of their positions and standing but also they are going to be the dominant *genre* in future. So, the outcome of this magnificent discovery is the glory of femaleness not only in every concept and view of the word but also in its broadest meaning.

The basic question. What is a female sex fixer? A sex fixer fixes the conception of a particular sexed offspring. And a mammalian one to that? There is the message. The nature has a message and that seems to be a very clear one. 'There will be more females'. There is a reading in holy *Koran* like this. There is nothing religious about the present discussion but the deciding of more production of females in mammals has the approval of the ultimate master of the universe, the Mother Nature; is the message. The reason for this may lie anywhere but the impact of this will be centered on the expected target.

Why more females?

Because, they are needed.

And, the need?

They were ignored and that also for thousands of years. A balance for this has somewhere to be struck.

And the time for this has come.

And the suppression.

How long will it continue?

Not so long. Seems to be the answer.

But why?

Why and how?

Inquisitive mind may ask.

The answer is not difficult to find.

The tide has turned. The time is now with the female folks. They have won the sympathy of the biggest influential personality in the universe. And that personality is the Mother Nature herself. The Mother Nature has come to their rescue finally and that is the verdict of time. And the verdict seems to be a stronger one.

Stronger and holding.

There may be other reasons as well. But the things are not that much easier to understand. The world really may need more females. The females of the dairy bovine species provide mankind with milk. The world today is already on the verge of a bigger food crisis of colossal dimensions. More female cattle will certainly save mankind from hunger and starvation. And as is the case in humans, the sex ratios have fallen drastically against the fairer sex. There should an equalizer. So a mammalian sex fixer. The equation fits well everywhere. The end product is the need for a sex fixer in favor of the females of all types in mammalian world. And that is exactly what we have in the form of discovery of Aulprofem. So, the nature decides by itself the happening of the discovery of female mammalian sex fixer and that is the inference of the discussion in this regard. And there should not be a questioning of any type in the derivations that evolve from this. The female sex has won by the grace and blessings of all powerful Mother Nature.

The women will be a liberated lot. They will be salvaged from their present miserable environments to those of the pure and cherished ones. This is clearly the greatest victory of Aulprofem. The social upliftment of womanhood, their social liberation and strengthening and the acceptance of their talents and the recognition of their positions of being creators and nourishers of mankind; is the obvious and clear message of the one of all times' great discoveries *i.e.* the discovery of Aulprofem. This message should drive straight through the minds of all and one. This if happened, will no doubt be the great contribution of a scientific discovery to the ultimate development of mankind in thought and action and the emancipation of womanhood in all times; past, present and to come.

Regretting and undoing the injustice done to them in the past. Acknowledging their rightful position and standing in the real activities and annals of world history.

Giving their rightful and just due in the practical working conditions in the present world.

Ensuring the proper equal and un prejudicial future development and flourishing of the female folk in the times to come.

Let the real message of the ultimate victory of female hood that has expressed itself in the victory of female sex in the form of the discovery of Aulprofem; prevail upon one and all and lead the future generations to have a perfect, harmonious and balanced growth and development; completely fair and unbiased, towards any of the sexes.

Let the Mother Nature bless the human race with such lofty ideas, intentions and deeds!

Amen!!

Economic Recession and the Way Out

The world economy was well in a slowdown of gigantic proportions in 1991-92.

The General Motors had declared heavy losses and so had the Ford. The Chrysler (see Appendix III) was already in the grip of a big crisis. Under these conditions the famous induction of Lea Iacocca (see Appendix I) as the general manager in Chrysler, happened.

These are old stories now.

Nobody seems to remember anything about them. But the alarming bells were ringing very high at that time. The entire economic system was trembling with fear of collapse and the fall was expected any moment. But a surprise rescue occurred. Two things had happened.

First was the fall of the mighty Soviet Union. The enemy of the free markets and the business world, was dead at last.

This gave industry the much needed hype and energy and boost. There was a sense of jubilation and exhilaration all around. The system on the verge of death started showing signs of revival. There was some recovery but it was not long before the system again started gasping for breath. It certainly needed more oxygen.

It is a fact that economics are economics.

They need something of real nature.

Soon the stagnation was there all around.

Now the markets needed a messiah. A savior of the entire world of economics. All the world was praying for it to arrive and arrive it did.

Yes, this time it really came.

It was the IT.

By IT, it meant the information technology. Some people also called it IT revolution.

The world had discovered computers just a couple of years back. They were wonderful machines but the world was still to acknowledge their utility in the day to day life of people, the business, professional and other circles and these new devices were struggling to find a due and honored place in the minds and ordinary practical lifestyles of people. Of course they were a successful entry but not so much. There were hindrances combined with this technology; of being costlier, of tough practicalities and many more.

Those days the computers were an elite lot. They were habitual to a luxurious life style of air-conditioned environments. On a cumulative estimation, there were obviously many obstacles in the way of computers being a common man's device, choice or infatuation.

Then one fine day the Pentagon (see Appendix I) found a unique innovative application of the computers. They connected many computers of a single department related with defense data, with one another. The intended motive was a real good one. They in fact wanted the easy transfer of data across many computers or probably across many offices, if possible. The other motive was to pile and store more data and use it across the many computers, connected this way.

The experiment was a success.

This was the computers networking and it was aptly called 'intranet'.

Soon the intranet came out of the boundaries of Pentagon's office.

The pampered child of the pentagon was soon the talk of the town, rather the talk of the entire world. Everybody wanted to have a glimpse of it, to hold it or to cajole it.

It was such a lovely child.

It was soon neo christened as 'internet'.

Then suddenly many people in the universities, research institutes, data organizations, discovered many useful and practical benefits of it.

There was a roar of this new discovery in the media and the fiction world. Even movies were conceived and made centered on this subject. Soon everybody, who was even a little literate was busy talking about this. Numerous conferences, seminars and debates were organized the world over to teach and guide the people of all continents about the benefits of this new development. People were really enlightened and they demanded de-control over this most handsome pet of the Pentagon and the rest is history!

The wise government at Washington soon approved its use outside the defense boundaries and this was the embarkment of a real revolution. The internet really spread very fast to the entire of outer world, to the academics in universities, to the data centers in research institutes, to the business organizations and many more.

Then there came the thought of combining the net working of the entire world's computers. This was the discovering of the right meaning of the word the 'internet'.

The internet was now a real blast.

More than its applications it captured the imaginations of the people. It was real beneficial, not like the false cloning discovery which until now is a real hollow mired in fantasies and is still based on the imaginations of the non scientific fiction and media world.

The internet was welcome everywhere. Thanks to the companies like Microsoft, Intel, Google and many more, the internet was a grand success.

It is not without saying that the internet was also popular with the fiction and literary world. It had rich exploits in these domains also.

It gave rise to the writing of countless books on the subject. Other than this also, it was a favorite topic in the media; both electronic and paper, in the entertainment world and many more. Who can forget big screen movies like Matrix, Speed etc where the applications and scope of internet were displayed and elaborated? The internet was by now the true netting of the entire brains of the world populations.

The IT revolution manifested itself in still an other *avatar*.

Another research connected with the IT was the rise of mobile telephony. The mobiles are one of the best manifestations of the electronic world's Pandora box of discoveries. They are really useful devices. They had their predecessors in the form of stationary telephones. They removed the biggest drawback of the stationery telephones *i.e.* the location specificity. The stationary telephones can not move with their user. There connectivity is based on the network of cables that connect these devices to one another through nodal centers and telephone exchanges. This was the biggest drawback with these time tested friends of man. A drawback, the overcoming of which was the need of the hour. A little before also, the overcoming of this drawback had popularized the pager phones at one time and this time now, this utility feature became the publicity point of the mobile phones as well. Now mobile phones had proved their worth. It was not long that they captured the minds of the entire world and became very popular. First they became the dream of everybody and very soon this dream was converted into the need of everybody in the world. A lot of mobile companies like Nokia, Motorola, Sony Erickson and many more contributed in the development and popularization of mobile phones throughout the world. The mobile phones are also a part of the IT boom but we will come back to it only after some time. Now we concentrate on the emergence of an industrial boom with the development of IT sector connected with computers and internet.

An industrial boom starts: a financial crisis is averted

The IT boom did herald a new era in the arena of mass information sharing and pubic connectivity. The use of computers on larger scales virtually made them the household tools. A day was reached very soon when they made their appearance in almost every school, college, business office and even in the houses of well to do people. Very soon, they became a tool in the hands of highly educated and professional people and then the students and even the ordinary not so rich and middle class people, made places for these wonderful machines of development, in their houses. The sales ran into tens and then hundreds of millions in numbers of the machines sold. The revenue in billions of dollars got generated and made new billionaires in the world like Bill Gates (see Appendix I) and so many other such people. Bill Gates became the richest man of the world. New companies in the software and hardware business, earned big profits and their shares in the stock exchanges rose to sky high prices and bands. In India more than half of the billionaires belonged to IT business from middle of the last decade onwards in the twentieth century with Premji Azim of Wipro (see Appendix I) group, heading the list.

The world economy which was in the grip of an industrial slowdown and was fast heading towards a recession; started showing signs of recovery with these developments. These newly became billionaire companies and their bosses went for heavy investments in the manufacturing sectors and trade soared with that. The employment was generated on mega scales and there was bountifulness in plenty. The new industrial settlements came into being and IT parks, IT hubs, IT townships and even IT cities were established. The Silicon Valley in USA (see Appendix I) and electronic city near Bangalore in India, are examples of this. The cash rich companies invested in the real estate and purchased large tracts of both fertile and non fertile land and settled hi-tech satellite townships there. The business process outsourcing (BPO) industry came into being and employed millions of highly paid young boys and girls in this profession. This led to the richie-rich life styles to emerge among these young people even in poor countries. In stead of these well to do people, the IT revolution before had also created many rich and successful professionals, entrepreneurs, businessmen and technical experts etc. The people, who few months back were hopeless struggling paupers, now became persons with status and money. These neo rich who had earned enough money, now wanted to have other amenities in life also. The first thing that a man wants to own is the house of his dreams. These rich people wanted to buy homes, so they decided to invest in real estate. The property became much in demand. The governments the world over, saw an opportunity in this and eased the norms for public lending and interest rates and cash reserve ratios for the banks were lowered. The banks in turn cut interests to considerable margins and this further made liquidity an easy access especially for the people who went in for real estate investments. This was the golden time for every body especially the neo rich in the IT sector who had just started earning decently to realize their dream of having a 'home sweat home' of their own. The real estate now had a real time to fly and fly it did. The prices of land simply sky rocketed. There was another boom associated with this. This was in the sectors connected with the real estate such as the sector of civil engineering construction. Many companies in the field of houses, apartments and building construction came into being and recorded super successes and the companies already working in these sectors also had a gala time. This was the real boom time. People had money and they went out in a frenzy to invest and the demand for property in the real estate sector seemed insatiable. As the demand for houses grew, many other related demands also grew. The demand for cement grew and so did the demand for steel and bricks and other related materials like timber and hardware etc. The interior decoration, furniture and numerous such architecture were also in big demand. Some phoney businesses like *Vastu* and *Reiki* and such others also devoured a boom time. The flow of money was such that there appeared thousands of Ganges flowing full with money in the world. The share markets appeared to have rocket fuels in their tanks and bulls annihilated anything that came in their way, leading the trading in the stocks to heights of speculation and culturing and cultivating many scandals in their big bellies of selfish greediness. Everybody was busy minting money, which seemed to remain the only business in the world. Hype and jubilation seemed to shine in the neon lights of five star hotels, resorts and clubs in metropolis and other mega cities. Everybody was simply exhilarated.

The governments were happy and extremely well satisfied.

The parliaments mentioned with satisfaction and pride the exploits of the speculators in the stock exchanges. As long as it was shining well to glitter, everybody seemed to take it for gold.

The finance ministers the world over, made references to the bullish developments in the share markets in their speeches.

Nobody wanted to think that there may be scandals brewing.

Nobody did think also. Everybody was busy in making fortunes.

We have to catch another story least we forget.

Another story that we discussed a little while ago. The story of mobile phones.

Mobile phones are also one of the finest discoveries of science ever made on earth.

They are also a part if the grand IT revolution.

They also generated basis and infrastructure for industry and industry in turn generated companies and business establishments. These companies generated employment and development and created industrial hubs and cities and the share markets trembled to the roar of their successes and jettisoned high with their exploits on the fronts of commerce and industry. They generated many billionaire companies and their bosses. They also needed to invest in the real estate. They also contributed to a larger extent towards the emergence of the real estate into a big flourishing enterprise.

This was also one of the causes of the real estate boom.

One more thing happened with the IT boom. Let's have a look.

When a person buys a home, he wants to acquire certain other things that add distinction to his home or are the essential and decorative part of it. This happened in the present case also.

The neo rich and saturated rich wanted to have cars and some of them wanted to have real big ones of that. They all went into the Car-Shoppe's and showrooms and bought them. The automobile industry certainly got a flip with this. Not a flip only but it was a real big boost.

The car makers had a wink of satisfaction which got soon converted into happiness. In fact they were real happy.

The General Motors were happy.

The Ford was happy and so was the Chrysler.

Even the Hyundai and Suzuki and Honda and BMW and Toyota (see Appendix III) were happy.

All had a gleam time.

All had real big sales.

The newly cash rich people kept coming to such showrooms and kept on buying cars after cars.

The slowdown had gone and so were the fears of recession. Their shares also roared. Roared with thunders like the light splashing dark clouds in the rainy season. This thunder splash became the cause for more speculation in the share broking circles. The casino economics had emerged fully well in the stock exchanges. The share markets had donned the mantle of dignified gamble houses. When some stocks roared with success, they also cut heavy harvests due to speculation in the stock markets.

All were happy. There was prosperity at display. Prosperity that glittered and radiated and hypnotized. Everybody was simply left spellbound and gaping in this running road show of splendid exploits of the Indiana Jones' of speculation. The circus of glitter displayed its antics everywhere in full stream.

Some more points we ought not to miss.

The boom had also the raising effect on other markets as well. When people had money, they spent heavily on other fronts like garments, consumer commodities, cosmetics, hotels and travel etc. Every market now had a real chance to boom. In fact a genuine chance.

There was a boost in the cloth markets. There was a boost in the tourism markets. So there was a boost in every market. Consumerism grew high in stature and strength.

The sale of refrigerators was recorded in record breaking terms and so were the sales of washing machines and video cameras and plasma TV's. Everything was sold to new record levels set high. Even the hotels recorded high turn overs.

Festivities began.

When a person is real happy, he wants to have one of the two companies, either the basketful of wine or a beautiful mistress. The markets made these things also available. There were businesses escorts who had jumped into the market and were having a gala reap. Their old sisters and predecessors, the call girls also had a nice business flowing in. The prostitution grew. The live flesh was in demand everywhere.

Almost everything connected with this magnificent world of capitalism grew beyond limits, in those splendid years of the end of the last century and the start of this twenty first century.

'Make hay while the sun shines' is an old proverb and which still holds good like other universal truths of all times.

So, the wise people in those times tried to dry as much hay as they could. And this happened to be the most unwise thing as it later turned out. 'Man proposes and God disposes' is another proverb which also holds well some times. The hay driers seemed to have forgotten about this totally. The hay turned damper and damper and even damper. Here is a story. Let's give a slight ear to it.

There are some wise men of all times.

Adam Smith was one among them. Another was Karl Marx and one more was known as J.M. Keynes.

They had worked really hard and formulated certain formidable principles of economics.

But when lot of money is in wallet, even a fool becomes wise.

And in those magnificent years of IT, every fool had a lot of money.

And everybody claimed to be wise. So, nobody listened.

Nobody dam cared. Nobody cared for what was transcripted in the holy books of economics, written by these great men.

And when the fateful happened, it was a bigger explosion than the Hindenburg or the Nagasaki. All the hydrogen and fissile matter had exploded and burnt.

The banks and financial institutions had made the maximum hay. So, they happened to traveling on board the Hindenburg.

They had really overplayed their roles.

They sanctioned as many loans for the real estate as they could. They were safe as they were happy. They in fact played much shrewd. All the properties were mortgaged with them only. They spent as much liquidity as their cash reserve ratios allowed them. But the markets were still hungry for cash. So they wanted to generate more cash. They played an unethical game. They further mortgaged the properties that were hypothecated to them, to financial institutions and generated more liquidity. But the real estate markets that had tasted and fed upon salt of shameless profit-mongering, cried for more. They demanded more cash. Those financial institutions further played the same dirty game that the banks had played in the first instance. The markets could not be satiated even then. They behaved like lusty flies of Othello of Shakespeare. They displayed more lust for cash. Those financial institutions in turn did the same maneuver again. The simple unethical dirty tactic got replicated many times in the most shameless act of profit-mongering. This is a story in itself and this is also the moral of the story that too much selfishness is a very bad thing.

But what to do when profit happens to be the most pious word in the entire literature of business and capitalism.

So, it was a fine spring time for the entire business world that one fine day when everybody was so intentionally listening to the finest flutes of the share market performances; the bubble got exploded.

It was Friday and there was something that christened it as the 'Black Friday'.

The share markets had crashed.

It was unbelievable!

How could it be?

How can the entire world of capitalism turn suddenly so bad?

But the truth is a truth after all.

Everybody had a little time in recovering.

But every body had a lot of trouble in recovering.

It was a real mega shock of a whole time nature for most of the citizens across the globe.

It was a slowdown, Bush claimed. The economists cried it was a recession.

Whatever it was, the economy was already in peril.

Everybody blamed the others. The banks and financial institutions were an easy target.

The sub prime crisis was point for talk, everywhere, in every corner of the world. The basis for the crisis was the very simple principle of economics. The demand and supply equation of Adam smith. When there was enough liquidity in the markets and banks lowered the lending rates and the IT boom had already enabled the many people to go heading in realization of their ultimate dream of having a 'home sweat home'; the banks extended magnanimous loans. The property that was prime and unsold at that time achieved buyers and was transacted sold. This simply worked on the bigger principles of economics without anybody's knowledge. The supply principle had been disturbed due to obliteration of it. When there is decreased supply in the true scientific sense of economics, the demand being already there, the prices are only expected to soar. The biggest question in this entire scenario is that the purchaser thrusted upon in this abnormal state of affairs was the one generated by the IT boom. He was not there in the first instance. He was IT generated buyer and he walked away with all the available supply leaving the market totally devoid of any available realtor property. This disturbed the otherwise natural equilibrium existing at that time. Whatever demand remained was met with zero supply as the entire supply had been consumed by now. The end result was that a false enhancement in the demand was created with this development. The possessors of the property that was unsold could demand somewhat increased prices for their property and the people who had to buy were left with no other option but to buy the remaining property at these increased prices. There was no other alternative left. Especially the bargaining point in the usual sale and purchase deal was gone as the property remaining for sale was very limited and the desperate purchaser was left with no other choice except to buy it under all conditions. This bargaining point by its very nature tends to push the prices down. This pushdown may be of a bigger magnitude or may be a smaller and even may be of a negligible one but by its very nature it tries to put a hypo-price effect. This hypo-price syndrome is simply inherent in the very nature of this bargaining point. Now the picture remaining was very abnormal. There was no spare property remaining. The demand being the same or even decreased a little, the buyers had to go in for the property that remained now. They were tempted, rather forced to settle on demanded conditionalities by the sellers. The sellers were in commanding positions now. The whole scenario had created too many contenders for the few remaining properties. The number of properties remaining fewer, the people wanting to buy them being the same, there was a proportional tilt in favor of the buyers being relatively more as per the contentions for the unit property remaining for sale concerned. So per unit property remaining for sale there were now more buyers. So, they could not miss the deal. They could not bargain also. They had to shell out anything that the sellers demanded. Otherwise they could end up losers as some other contender for buying, would step in and get away with the bounty of deal. So the bargaining point being totally or almost totally absent; the prices simply rose to the levels, slightly higher than the ones existing before. The end result was that the prices of the property rose.

The rise in property once started, continued being so due to many reasons. First the IT boom that had created the buyers in the first instance, continued doing so by creating more moneyed people in the process of creating employment and liquidity that it had done in the first instance. There was one more reason and that reason is a bigger one. Let's discuss a little.

The speculators came on the scene. They are powerful deities. They can play bigger games with versatile maneuverability. They can exhibit many traits that are particular only to their species, discussed just down under:

They are masters of rumors. Anybody familiar with the share markets functioning is much aware of this feature. Speculators usually run riots in share markets. When they came in the realtor markets, they simply played havoc in this field of trade also which already has a considerable rumor quotient attached with it.

The prices were going up and there was no land left for purchasing, was the news in the circles of property trade. This was happening when there was not much purchasing activity taking place in reality in the realtor markets. A type of fear psychosis was created in the minds of the potential buyers and the end result was only as expected. The prospective buyers were terrified and they rushed for actual sealing of purchase deals. The delay factor was eliminated which otherwise usually acts as a hindrance on the increase in price tags. The hush and bush in the real estate markets soon got converted into a chaos. Everybody was crying that there was no property left for sale when in reality there was lot of property remaining for sale. The deals were actuated on high prices, usually much higher than the prices that otherwise could have been. The end result was exactly as expected. The real estate prices almost quadrupled in one year and this trend continued for many years.

There is one more way by which the speculators played dirty and this was the method of risk marketing. Let's discuss.

The speculators when they came on the scene, wanted to achieve something really big and they found a way for it. Actually it was not a new discovery but ages old and well tried out methodology. They actually purchased property. Almost all the property that remained from selling was identified and bought by them. In addition to this; the property that was not for immediate sale but that formed a futuristic alternative to the land that could be offered for sale was also purchased by them. This left the realtor markets totally devoid of any worthwhile property that could be offered for sale. Now the situation had changed drastically. The realtors usually being high profile people, did never had a problem in approaching the banks and financial institutions and making them agreeing to their proposals. Actually banks and these institutions are normally more than happy in extending credits to such high placed and rich people. So, the liberal loans usually on lower lending rates were extended to them. The big tracts of land that formed the prime property and even the property that saw no prospects of being sold as a worthwhile alternative to prime property was also acquired by these speculators and by the time actual buyers came on the scene, there was no property remaining for purchase. Whatever property that remained for sale was being quoted on higher prices. So these actual buyers were forced to buy the same property on higher prices that was beforehand available on relatively or considerably lower prices. The trend of price increase had started. And it seemed irreversible, at least for the time being. The neo rich people of the IT revolution had a lot of cash and this was a big problem for them and it had already started to itch them hardly on their buttocks. The itch became severe when they could not buy any property for their ultimate dream of life of their 'home sweat home' of the finest dreams of an ordinary human being. And when this ordinary human being earns a lot of money, he starts thinking that he has attained a special status and now he is on a much higher pedestal than the other ordinary creatures of low ranks. A lot of money after all, is a big problem. And the problem becomes even more pronounced when this extra pile of money cannot buy anything and turns out futile. Just like Mr. Paul in 'The Capital' of Marx. They wanted to get this money converted into some pride possession. And the most worthy possession in a capitalist ordinary world is the acquiring of a dream habitation.

After all what is the use of money if it cannot buy anything worthwhile?

They wanted to seal some deal on a house. This was the time and they desperately wanted to do so now.

And this is where the biggest problem of the entire practical material universe started to germinate.

The seeds of the biggest crisis of capitalism of all times, had been sown.

The speculators had made a ground for this. In fact they were the driving force behind this all.

They had a golden opportunity to make heavy profits and they did not want to lose this. In fact they made super profits. This is a universal truth that profit is the most desired and pious word in the entire dictionary of business in the entire commercial world of capitalism. They opted to go for the noblest cause of business. They went to make heavy profits. After all this was a moment of theirs.

A moment to cherish and attain.

To attain glory and achievement and accomplishment and to reach a state of fulfillment.

To be a perfect man. Just like the ad of Raymond's.

To be a perfect man, is the desire of all men of all creeds, places and generations.

The speculators had a chance.

A chance to achieve the ultimate in life.

The ultimate success in business to make heaviest profits.

A success achieved, satisfies the ego of the man who does it.

So, these exploits satisfied the speculators as well as their egos and the society also seemed to acknowledge this. The society usually acknowledges everyone who progresses especially by stepping on the shoulders of others and using them as stepping stones and in this particular case, the speculators used the entire societies, in fact the entire nations as their stepping stones by robbing them all, of the worthwhile property that the society and nations had and in the process converting this property into their prized and personal possession.

The end result was magnificent. The speculators emerged the winners in life as they usually do. Winners in the sense that they were successful accomplishers now that duly rose on the higher pedestals in societal life.

The big men.

All great men! Everybody seemed to acknowledge.

The society suffered at large. But the property markets boomed. Boomed and booed.

Booed with joy and escalation.

The prices of property had gone up.

Now starts the second act of the play.

The secondary buyer arrives on the scene.

Who is he, is not difficult to recognize?

He is the person who had missed the bus in the first instance or the person who jumped the bandwagon of IT revolution in a later stage or the otherwise needy person who had to buy a house but he might have nothing to be related with this boom in the IT, except that he was made a victim by shelling out more for the same property that could have been available otherwise on much cheaper rates. He was forced to buy the same property on higher prices. There was no alternative to it. All the property was in the hands of speculative supermen and they won't allow it the indignation of any sort by letting it to be sold on decreased prices. So, the picture is very clear.

The late arrivals on the IT bandwagon and people otherwise, were made victims to pay more for the same old, low priced commodity. But this is destiny. They had to do this otherwise they might have faced the fear of further losing because the property was on the continuous trend to rise. And they did not want to be losers. They went in for the purchases. They had to, there being no other alternative and way out.

The result was out. The property registered the hike in value.

This process went on for many repeats.

Many acts of the same nature got repeated one after the other. The property prices being hiked after each such act. So, it turned out that after one year or so, the prices of the property soared. The nice reasons for this have already been explored and told above.

Some profit mongers made super exploits in the process.

The society at large was held hostage to the greed of a few men.

The banks, the financial institutions and the governments, all had combined with one another to facilitate all this and the most glorious days of the market economies and the share markets were realized in actuality, this way. The property soared. It had grown wings.

The share markets soared. The governments were happy.

Till one fine day the bubble exploded.

But there are still some more things to be discussed.

The lower lending rates were ensuring lending on bigger scales and the first purchases by the speculators on the lowest of the rates depending upon the positions of these people in society and financial circles. Then the primary buyers came in the first phase and these deals were also financed by banks on slightly higher rates. The banks or the financial institutions have to finance a deal otherwise this much large money is usually not available in the markets. Then the secondary sales happened, slightly on little more increased rates and these deals were also financed by the banks and financial institutions. Similar was the outcome with tertiary, quaternary and later deals. In fact the banks and financial institutions had supported and financed these transactions as they usually do in normal course of business. Financing is their business and if they do not do this, then what commercial activity is left for them? But here in this case they moved a step further. They had combined with the speculators to facilitate this entire operation virtually on a win-win situation for everyone except the real needy buyers who were forced to pay more for the same old low priced commodity. Every speculator and realtor had the blessings and shelter of a bank or sometimes of many banks or financial institutions in those meaningful days. In fact, the banks and financial institutions competed with one another to get in the good books of the realtors and speculators. Each time a buyer or owner was changed for the same house, a bank or financial institution was more than eager to step in as the financer. The same house got financed many times; every time it was sold again usually on a higher price and the situation was such that in these circumstances, the same house got sold on a price much higher than the old and original one because the prices in the vicinity had gone up. The banks financed such deals on bigger price magnitudes simply because the prices were up by now. The tertiary or quaternary or any other buyer later than this purchased this property at the prices much higher, simply on the pretext that overall prices in the locality were up. So, the situation was justified fundamentally. Nobody had a problem in understanding this. The implementation was also easy as the matter is quite digestible to *all and sundry*.

The truth that was true in purchase of houses was also true in case of apartments, country houses; farm houses etc. The prices of all these things went similarly up exactly on the same lines.

The companies that were marketing houses, and plots were also marketing country houses, apartments etc.

There was an overall boom in the construction sector. The companies in the real estate and construction business were simply making hay and they in fact were able to dry as much hay as could have never happened in otherwise normal situations without an iota of speculation, attached to it.

The nations saw the emergence of mega construction companies. These companies were in fact big conglomerates of speculators. Wishing or unwishing, knowing or unknowing, they were into this.

And this everything played havoc in the markets. The emergence of mega construction and real estate companies soon registered a very conspicuous presence in the share markets. There came a stage when the construction business overshadowed the performance of the manufacturing sectors and it started playing a dominant role in the national economies. In India, the otherwise industrial production was never much above five percent but with the construction and real estate boom, the national growth rate was somewhat near nine percent. That was the power of the construction and real estate companies. And not surprisingly, this power was soon felt in the share markets with these companies registering strong presence there and the shares of these companies fetching bigger values in share market's trading and brokerage. These companies started playing bigger roles in the bullish behavior of stock exchanges.

Combined with the original emergence of IT companies, the share markets simply roared with the run over, that these realtor and construction companies displayed. The indices in the stock exchanges went upwards, sky rocketing.

Every thing was going fine till one day, a universal truth came to tell that every coin has two things. One heads and other tails. When heads side looks sky side, it is not essential that the tails side will also follow it. In fact it looks exactly the other side then, which is the side down below. This is exactly what happened and it was the other side of one of the most magnificent and exemplary success stories ever told in the golden books of trade and capital. Let's go through it.

The markets are ruled by the simple and fundamental principles of economics. It is a different story that the Nobel Committee gives a Nobel Prize in Economics every year but still the market economics are guided by the eternal principles founded by the three great men, Adam Smith, Karl Marx and J.M. Keynes (see Appendix I). The most fundamental principle of the economics is the demand and supply curve as formulated by Adam Smith. Then there is the principle of overproduction of commodities and the law of the inevitability of economic crisis called recession in technical language. The theory of J.M. Keynes elaborates and tells about the need for state intervention and government regulation in the market economies. These principles clearly forecast that a recession was in the offing. First of all J.M. Keynes' theory told that when markets are bullish with extraordinary features, under normal parameters for development; certainly there is something, brewing foul. Secondly the theory of Adam Smith told that when there is overproduction and hence oversupply of the commodities in the markets; the demand being lesser than the off take, the prices are definitely bound to fall. Third great man Karl Marx' theory clearly demonstrated in full details about how a market competition in a period of favorable market growth turns into a period of commodity overproduction automatically, thereby, decimating the margins of profit, the pivotal point and life center in the functioning of a market economy. This state very soon gets converted into a market glutted with commodities and profit margins almost disappearing and the economy lurching into a crisis. These are the very basic things in the working of a market economy and one needs not to be a prophet, to forecast them.

So, the real estate markets were full of real estate companies that came hurling, in search of more and more of profits. It is a basic knowledge in humans that if one person is successful in a field, he attracts many takers into this field. When one company was very successful in the business of real estate, it naturally attracted many a companies into that foray, of successful enterprise. With more and more of such companies jumping into the bandwagon of realtor business, there was naturally much activity on this front of development of real estate property and homes and apartments construction. By the start of year 2008 there were more than three hundred thousand apartments waiting for their prospective buyers in Bangalore city of India only. But the wait lingered. The buyers never came. The people, who could buy them, had done so much earlier. Now they did not want to buy a house any more as they already had it or they were totally drained of the money that they could pile upon with their first purchases only. The other people who remained from buying were probably not so much well to do. So the markets were devoid of buyers and commodities, that is plots, houses and apartments; were too many.

And no purchasers came.

The sellers had to sell their products, so they started marketing efforts aimed at increasing sales. These efforts may include advertisement, publicity and incentives for purchasers and so on. The incentives also included gifts in cash and kind and also the decrease in prices. But these efforts were not successful. So, in certain cases prices were slashed. Slashed to bigger margins. This worked some where or did not work some other places but the end result was that the prices were considerably cut, in most of the places and cases. But even these remedies could not bring the buyers in.

Meanwhile another thing happened. Some people who had bought property on exorbitant prices, found it very difficult to pay their monthly or bimonthly or yearly or whatsoever, installments in time. The reasons behind this might have been many. Some people might have lost their jobs and some might have befallen prey to bad luck. Who knows? But the end result was that in many cases the defaulting on part of installment payments, happened. The magnitude of this was small or big but the banks and financial institutions reacted very sharply.

It is true that the entire business of banks and financial institutions is based on the amount of interest that is accrued upon the sums that these institutions extend to people and organizations, private or public as loans. If all or some of these people or organizations turn bad debtors, then from where, will these banks and financial institutions recover their loans back and naturally they themselves, will run into bad debts or losses? The banks have to recover their loans back. So, they reacted very fast. They started attaching these properties. Very soon the auctions were announced. But a strange thing happened. No buyers came for these auctioned properties. Now a real bad time had started for the banks. When all this was happening, another aspect of this problem came to light and that is not less interesting. Let's have a look.

The properties in the boom time were overpriced and banks had extended liberal loans in those heydays. But now the buyers being absent from the markets, the prices started to come down to the original or near to the original values. Now if one has to attract customers, he has to bring prices to the levels as lower and sometimes nearer to the original ones, before the boom. But how could banks do that when they had financed on higher rates?

So the banks were in soup now. What will they do with the property which has no customers? After all they are financial institutions and not property dealers. The financial institutions who had further mortgaged this property were also in equal soup. Everybody was devoid of cash and liquidity. Now there started a crisis that came to be known as the sub-prime crisis. It had some of the very interesting features.

The people who were defaulters, had no money.

The banks that had financed, had no money.

The financial institutions that had mortgaged, had no money.

The companies who had planned and developed colonies, constructed houses and apartments, had no money.

The worst affected were the companies which had started their projects and the projects were mid way in construction and incomplete. They also had no money. To add insult to injury, the banks refused to provide them with further cash that was so necessary for their projects to be completed. These projects simply lurched in uncertainty and their fate hanged in imbalance of ill luck.

The end result was that everybody was robbed of cash and liquidity.

A crash had started.

When real estate fell, it made the steel sector to fall with it. Then followed the cement. Then followed the hardware and timber and transport was also affected. Then a chain started. The people, who had become rich in these sectors, needed no more luxury or otherwise related amenities and cars and air-conditions and lot more commodities of elite living.

So, all these sectors fell.

The automobiles fell, the air-conditioning sector fell and also the consumer goods fell with it.

All companies manufacturing these magnificent items were listed in the share markets. So their shares fell in turn and there was a rout in the stock exchanges.

And the share markets fell.

This was the story behind the 'Black Friday'.

When the fall started, nobody wished to be its companion. Everybody distanced himself from it. The intellectuals and wise people appeared on the scene who started explaining and blaming others for the fiasco. They in the process forgot this also that they themselves were the biggest advocates of all the misdeeds and happenings, that lay foundations for or were responsible for this mega crash.

Nobody wanted to be proclaimed or called as the 'devil's advocate'.

The writing on the wall was quite clear to all.

The doom appeared close and in sight.

A full-fledged recession had started.

The summary

The summary to all this discussion is that the world economy was well on the verge of a recession in 1991-92 but the fall of Soviet Union gave it a support and saved it from falling into a recession. But very soon it became clear that it was not enough and then the IT boom came and it was such a revolution and it clearly wiped all gloom from the markets. The two aspects of this IT boom were the computers with their internet and mobile telephony. These two formidable engines of the IT revolution virtually gave birth to another industrial revolution and the world economies flourished. Even the economic slowdown in Russia in 1997-98 could not prove to, a stopping point for this IT revolution. But by the 2005-06 the signs of a slowdown had started to appear and 2007 saw a full blooded slowdown and by the start of 2008, the world economy was well into an economic recession of mammoth proportions. The recessions of 1911 and 1929 had witnessed the rise of ideologies associated with blind nationalism, religious, regional and linguistic chauvinism and racial discriminatory venomous propaganda and such strategies that ultimately plunged the mankind in a period of utter desperation and darkness and resulted in world wars one and two but the world simply cannot afford a world war three at this stage in 2008, to come out of the economic recession. The lessons of history are very bitter and we do not want to tread again on those paths of destruction, designed and thrust upon the world by evil forces like fascism and Nazism. The humanity has already paid a heavy price for that. Apart from the material losses and halting of progress for many years, the mankind has sustained enough damage in the form of five crore and seven lakh human lives that died due to sinister designs of some of the worst demons in history books like Hitler, Mussolini, Tojo and Franco. Now we pray and hope for some fruitful, amicable and constructive solution out of this economic crisis. May the world is able to come out of it with some better thought and thinking and steps in the brightest of future. All hopes at this point of time are centered on this in all corners of the world and may these hopes be fulfilled and honored.

Amen!

The way out of the recession with Aulprofem

The recession is damage on economy.

The industrial outputs are decreased to the minimal levels and industrial growth becomes not only sluggish but many times it reaches almost to zero levels. In many countries in Europe including Germany, it has recorded negative values. The United States economy is one among the worst hit. The Lehman Brothers have gone bankrupt. Sixteen more banks have fallen. The General Motors, Ford and Chrysler have asked for a financial bailout. The federal government has already provided a relief of seven hundred billion dollars but still the near dead economy has displayed no signs of coming to life.

The position on the employment front is equally bad. The City Bank group has sanctioned a sack of over fifty thousand employees all over. The HSBC is in the process of doing so. More than a million people have already lost their jobs and livelihood and more are on the process to lose.

The situation on other fronts is equally abysmal. The energy demand in USA and other countries, has gone down. The crude prices that were soaring on dollar 160 a barrel in April-May 2008; have touched levels as low as near to dollar 50 a barrel in November, 2008. The value of world currencies including Euro, Rouble, Yen and Rupee etc is daily touching new lows.

The effect of a recession on the day to day lives of people, has already been started to be felt. Many people have committed suicides due to the financial equations suddenly turning bad to worse. There have been a drastic cut in the spending. The markets are glutted with the products that are out of the reach of people's buying powers. Moreover, people are out of jobs.

The unemployed people do not earn, is a simple truth.

The unearning people do not buy much.

Sometimes, they do not buy, at all.

If they do not buy, then what will they eat?

They eat less, is the answer.

Sometimes they do not eat also.

Then how do they live?

They live on state subsidies and rations, is the answer.

But how long?

There is no answer.

The recession cuts sharply on the pockets of the people which lose weight very soon.

Sometimes, it cuts on their bellies also.

They also shrink very soon.

But it never cuts on the brains of the policy planners.

They become even more intelligent.

They think and come out with newer and newer policies.

And every policy strangely enough, makes the poor poorer and the rich richer.

This is the beauty behind policy planning.

It plays wonders always.

When the recession will be over............

When good days will come..........

When there will be blossom everywhere.........

The rich will emerge wealthier by a strange magic.

And the poor will become paupers by the same magic, it seems.

And the policy planners will be wiser by no magic, it seems.

Many of them will be nominated for Nobel laurels and many will probably bag it also.

The poor are strange creatures.

But they should be fed. Let's see how?

Because if they are not fed, they may cry and create restlessness.

They may create problems also.

Like what?

Like forcing the governments to gain control over the financial institutions and markets.

And which government is such a silly buffoon to intervene in the functioning of the financial institutions and in markets........?

No..... Never.... is the answer.

So, let's think over as how to feed them.

The people want three basic amenities.

The right to exist and right to live is mainly expressed in the ability to live with proper shelter, clothing and food and these are the main forces and necessities for life.

Can Aulprofem prove a help in this?

Aulprofem is a mammoth project. This can play a bigger role in the saga of national developments of nations in much better way. This can be a magnificent tool in increasing milk productions and the development of dairy sectors in agrarian economies of the developing nations. Also it can have a boosting effect on the developed and organized dairy sectors in countries like USA, Australia, European countries and so on in many ways. A female sex fixer for mammals can be helpful to dairy sector, is not a point to debate. It is well understood and rightly taken in the professional circles, related to agriculture and dairy. There are some other points to consider. Let's have a look at them.

The milk is still one of the main foods all across the globe. The use of Aulprofem to increase milk productions can be very beneficial in feeding the people who have no access to proper food and other resources, necessary for sustaining proper life.

As recession is going to progress, it will have a negative effect on the industrial productions. As it progresses further, it will definitely be acting in the form of a big hindrance on the agricultural outputs also. Many agricultural productions are dependant on the industrial performances like the productions of fertilizers, insecticides, etc.

The recession in the process of disturbing industrial outputs, brings those industries engaged in the manufacture of products such as fertilizers, insecticides etc in the category of low performing industries. This will adversely affect the agricultural productions and growth also. So, it is well understood that agricultural outputs are also most likely to come down during an economic recession. This scenario is also clubbed with another phenomenon of dairy and livestock industry. The dairy industry and livestock, are also the sufferers under such conditions. Poor agriculture will have a determining effect on the production of quality fodder and feed productions. The veterinary industry will also be affected. There will be direct impact on the health of the animals in dairy sectors. In addition to this, many other things necessary for normal healthy functioning of dairy and livestock sectors will come under pressure. The quality and criteria for sheds, animal houses and related infrastructures are also affected. This all combined, leads to the poor and bad health and hygiene of the animals. The poor health of the animals puts negative effect on the milk yields and the quality of the milk also goes down. There is no refuting the fact that milk productions as well as dairy industry performances are bound to suffer in these conditions.

But with Aulprofem in hand, the milk productions will definitely increase. This will have a direct boosting effect on the dairy industry. With lots of milk productions, we can expect dairy industry to pick up.

The increase in milk outputs can be well understood on national as well global levels after the Aulprofem use is resorted to. Let us consider the example of India.

The animal wealth of India is 300 million dairy bovines. If only a 10% of them are taken for the application of Aulprofem for a short period of 10 years; then the annual increase in milk production will be nearly 160 million tonnes per annum, giving about 400 grams of milk for every Indian, per day, in addition to saving nearly 70-80 million tonnes of food grains, 10 million tonnes of pulses and nearly 100 million tonnes of vegetables per annum.

So, the Aulprofem can prove a milestone tool in providing better food to vast populations passing through stress phases of an economic recession. 160 million tonnes of milk calculated at the rate of rupees 20 per liter means an annual increase in annual GDP of India by 3, 20,000 crores in rupees. This money will be passed into the hands of the people who constitute the biggest chunk of the population and form the agrarian backbone of this vast nation and are scattered throughout the length and breadth of India to the distant hutments and villages in plains, jungles, mountains and all possible locations in the country. The money passed into the hands of such vast majority of populace can rightly be termed as the national money or the national asset or even more appropriately, the national saving. A point to be noted here is that this much resources and materials are to be generated not out of new assets but from the assets that do exist before and are being wasted on the unproductive male progenies and adults of the cattle and buffaloes. This increase in GDP can be considered a national saving also. Every saving has two faces. In changed conditions, this can turn into investment concern. In other words, this can be a national investment also. The poor and destitute people on the verge of penury and starvation definitely need some amount of finances to be spent on feeding and sustaining themselves. This money that will be directly passed into the hands of the common masses in poor and rural localities and this will definitely be increasing their purchasing powers. The increased purchasing powers will directly be converted in enhanced buying and there will be a swaying effect in the otherwise sluggish markets. There will be movements of commodities into the hands of the consumers who will definitely consume them up. The cash will flow into the business circles through the dealers and retailers who will sell these commodities to the people. The commodities once consumed will ask for fresh supplies. A need to demand to replenish the depleted markets will be felt. This will create more demand. There will be placed more orders for more of the commodities to be bought by the dealers targeted for the ultimate use by the consumers. The industries will get more orders as well as the much needed funds. The liquidity will start flowing. There will be proportional movement of materials and commodities in the markets. This will constitute an increased economic activity.

One more aspect has to be considered. The increase in milk productions will result into the saving of millions of tonnes of food grains, vegetables, pulses, cooking oils, fuels and energy. This in itself will be a saving booty. These materials can be exported, stored or used in the production of more feed and other materials for the dairy and livestock. On the other hand, this will help in undoing or unmitigating the bad affects of the recession on economy and on the lives of the people. The common masses with surplus supplies to eat and sustain will be much better off in stark contrast to the starved and misery struck people faced with recession and lay-offs. These many resources and food stocks as told above can act a rescue to undo the starving and poverty and misery generating affects of the economic recession.

But with Aulprofem, the picture is entirely different.

The scenario will get changed to a better one in many ways. Let's discuss.

The sluggish markets are characterized by negligible sales and hyper glutting of the commodities in the stores. The commodities have already passed out of the industries for which the industry owners have already made payments in the form of raw material purchases. The banks have already issued funds for such purchases. The money to the banks has to be paid back. The goods pass on to the dealers in the markets who may or may not make advance payments. If they make advance payments then the industry owners will get back the money and they can pay the banks back. But on the other hand if the distributors make payments, they themselves might have taken the money from the banks and this has to be paid back to the banks. This money they can raise from the retail dealers who in turn may buy by cash or credits. In case of cash sales this money is recovered by the distributors and passed on to the banks but in case of credit sales, the goods are passed on to the retail stores without the money being passed to the dealers. Now if these goods are not sold to the ultimate consumers, there will be no flow of funds to the retailers and the dealers will not be paid also. Another aspect of these types of transactions is that the entire circle of transactions runs on credit. The money blocked on consumer level is the money blocked to the retailer and in the same operation this is the money blocked for the distributor, for the industrialist and for the raw material provider up to the bank which has financed this raw material procurement. A full chain is blocked if the consumer does not turn up for purchases.

During such circumstances, the market right from the industrial producer to the retailer is glutted with the goods for which there are no customers. The customers, who are the common people under these conditions of economic slowdowns and recessions, do not have much money in their pockets. They cannot go on for more purchases. The goods remain unsold. There is a big glut in the markets that remains unsold.

There is only one way that this glut and economic inaction can be broken in such circumstances. We have to put money in the hands of the people who have become cashless in times of recession. This can be done in the form of putting people back to employment but this is not easy as the recession has already eaten upon the employment sources. There are bigger layoffs in industry in such times and it is foolish and childish to think of creating more employment opportunities. The economic recession in the first phase, is a big layoff creator and employment eater. But in the present case the picture is different. Now we have the power of Aulprofem in our hands and let us find what we can do with this.

We have huge funds to the tune of 3, 20, 00 crores of rupees already flown into the hands of common citizens of the nation and this can play a bigger part, much larger in scale to the mischievous plot hatched by the Indian government that it did to bolster the dignified gamble of stock markets by announcing the fifth pay commission in April, 2008. This really played the trick and stock markets improved albeit for a smaller period and also there was a short term recovery in the commercial markets. During August to October, 2008 there was increased sale of cars due to this pay commission hike only. But the ploy did not work for a longer time and everything had proved futile by the end of November. The magnitude of money for this exercise was close to an annual increase by about 50,000 crores. If a small amount of 50,000 crores of rupees, can do the trick may be for a small period by pushing the cash into the hands of well to do government employees who form a bigger chunk of the share markets gamblers because they have too much money and they do not properly know what to do with this. They usually go to gambling in the share markets and do all types of cheap and third class immoral and indecent activities in the real public life also. In a country where nearly seventy percent of the masses live with a monthly income of less than six hundred rupees per month for a family of more than five to six members, on a daily income of only twenty rupees, the shameless ruling junta has no shame left, in increasing the salaries of the government employees in a country where even a peon, the lowest rank in public life, is paid more than rupees ten thousand a month. But we are talking of bigger economics and of Aulprofem. Let's continue.

The real good money to the tune of 3,20,000 crores when passed on the poorest of the poor, will find a flow in the markets and will be immediately passed to the higher levels of commodity and industrial trade. This money will not be lost in the share markets gamble but will form the real strength and play a backbone of the economic and industrial activity. Liquidity will flow. The markets will be buzz with activity. The cash and payment crunch will diminish and subsequently disappear and every where, there will be movement of funds, finances, commodities and materials. The banks will get back their loans and they will finance more. There will not be dearth of funds for the raw material purchases, neither for the process costs in the industry. The dealers and retailers all will be happy. The masses that are the usual inhabitants of a country, will be happy. They will be having cash and liquidity in their hands, along with the employment that Aulprofem is going to generate for them in the form of increased females of dairy animals that are the biggest asset for a small and marginalized farmer. The increased productive animal wealth will be an increase in the national assets.

This methodology can be applied in other countries of the world as well; especially the countries which are agro based and have considerable animal wealth that accounts for an appreciable proportion or contribution in the national gross domestic produce. All nations can benefit from it. The nations and the people of all nationalities, castes and creeds and religions. All alike.

These people as well as the nation will pass into a newer phase of prosperity and happiness and there will be no more recession.

And no more unemployment.

No more economic sluggishness.

No more gluts in the markets.

No more layoffs.

No more problems and obstacles for the common people.

No more tears.

No more sorrows.

Happiness all around!

Festivities and celebrations everywhere!!

And certainly the world will not remain a bowl of woes.

Fighting Terrorism – A National Salvation

Terrorists are the intra terrestrial aliens, from planet blood.

They have no concern with this world except destruction.

They have their vested interests.

Vested interests of separatism, linguism, fascism, hooliganism, vanadalism etc.

They give a damn caring to anything except their vested interests.

They display a stout fascination to achieve their interests.

In a practical capitalist world the biggest fascination is profit.

In this way they are hundred percent like capitalists who are often ready to do anything to procure their most pious fascination, the profit.

Or just like communists who are ready to do anything to procure their most pious fascination, the dictatorship of the proletariat.

The other questions are secondary.

Like capitalists do not strive for profit but in reality they work to satiate their personal selfish goals of individual betterment at the cost of others and indulge in deeds often aimed at exploitation of general masses and society.

And the communists do not strive to establish the dictatorship of the proletariat but in reality they work to satiate their personal selfish goals of individual glorification at the cost of people and usually resort to ruthless subordination of general masses and society, especially the proletariat in whose name they try to justify their misdeeds.

In this way, they all are same; the capitalists, communists and the terrorists.

All have their set and fixed goals.

All are one, in their most prominent characteristic.

All are breeders of misery.

Misery and poverty.

Poverty is a wonderful thing. In fact it is a multifaceted personality.

Let us have a small discussion.

Poverty is the biggest breeder of problems. It creates unevenness in the society or on the contrary, it itself is created by the unevenness in the society. It creates two worlds in the otherwise only present single world. It creates two societies in the single society of humans. The haves and the have-nots. In fact the have-nots are produced by the haves.

And the poverty is also produced by the haves.

And poverty produces a lot of other master pieces.

One among them is human suffering.

Human suffering and misery in the face of plentifulness, is produced in the immediate and opposite neighborhood.

Two living conditions, two societies inhabited by two types of humans; the rich and the poor, the haves and the have-nots.

And a common bonding between them, the mutual disbelief.

Disbelief and disregard.

The rich do not like the poor; is a common knowledge.

In fact they consider them even inferior to the animal folks.

They treat them likely also.

The poor do not like the rich; is also a common knowledge.

They in turn despise and disapprove of them.

But they cannot treat them like animals.

There is a river of distrust flowing between the two communities.

In fact it is like the railway tracks that never meet.

Distrust and disbelief are the glues of a modern society formed of haves and have-nots.

The glue binds them together but produces certain chemical reactions based on its nature.

There are certain by-products also.

The hatred is a major by-product produced by disbelief.

This in turn is the manufactory of all sorts of magnificent equipments.

Suspicion, conflict, intrigue and conspiracy are some of the few to be named.

Sabotage, agitation, violence are the usual results, worked upon by above.

End results are the crisis.

Crisis of many types. Let us consider a few points.

The world today is a hotbed of crisis of various types. The most prominent and perhaps the worst of them is the menace of terrorism which has virtually taken the entire world in its ugly garb. The shortcut to political struggles and the way to an immediate end to force even the powerful political enemy to its knees; it has found innumerable takers in the weird world of political destructionists. Terrorism is generally the destructive blackmailing. Unfathomable damage is done to the political enemy who generally may be the stronger opponent who cannot be taken head on in the ordinary course of events. The weaker points in his defense are found, studied and then attacked. The enemy is inflicted with severe unexpected destruction and damage. He is attacked on the points where he is the weakest. The humanitarian and internationally accepted norms, so necessarily needed for the usual human life to be conducted smoothly, are all thrown to the winds. The morality, the sense, the logic, all are meaningless words in the dictionary of advocates and theoreticians of such mean and inhuman political thoughts and movements.

The world has seen the outcome on many an occasions.

A conspicuous example was witnessed in the United States recently.

The 9/11 incident (see Appendix III) is a clear testimony to the level of destruction that can accrue of, as a result of such a cruel terrorist design, coming true. The start of First World War was an act of a terrorist attack on the life of the Austrian prince, Ferdinand which in turn took millions of human lives in the process and resulted in the complete destruction of countries like Germany, Austria and Italy, in addition to causing tremendous damage to the world economy. The world economy could not recover from the devastation as late as till the end of thirties in the last century.

The bombing of U.S. embassy building in Beirut by the Khomeini backed terrorists is a vivid example of the measure of damage that can be done to the civilized society by this monster of terrorism. The Chechnya, the Kashmir, the Palestine, the Lebanon; all are examples of the ferocious destructive power of the force of terrorism. We have come to such a position that the picture of the today's world is incomplete without the horrible face of terrorism not showing its frightful presence in any part of the world. This is the most horrible but the true reality about the present times.

What are the various causes of terrorism? There may be different views as to the existence of various political, religious, geographical, racial, linguistic, economical, social etc causes for it and it may be the subject of a thousand intensive discussions on the subject. But the most basic question is that why the people and usually the young ones are attracted towards such destructive ideologies and approaches of short sighted political shortcuts? Are they fascinated by the ideologies? Are they the victims of the systems prevalent in their countries of origin or inhabitation? Are they really the loathers of better things for the people of the countries of their birth or living? What makes them to adopt the destructive ways and means to achieve their desired goals? These are some of the basic questions that need answers to find a way out of this much entangled puzzle of terrorism.

This may be beyond the scope of the present write-up to elaborate fully well on the bigger question of terrorism but the mainstay of terrorism is the entry of people and that also of the young ones into this.

A sea is made because many rivers flow into it.

An army is made when countless people are recruited in.

A hell is conceived when many devils join it.

Why innocent young people are pulled into the profession of killing fellow brethren; is the mother of all questions?

The human being as usual, is a peaceful creature. He prefers to live by rules and principles. He is not perturbed unless he sees some sort of injustice or suppression of immoral origin, in society. If everything goes all right, he prefers to live in peace and harmony. But these ideal conditions arc rarely seen in most parts of the world. The world as rightly said is the hotbed of problems. There are haves and hvenots. There are oppressors and oppressed. There are tormentors and tormented. There are satisfied and there are unsatisfied ones. There are achievers and there are desirers of achieving. There are cunning and there are innocent ones. The world is made of people of all colors and hues.

Some people revolt when they see injustice of some type in society and they wage violent or non violent struggles against the inequalities and injustice. But there are people who are very shrewd and they keep on searching for the opportunities prevalent in the usual societal set up and when they find them, they try to make maximum out of that for their personal glory and gains. Such people usually are the starters of various territorial, racial, religious, linguistic, etc movements and struggles that usually benefit no one, but these people only. Politically they can be called fascists also. Hitler was one big example. He unleashed terror of highest proportions on the Jews. He killed countless number of Jews. But more than the Jews he killed; he caused Christians to death on whose name he was doing the godly act of eliminating Jews from their very existence on earth. For every Jew killed, there were nearly ten

Christians put to death, in the final scores. He along with his team of notorious hatred mongers like Goebels, Himmler, Rohnne etc; started the most successful terrorist campaign in the entire history of mankind. In course of that, he even used the government machinery and power to wage a relentless war of hatred and intrigue. In spite of being in government, he even burnt the parliament building down, simply to put the blame on communists who were a threat for his much touted and phoney patriotism. His terrorist and extremist ways did not end there only. He organized the massive genocide of Jews on a global level after coming to power and subjecting many countries to surrender and defeat. But the evil designs of him were put to an end when his massive armies were defeated in the Second World War.

A person so hated and despised today; he found his way to success and glory so fast and easy in his time. His rise to power was simply meteoric, unimaginable and charismatic. In fact he captured the imaginations of many people in his times and even continues to act a role model for countless traders of political sabotage and hate. After that many people across the globe have tried to emulate Hitler, in acquiring power and political success. The prominent among them are KKK of U.S., Khomeini of Iran, Taliban of Afghanistan, LTTE of Sri Lanka, Khmer Rouge of Cambodia, Hezbollah of Lebanon, Kashmir and Pakistan (see Appendix III) etc and many others. Some of them have been successful like Khomeini, Taliban etc. and some have not been successful like Hezbollah of Pakistan, Kashmir etc. These days a big organization called Al-Queda is working in many countries of the world like Afghanistan, Pakistan, Philippines, Iraq, Sudan, Uzbekistan etc. This organization which was responsible for the wreckful attack on World Trade Centre in New York in 2001, is believed to have the biggest terrorist network in the world today. There are countless other terrorist organizations working in different parts of the world. Some are big and some are not of much influence and impact. But it is a stark reality that there is barely a part of the world today that is untouched by this deadly menace of terrorism.

Again we come to the basic question as to why people are attracted into joining the terrorist organizations. The people who are otherwise basically peace loving and gentle; then why do they find the infatuation to become violent and destructive in motives and action? There may be the possibility that they are forced into joining these organizations. One fact, that the people who are generally well to do and belong to the upper strata of society; usually show a degree of skewness towards joining theses organizations; tends to prove this point. Also the people in developed nations and societies are not usually keen to support terrorism.

The main nurseries of terrorism these days are the poor and under developed countries like Afghanistan, Sudan, Bangladesh, Ceylon, India,

Indo China, Africa and Latin America etc. The biggest problem with the people in these countries is that they are living in conditions of extreme poverty and resourcelessness. They find it very difficult to make both ends meet. Many of them have to go to sleep with empty bellies. To add insult to injury, there is society who despises such people who are have-nots. Such people usually live in mental and psychological turmoil, always lamenting their bad conditions of existence. The main point in character and thinking of such people is that they always want to come to the equal levels of living with the well to do inhabitants of society so that they should also get a sort of respect and recognition. These are the people who are the easiest targets of terrorist organizations and their ideologues. They are offered promises of financial salvation and high prospective positions in society and ruling hierarchy that is supposed to be the end result of such terrorist designs coming true; and they are attracted towards it. Then they are brainwashed and thrown into the destructive destinations of terrorism. In the process they are subjected to various processes like emotional blackmailing, intimidation, false promises and what not. In the end, such people are generally sacrificed as a rule to bring unfathomable damage to them and to the society. The course of events generally is much fast packed and confusing in such a sequence of terrorist working that nobody is allowed to use his brain or think rationally. Anyone who talks sense and logic in a terrorist organization, is considered an enemy and is thrown out or executed without showing an act of mercy. The people go on being sacrificed to bring countless miseries to themselves, their families and to the general public but the cruel cycle of terrorism starting from deprivation of poor citizens to attracting their entry into the terrorist outfits and the ultimate destruction of them and the society; goes on and in the process, the terrorism mongers; the people behind these nefarious designs and outfits, keep on flourishing. The nursery of terrorism *i.e.* the extreme and abject poverty and destitute living conditions in poor countries and societies keep on existing and so does the brutal cycle of terrorism. The fate of such people and countries seems to be having the same vicious run, going full throttle, for at least the time being and also in the immediate future to come.

But there are certain benefits of terrorism; a point can be argued like this, and there are many takers for this in the world. In fact people are emotional advocates these days in favor of terrorism. This is the imminent fallout when they are stung by the deadly *mamba* of fascist ideologies. Usual support arguments put forward by them are as:

What should people do?

Where should they go?

What options are left open for them?

By word people they mean 'terrorists'.

They believe that when people are unemployed, poor and without work, they are free to go for destructive and subversive activities. At least, they get justification to do so because a human being by nature should be a dynamic entity and not a passive creation. It is a different question that these activities are oriented in creative or destructive outcomes. Such people find devastating options just right for such acts.

The Aulprofem can be a salvager in this regard. Because basically Aulprofem is poverty fighter and poverty reliever and can do well the role of an economic liberator and as well a social consolidator. So, it can block the recruitment of people into the terrorist organizations. The well to do people do not usually fall a prey to the bad motives and designs of the terrorist agents and ideologues. So, the Aulprofem should be promoted to counter the growing menace of terrorism in the world. This message should be taken well because the world with terrorism has become a worst place to live and everything possible should be done to bring this deadly menace of terrorism to an end. And this will be a bigger contribution of such a mega discovery like Aulprofem to the cause of maintaining peace and harmony in the world that of today is so torn in the ugly fight of terrorism in fragments so innumerable; each fragment bleeding to the core with the blood of the innocents, suffering in this brutal savagery of destruction.

May Allah, the Almighty bless the human race with a rid from this one of the worst scourges of entire history of mankind.

And with this we hope for the best.

Amen!

A Strong Case Against Narcotics and Addiction

The golden triangle is a famous word.

In fact it has a meaning also as all words have meanings.

But sometimes even the words lose their meanings.

Like, a chronic cancer patient loses the feeling of pain and when the pain is unbearable, he loses all feelings and fades into senseless unconsciousness.

Like a smack smoker or cocaine user or morphine injector or a heroin addict loses all meanings attached with the dictionaries of aesthetics, taste and sense.

They remember only one thing.

Addiction, addiction............more and more.

They even forget their families; parents, children, friends and all.

Euphoria, is a state of mind.

The word has place in medical dictionaries. An intake of alcohol is the easiest way to feel this state and millions; in fact hundreds of millions of people across the world, ingest alcohol in various forms every day.

But sometimes even words are raped.

The euphoria is such a victim of a word.

It is raped countless times on daily basis by millions of cocaine, smack and heroin addicts.

It is raped by the drug lords and their syndicates.

It is raped by the cocaine governments in the Caribbean and the opium based ministries in Burma, Laos and Afghanistan. There are countries where governments are formed with the will and mercy of the cartels engaged in the narcotic trade. Even thee presidents of these countries happen to be "bastards", involved in these nefarious activities. The people of these countries reserve the use of word 'bastard' for the presidents or heads of the state in these countries. This is the natural respect that they shower upon the rulers of their nations.

They have no other choice and they vent their anger and frustration this way at the state of political and administrative affairs in their respective countries.

Sometimes the words offer solace.

Solace and peace of mind.

If only anybody can have them in real terms!

But the days when time himself becomes the biggest enemy of the oneself of someone; then where does remain the comforting way out for the minds and souls?

The golden triangle is famous the world over.

It is a big piece of land spread over the hill tracts and valleys in three adjoining nations of Burma, Laos and Malaysia.

This is famous for the cultivation of a beautiful crop with big, colorful and conspicuous flowers and delicate, charming and lush green weeds.

A perfect picture of beauty!

The ill omened and cursed beauty. In fact the beauty is destined to be cursed, most of the times.

This crop is the only source of livelihood for the inhabitants of these terrains and localities.

The people of these tracts across the three states have a common distinguishing bond that unites them with one another and builds a strange and shared fraternity and brotherhood between them.

They are very poor.

This is the bond of commonhood between them that unites them as one and at the same time, with a strange and peculiar action cuts them off from the other people of the world.

They cultivate poppy.

Poppy is the source of opium and the people of these areas rear and raise this crop and harvest opium and their duty ends here only.

The opium is collected by the local lords who interestingly happen to be war lords also and they decide the levels of compensation to be paid to the farmers for the exchange of opium that is produced by them.

And again interestingly enough, this exchange value is very less and it is meagerly sufficient to keep those farmers in the permanent yoke of poverty and dependence over the local war lords for sustenance of life's bare essentials and necessities.

A good question arises here of the use of the opium that these local war lords collect this way from the farmers. But the answer is very strange and illusioning.

The war lords do not do anything of the opium.

In fact the opium is useless for them. They have invented a strange application of it in the research laboratories of evil mind setting in a denigrated societal structure embedded in selfishness and immoral human thinking and behavior.

These war lords are the Midas'.

They have the Midas touch with them also.

They can transform anything that they touch, into gold. In fact they can convert anything into substances even more precious than the gold itself.

They convert the opium into morphine.

And with the same deft craftsmanship, they convert morphine into heroin with a very simple and old fashioned methodology called double acetylation. This entire process is completed in very simple and poorly equipped make shift manufactories by men with ordinary skills and no formal training.

The only training that they need to have, is the existence in their minds of a strange and selfish lust for quick money. All other techniques and skill come automatically. One needs not to go to specialized training centers or universities to study and master this professional acumen. Neither some specialized degrees or diplomas are required to be earned to perform such practices. The only condition required is that you should be more than ready to sell your soul to the *Satan* and once this is done, he gets all other formalities completed himself with the display of same shrewdness, efficiency and outcome that he demonstrated once when he plotted to get the *Adam and Eve* out of the gates of the God's kingdom. He still owns and maintains that ability and cunning wickedness and he amply demonstrates it as well whenever he pushes a new entrant into the devilish, ugly and gory world of drug underworld.

One more destination decided by the *Lucifer* is the barren land of Afghanistan devastated with war and turmoil and pushed to conditions of abject poverty. This country has a long shining history of defying tyrants and foreign invaders. Otherwise a hilly country, having a plenty of land not fit for cultivation; the valleys of Afghanistan also have large stretches of good agricultural fertile land. Usually the farmers grew wheat and vegetables in these tracts of land. Even Afghanistan is famous for its production of dry fruits. But the twist of the time in favor of religion based dictatorship turned the picture upside down. The land and its people were subjected to the agony of war due to the rivalry between the two super powers of the then world; the USA and the Soviet Union. The country saw many upheavals after that. The political instability and the outcome of fascism later on, in that land added insult to injury. The neo Nazi ideology with a newly christened name of Taliban was established in that country. The Taliban rulers imposed ruthless suppression tactics in the name of Islamic *shariat* laws and a land once with glorious traditions of daring and free spirit, thinking and existence; turned into the likes of Nazi occupied territories in Europe, during the infamous rule of Adolph Hitler. The poor farmers of that country were forced by the new Taliban leaders and war lords to start the poppy cultivation and this is how the beautiful country of Afghanistan turned into a haven for the international drug cartels. The poppy crop was grown on national scales and the poor and penniless farmers had no other option, except to yield to the new love of the rulers for the narcotic trade; and had to go for the rearing of the deadly crop of opium in their fields which once grew wheat and vegetables.

The Afghanistan very soon earned notoriety and disrepute for its place in the international drug trade. The destiny of a country was in the hands of a handful of the crusaders of hate, propagating terrorism in the name of religion and God. The illiterate and ignorant people were lured into growing poppy by terrorizing or offering a little more money. The poverty danced to full drums and the desperate people were left with no choice other than to co-operate with the infamous drug trade or die with the gun shots or starvation on refusing to co-operate. A full nation became a virtual factory of narcotic production.

The biggest needs of every human being in the world are still the food, shelter and clothing. Of these the food comes the first and prime most priority. The other two; *i.e.* clothing and shelter, are the immediate followers. Of course in modern times; medical aid, education, employment, right to participate in democratic functioning of the state, right to know, accountability on the part of governments etc, are the other priorities and needs of the people. But in a country, where the basic amenities of food, shelter and clothing are a scarce and luxury, who can think or even dream of these other amenities needed for an ordinary and suitable human life?

So, the Taliban imposed poverty and misery in that country. The poverty is a good weapon in the hands of the governments which are anti-people by their very nature. It helps the governments to subject the people to continuous worries regarding the earning of daily bread and the bare needs for a primitive and crude human sustenance on the lowest of the levels. Such people usually lose all interest in activities, other than the daily chores aimed to earn the meager living and they never think or dream of revolting against the rulers. The relationship of poverty with the anti-people nature of governments is an inseparable reality. If there is an anti-people government; then it for hundred percent for certain that the people of that country has to be poor and *vice versa*. The major countries in Asia, Africa and Latin America are the shining examples of this fact. The Afghanistan under the Taliban was one of the most appropriate examples of a country under such classification. The fascist rulers were a perfect picture of the anti-human suppression machinery in full action. The optionless farming community in that country had no choice except to yield and they became just pawns in the international drug mafia trade.

A poor man sometimes even sells his soul to *Satan,* to feed him and his family, is a universal truth.

This truth was well experienced and practiced by people of Afghanistan in those curseful days.

Another safe haven for the international drug mafia is the countries of the Caribbean. Cocaine is the national infatuation and a mega employment generator there, second only to the beauty and the related flesh trade, widely practiced in these countries. There exists a common bond between the ordinary people in these countries and the common masses residing in other countries mentioned above and engaged in the production of raw and finished materials for the international narcotic mafia syndicates.

This bond is called poverty.

Helplessness is a technical word used in essays on studies on psycho-analysis and evaluation of human behavior in applied human psychology.

But this word has some other applications as well.

In fact, this word has come to exist as meaning something, like a close relative of almost every common human being especially if he is poor. Helplessness usually inhabits the households of all poor people. It develops a very close relationship with each and every member of the family, belonging within the boundaries of poverty. The father, the mother, brother, sister, uncle, aunt and every possible human bond and binding between the humans is shared also by this particular moiety called helplessness. It is the mother at one time to a person, father at the other time and son, daughter, grandpa or any other possible human relation at all the times, to all the members of such families. It is such a faithful relative or a friend indeed who never betrays such people with whom it is attached. It develops a permanent attachment and cohesion. An example of a perfect cementing bond!

Let's discuss a few case studies:

An ordinary family of a daily wage worker husband and a domestic help wife, with two or more children in a poor country earns enough as to feed their family with the minimum bare essentials of life. They usually go on working day in and day out and the wheel of life drags on. But all days are not the same. There may not be work for the man some day and the wife may run out of the petty vessels cleaning or the sweeping job that is so valuable to her, in the running of her household. Sometimes, someone in the family may meet with an accident or may fall serious or terminally ill. Suppose a chronic disease develops a love relationship with the family. The doctors and other people in medical and paramedical professions in these countries, are the true representatives of the *Lucifer* himself and this is no hiding or surprise happening to anybody in these countries. In fact, it has come to exist as one of the most universal truths of human life of modern times under such conditions. If anything of the sort, told above happens; then one thing is for certain that the vehicle of household ceases to be driven. All life comes to a halt. But the needs of life never stop. In fact, it is in their nature to go on. On and on. The belly demands food. The ailment demands medicine. The house lord demands rent. The school demands fee *in lieu of* the provision of education to the children. All these needs are a big reality of life and they can be satiated only with the exchange commodity called money. And the money is the substance that is in scarcity, under such conditions. The earning hands in the family or either out of work or not in a position to undertake working.

The disease is a funny thing. It is like the governments. It taxes the person in the form of spending on treatment and it also inhibits him from doing his earning job by laying him on the bed and decimating his working capability by making him either weak or disabled just like the governments force harsh regulations on the working of the enterprises and force them to shut down business and on the other hand levy heavy taxing on them to make them unprofitable concerns.

The summary of the discussion is that under such conditions, the head of the family finds himself unable to run or maintain the family. But the family has to be maintained under all conditions and there is no escaping from this fact. So, he starts thinking and searching the possible way outs. The uncertainties of such a penniless life, start bearing on his mind and it is no surprise that he even starts to think or follow certain shortcuts in life to earn more and also earn quick. Any human being wants a minimum iota of social security in his life and in a capitalist society there is no better security than the capital itself. The capital in ordinary life manifests itself as cash or money. Everybody finds the best solace in the lap of money and this is the reason that almost everybody is after money, in such a social set up.

So, the person in question starts thinking and analyzing the ins and outs of the situation, if he indulges in such an act of short cut money making.

After all, money has to be made.

Social security has to be earned.

Life is to be lived and life without money and materials necessary for life, just means death and starvation. Even the materials necessary for life are earned with money getting spent on them. So, the money is the mover of life and hence the life demands that the money should be earned.

The thinking mind starts working out solutions. The solutions are sometimes daring in nature. They frighten also and display risk shades. But some risk has to be taken if a gain has to be made in life.

Here comes the moral preacher. He is hidden in every human mind. He is the path to justness. He is the watchman on the way to good doing. He is the goodness himself.

He cautions. He tries to intervene. In fact, he intervenes. He tells about the futility of acquired gains, at the cost of moral loss. He explains about the moral hollowness, connected with the material luxuries earned with evil ways and deceit. He even goes to the lengthiest details in explaining about the mind haunting and mental torture that are the usual results of such compromised approaches about life.

But helplessness is not a hollow word.

It has substance. Substance and value and....standing.

It exerts pressure and that pressure works most of the time. It works on the man. It works also on his mind.

A helpless person never listens to the voice of righteousness.

In fact he listens and then turns a deaf ear towards it.

The voice starts turning feebler. It starts appearing that the voice is coming from a far off place and then it feels that this stranger voice has nothing to do with ourselves and it is concerning some other people who are not close or known to us. It even appears that this voice is non existent and just an illusion of human thought and overworked mind.

Then the voice stops at all, all of a sudden. A strange hollow numbness appears to be spread all across the horizons. This is the stage for the wrongness to take an upper hand in the thought and reflections of an ordinary man. He ceases to walk on the path of righteousness and takes the path of the Devil.

Now it does not matter to him what he does.

It does not matter whether he sells the narcotics to the innocent youth and poisons not only their lives but also their future and careers.

It does not matter to him that the young men so betrayed and disguided by him are the children of someone. Some mothers are sitting in wait of their children and some fathers have pinned high hopes on the performances of their offsprings and all these waits and hopes are just annihilated due to the evil designs of a representative of the *Satan*.

Such a person becomes more than ready to play a representative of the *Satan*. He on the other hand, tries his best to become a faithful servant and ally of the king of all demons of the universe. He takes the side of the evil and starts disposing his new duties of immoral and beastly job description. He becomes a beast in human skin.

Such are the persons, hijacked by the international drug mafia syndicates. In fact their men everywhere, are already on the lookout for such persons who can easily be lured into their nefarious activities and they keep a constant vigil in search of those who are already in the need of some easy money making efforts.

This is how the recruitment is done in such devilish professions and the ordinary innocent people are lured into the drug running rackets. There may be some other people made of a different stuff who are born and brought up in sinful environments and being already in the thick of such anti social thinking and actions, they directly go for involvement in such criminal gangs and evil doing and there may also be some people who are fascinated by the privileges of luxurious life, usually associated with the world of money and glamour and it is a fact that money is the greatest attraction for people, to be pulled towards such nasty professions.

There is another fact that all the bad guys are brethren in blood. The same truth holds on bad professions as well. A bad profession is always the natural ally of the other or sometimes even an integral or contributing part of it. That is the reason that drug syndicates of today are run in collaboration with the terrorist gangs and terrorist gangs in turn are run hand in glove with smuggling syndicates or the international cartels running flesh trade. Almost all bad professions have some visible or hidden bond with one another. They are like the members within a same family. The brothels, the call girl rackets, the dance bars and other professions in the periphery of obscene and immoral trafficking or categories, are closely controlled by the *Mafiosi* or are attached with them. Even the socially accepted and to a certain degree respected professions like glamour, fashion and modeling world, are sometimes controlled by the bad guys or their syndicates. The comforts of money include costliest of the cars, magnificent farm houses, personal cruise liners, luxury yachts, the best and glamorous females, the dining and staying in costliest of hotels and what not? These are the credentials that generally appeal to people of all categories and classes in society. Even the celebrated intellectuals who are hollow within and socially up class people, are attracted to such moral bribing, the kind of which is listed above. But a fact remains that the general people attracted to dirty immoral trades are of the just ordinary and lower mental levels and most of the times they also belong to the lower ranks in society. They find these attractions a big infatuation and they are automatically pulled towards this risky and sometimes frighteningly dangerous profession in flocks like the carbonate reacted sperms, directed towards the ovum lying in the oviducts in the female genital organs of a mammal. This explains the much activity and hush and bush in the factory of the evil that is demonstrated by the presence of so many news and media covering, regarding the immoral and underworld professions these days.

The immoral and underworld professions are a blot on the face of the society. All of them are equally bad but some of them even fall under the class of heinous acts. The narcotic mafia trade falls under this category. This anti people misadventure, even falls under the category of anti humanity crimes.

'If there is an ailment, then there is treatment for it also'. A proverb in *Hindi* reads like this.

Another proverb in *Urdu* seems to give the exactly opposite message and runs like 'the condition of the disease kept on deteriorating as the treatment progressed'.

Ironically both the proverbs find a total fit on the state of affairs as regards the anti narcotic drive on national and international levels. It is true that there is a remedy for every disease in the world and a way out of the problem can be found if a sincere and intelligent dithering is done on the subject and suitable methodologies are worked out and executed to put the given problem to a permanent end.

It can't be said that the respective governments in many countries of the world have not spared any action on the subject and no appropriate steps have been taken in various parts of the world. Actually, it is the exact opposite of what is said above. The governments in many countries; especially the US government, has done great efforts in trying to control this menace, worldwide. It is said that in the days of president John F Kennedy, a big offensive was launched against the mafia lords, involved in the narcotic trade and Robert F Kennedy (see Appendix I), the younger brother of the president and at the same time the attorney general of the United States, personally looked into the matters relating to the crackdown on the drug underworld. That is why a strong reason has been cited for the mafia funding and involvement in the murder of the two brothers; the president and the Robert Kennedy. An efficient needle of doubt has periodically been raised towards the dons of the narcotic mafia cartels along with the right wing ideology outfits like Ku Klux Klan (see Appendix III). This goes without saying that US government has been one of the most serious governments in the world, as steps to curb this menace of international drug mafia are concerned. But it is a pity that even after such an transnational uproar on the matter and the media highlighting of the bad and unwanted effects of this menace with short and long term perspectives on the countries and their people at large and their young generations in particular; no concrete success has been reported either in any specific region or country or the entire world as a whole.

This is a very sad situation and as the time has passed, the situation has only turned from bad to worse.

The gradual growing in strength of the international drug mafia is a direct blow in the face for the efforts of the world governments and the international civilized society and world community at large. There are many reasons to this. First of all there is so much money in this illegal trade that it is sufficient to lure people into it and bribe them morally and materially to become ready to pay any price or to go to any length in risking themselves, their families and even their lives, to get involved into this heinous crime against humanity and in the process, thrusting countless misery and suffering on the society at large and also upon themselves. The people just so willingly come and join this filthy and mean profession, is very difficult to imagine. But when the amount of money involved is considered, we come to conclude that this may be the reason that makes them ready to sell even their conscience and souls directly to the Devil himself.

The second reason may be the loopholes in the structure and/or constitution of the organizations involved in such crackdown and the people employed or hired in their ranks. There may be some rules that are easy to be trespassed or manipulated to the advantage of such agents of the *Lucifer* involved in evil trade of drug mafia. Then there also may be some people in these very organizations meant to curb this menace, who are also lured with the show of greenery on the currency notes and they are morally and materially bribed to offer help and support to this, one of the dirtiest of the professions ever reported in the entire history of mankind. The third and not the least reason may the selective state sponsoring of such a heinous crime against humanity. The Taliban government in Afghanistan has been one such example which openly sponsored the poppy cultivation across the entire territory of that country. There have been more examples in the countries of the Caribbean where many dictators have been known to have drug mafia links and roots. Such people when they are sitting at the helm of the affairs in a country, can do havoc to the cause of goodness and welfare of the people. There have been arrests of the running and past presidents and rulers of many states in the Caribbean countries, by the anti narcotic federal agencies of the United States.

Will there be an end to this gory business, ever?

This is not a question but a cry of a humanitarian person or thinker, originated from the inner core of his mind and heart.

The history of mankind is a witness that all demons get defeated in the due course of time and all evildoings definitely come to an end today or tomorrow. This has been the most peccable lesson of history that the mankind has learned in its long struggles and agitations fought against the tyrants, dictators and the other tormentors of mankind. The written history has been a witness to the fact that the goodness and human welfare has always won whenever they have been pitched against the forces of destruction and evil doing aimed at subjecting the offsprings of the most merciful God in heavens to subordination, torment and enslavement. The forces of nature always play a part or intervene in the process so as to make the victory of forces on the side of human welfare and betterment, possible by the well established and time proven methodologies; known and devised by themselves only.

The discovery of sex fixing is such a development.

The technique of sex fixing has a hope in store for emancipation from this curse on the human race. The Aulprofem is a known poverty fighter and eradicator. This also generates employment. These credentials of it can play an effective tool in preventing those people from getting lured into the business of drug mafia who opt to join this nefarious trade in a state of helplessness. Such people can be made self employed and self reliant with the offer of employment opportunity with the help of Aulprofem, well before they become out of job and the means to sustenance of life. A good reasoning person will never go on the path of the evil if he is getting a respectable remuneration of what he is doing in society in a well acknowledged and good way of livelihood. If he is able to feed and maintain his family and at least maintain its necessary and basic needs, he will never go astray on the paths of destruction decided by the forces of evil.

The sex fixing has the potential and strength to play a foundation for the rapid progress for the people in the countries, with the agro based economies. So, the people who are on the edge of losing their quantum of minimum bare needs, necessary to keep and preserve a respectable human life, can find a place in the blissful share of prosperity generated by the constructive forces of nature in the form of the power of Aulprofem and once they decide not to go on the path of the evil and maintain a stand on the pedestal of goodness and human betterment and progress of society, made possible by the ever growing fields of scientific research and development; the devastating professions and ugly crimes against humanity; like the one of international drug mafia, can be put to a permanent end.

Alleviating Poverty – A Humanitarian Cause

A blur on society, a curse on the rulers and the worst word ever badmouthed; the poverty has existed since times immemorial and still torments the lives of billions of human inhabitants across the globe and going by the indications that exist, it seems to extend its ugly presence in the farthest of the future to come.

Poverty is a condition, ambiguous in defining but very vivid in judging. The conditions of living in general are a measure to ascertain the existence or absence of poverty in a society or in a section of people, living in society. The availability of clean drinking water or the absence of it, the accessibility of people to healthy and sufficient food, clothing and shelter etc, all are the indicators of this phenomenon existing or not. About 40% of Indian population lives under the poverty line. Let us take India at a glance. According to 2001 census there are nearly 192 million households in India. Out of them 24 millions have closed drainage system, 65 millions have open drainage system and 102 millions do not have a drainage system at all for the waste water outlet. About 34 million households use LPG as fuel in their houses.100 millions use firewood, 19 millions use crop residue, 19 millions use cow dung cake and 12.5 millions use kerosene as the fuel to cook. Electricity is available to 107 million households *i.e.* 54.7% of the population and remaining nearly 45% of the population, live still without electricity in their houses. Toilet facility is available to 69.8 million households *i.e.* 36% of the total population and remaining 64% people are still without it. About 68 million households avail banking services and remaining 65% are still without it. Car, jeep and van is owned by a little less than 5 million households and 22.4 million households have a scooter, motorcycle, moped etc. About 83.8 million households have a bicycle, 17.5 millions have a telephone, 60.6 millions have a TV set, 67.4 millions have a radio, transistor etc and to the horror of it, nearly 66 million households do not have any of the above mentioned assets in their possession. Nearly 30% people still have the walls of their houses built by mud or unburnt brick and nearly 10% have these of grass, thatch and bamboo etc. 21.4% have the material of roof made of grass, thatch, bamboo, wood, mud etc. Out of the total 220.3 million married couples in the country, only 133.9 millions have independent sleeping rooms, available to them. About 70 million households have drinking water from a tap and 68 millions have it from a hand pump and remaining people do not even have the access to clean drinking water. Nearly 50 million people go to sleep on empty stomachs daily. So, India a land of great geographical and cultural diversity, is also a land of remarkable depths of deprivation and lacking.

All this information, quoted above, is taken from India at a Glance 2004, published by Jagran Research Centre, Jagran Building, 2, Sarvodaya Nagar, Kanpur, India.

Aulprofem, can be a tool in benefiting the society. It can provide the food *i.e.* the milk, which by any standards is still the best nutritious food in the world to the half starved and hungry masses of India. It can increase the income levels of the poorer of the poorest by strengthening the dairy outputs in rural India. It can enhance the purchasing power of the farmers by increasing their income levels. The flow of money to rural India with Aulprofem can be tremendous. Nearly three lakh crores will be flowing to rural India every year. This money undoubtedly, will be used in raising the living standards of the people especially the poor ones, in villages.

So, the project Aulprofem is definitely bound to go to longer destinations in alleviating poverty and distress conditions in rural India and salvage the poor Indian masses from pangs of hunger, malnutrition, shortages and deprivation.

Delivering Designer Babies

Can fantasies be marketed?

Or can dreams be auctioned?

Here's news that offsprings will be designed on order and as wished.

The designer babies.

The world may call them with this christening.

The color of skin, the contour of the eyes, the curls in the hair, the design of a nose, the length of the neck and the extent of the height. All these things will be put in a prescription pill of some magic therapeutics and delivered to the would be mothers and fathers of course and they will be the proud parents of the children with the best of the traits, in terms of beauty, personality and built.

And the sex of the child?

Yes, we have Aulprofem to decide that and the only condition that makes this possible is that we ought to be living in an **era of sex fixing**.

Suppose a beautiful creation with long face and neck and similarly suitable pencil lips, sexy eyes and scintillating looks with long legs, thin waist and sharp curves and angles and a cat walk... but the sex......?

What if it is male?

No comments......sshhhh......

No? Anybody can cry that. It has to be a female only.

But if it is male, then the artist that envisioned and then crafted it in practical reality; should be rightfully hanged.

The entire process of producing such a crafted and made to order progeny can go into a full waste if the necessary and suitable genital instrument of choice is not fitted and if it is wrongly fitted or is not in a working order and fluctuating in between the two ends; then the whole outcome is definitely a blasphemy.

Similarly we cannot expect to afford a girl child with big and strong built, dark color, high forehead, broad shoulders, broad chest and muscles bulging.

If we can design order such a child and the sex part only is not relevant and as wished, then the whole exercise will turn into a big futile ending.

So, the designer baby is a hollow criterion without the fixing of the sex part. The sex part has to be as wished and this desire can not be left unfulfilled. Otherwise the whole exercise of going into it and getting the outcome is a wasteful concern.

In fact, getting a child of desired sex is almost as getting the designer baby. Even if the effort does not yield the necessary features of a hundred percent desired child, it certainly fulfills the conditions of at least getting a ninety percent success in getting the child of desires and fancies.

The very fixing of sex in the offspring, delivers the outcome of a designer baby to almost near hundred percent values or at least to a more than ninety percent.

And this is the biggest in truth about the process, ideology and thought about the wonderful phenomenon of designer babies.

A designer baby is certainly non-existent without the proper fixing of the sex and genital parts, and spares and even if the child is a made to order one and exactly formatted and formulated as wished, it is a full exercise into nonesense without the proper instruments of masculine or feminine performance and efficiency.

This is the most bitter and real truth about designer ordering a baby and then having it.

There can be a designer baby without the proper phenotypically engineered accessories and aesthetic finishes but there can never be a designer baby without the proper fixing of the sexuality and genital organs in it and this is the deciding guideline and the binding principle in whole of the thinking and philosophy behind the concept of having the designer babies in the world of human life and existence.

Basically, the idea to have designer babies applies to human beings only and the idea has quite fascinated the authors and intellectuals in the literary and fiction worlds, throughout the world. The rapid progress that science made in nineteenth and twentieth centuries fired many a passions and desires in the minds of inquisitive and curious inhabitants of the world citizens who at times became restless and impatient and demonstrated these traits often in the open and wanted to have what the science offered and what it could offer and at the same time desired for that also which science promised to offer or seemed just to offer in the near or distant futures.

Robots were one such desire but rather than being fulfilled, this desire is still a dive into the fantasy. Clones were another object in this category but they also seem to lie in a far fetching lap of future happenings. It is a known fact that during the second anaphase in meiosis, genetic transfer takes place in the form of exchange of chromatin material between adjacent chromosomes, paired opposite to each other thus knocking down all possibility of the existence of the same DNA or chromatin in the daughter cells; a foundation of the all cloning theory or practice. This simply turns the very concept of cloning a hollow imagination, without even a trace of scientific realism in it. The genetic transfer is a phenomenon, perfected by nature to create and maintain the biodiversity so vivid and present in the world of life and live organisms and this is the process by which all biological evolutionary practice has taken place in the universe, thereby resulting in the evolvement of millions of organic species in the plant, animal and prokaryotic worlds. One more thing and message that nature has tried to convey with the existence of this phenomenon of genetic transfer is that nature likes the change and not the repetition in the life and life forms and thus it strongly disapproves of any concept or principle of maintaining a stagnation or *status quo* of any type; maybe it is cloning of any type. Thus nature doesn't like the very concept of cloning or keeping of organic species in the unaltered and former conditions.Getting a panacea to treat all diseases is another such imagination wish and the magic potion to bless every human with eternal youth is another. The latest arrival on this wish list is the concept to have and own designer ordered and accordingly crafted babies.

This concept has also been advocated in animal and livestock world of late. There is no harm in getting a cow which can yield hundred liters of milk per day and eat half as much as the other ordinary cattle or have the best swine that can grow to give much pork in a lesser period with lesser of the diet consumed, in so getting. But what if we get the male progenies instead of getting the female ones in case of cattle and similar exercises result in getting only male herds instead of females when we opt to increase the numbers of animals in the cow herds or pig farms by preferentially producing more females of their species.

The sexed babies can be a boon of science and fortunately enough with the discovery of Aulprofem, we have already marched in the magnificent **era of sex fixing**. The conclusion of the story is that designer babies can be a boon of science but without the success and provision of sex fixing activity in them, the entire process and objective of getting them runs into the danger of going into a full waste.

The sex fixing activity if combined with the latest advances in the science or the developments that the science has to record and make in the near or distant futures; can make this wish of the humans to have designer babies fully granted and accorded. But at present the performance and achievement levels are near to zero except for the discovery of Aulprofem which is a very recent entrant in the world of scientific discoveries.

Without sex fixing, there will never be a single designer baby but with sex fixing at the disposal of man, the designer baby concept and criterion is just like almost covered and accomplished.

And this is the crux of the story in this regard.

Regarding more elaboration on this subject, I will like to take this matter of having the designer babies in the light of sex fixing development of concept and discovery and the possible benefits and impact of the advances in the field of sex fixing on the scope and possibility to have designer babies of choice in my forthcoming book on sex fixing entitled "Sex Fixing & The Human Development".

So, we close the matter here itself and conclude that sex fixing is a must if one ought to opt to have designer babies and without sex fixing, the concept of having designer babies is a hollow imagination and nothing else more than it and on the other hand, having a sex fixed baby is just like having a designer baby to almost near hundred percent extent and whatever the shortcomings remain to be bridged, in order to make this effort hundred percent successful; can be borrowed from traditional scientific advances in science in the domains of genetics and molecular engineering etc but their authenticity and effectiveness is surely a thing belonging to future research and development as of today the greatest achievement and satisfaction that we have is that we have already recorded success on the front of sex fixing and this should make us very, very content and near to having almost complete designer babies in future.

And this sums up the discussion.

And we hope for the better things to happen in future.

And with this good wishes and.......

May God bless man .... With designer babies.

And ...... Amen!

& *Adieu* !!

PART - F

Clinical Information

- Clinical Information
- Pharmacodynamics
- Pharmacokinematics

Clinical Information

This is discussed as under:

Special Populations

Renal

Impaired renal function does not result in a decreased rate and extent of excretion of Aulprofem active constituents and their metabolites. In animals with creatinine clearance of less than normal value, however, adjustment of the dosing regimen is recommended. The total amount of Aulprofem active constituents and metabolites removed during a 4 hour dialysis period is insignificant.

Hepatic

Metabolism of Aulprofem active constituents and its metabolites is reduced in animals with advanced cirrhosis of the liver, resulting in both a larger area under the concentration time curve for Aulprofem with respect to its active constituents and metabolites elimination half lives (plus 26 hours). In cirrhotic animals, adjustment of the dosing regimen is recommended.

Age

Healthy elderly animals aged 15 to 20 have plasma concentrations of Aulprofem active constituents and elimination half lives comparable to those observed in healthy animals less than 15 years in age. In animals over 20 years in age, maximum serum concentrations are slightly elevated and the elimination half lives slightly prolonged (18 vs. 16 hours) compared to animals 15 to 20 years of age. Adjustment of the dose is recommended for animals older than 20 years in age.

Gender

The absolute bioavailability of Aulprofem was 81% in males and 86% in females. The plasma clearance per ml per kg body weight for males and females was of the corresponding values. Following a single oral dose and after adjusting the body weight, females had a 12% higher peak Aulprofem concentration and a 35% higher area under the concentration time curve compared to males. The clinical significance of difference is unknown as Aulprofem as a sex fixer is meant to be administered to females only.

Clinical Studies

Aulprofem has been given in 100, 150, 200, 250, 300 ml single oral dose to animals meant for sex fixed for the offsprings. The animals were subjected to the sex fixing trials for offsprings after administration of Aulprofem on mating or AI.

In models with 100 ml dose the sex ratio manipulation to produce female offsprings was demonstrated at inappreciable levels. A dose of 150 ml provided sex fixing of 80 percent manipulation in sex ratio towards the side of the female sex resulting in the deliveries of female calves. A dose of 200 ml provided sex ratio manipulation of superior levels, up to a change in sex ratio of plus 90 percent to produce female calves. The results of 250 ml dose were comparable with the 200 ml dose and the sex ratio manipulation towards the female side was again plus 90 percent. The sex fixing trials could not be comparable with any other sex fixer drug because the Aulprofem is the only one and the first sex fixer drug of the world, known in medical history so far. The dose of 300 ml produce sex fixing but the animals showed mild signs of uneasiness and a nausea type feeling behavior and this trend in the animal behavior was more demonstrable with dose of 400 ml. But these doses were not at all fatal and the recovery of unwanted effects was soon recovered with the administration of simple antacids like aluminium hydroxide and magnesium hydroxide etc.

Titration Trials

In randomized clinical studies with 25 to 30 animals per group, done with a standard dose of 225 ml per animal, the female sex fixing activity with respect to the offspring was demonstrated in a success rate of plus 90 percent. In animals with trials with doses of 300 ml, 400 ml or more the signs of uneasiness or a discomfort to the animal were demonstrated but the trials within the dose regimen of 150 ml to 250 ml did not show any of these signs and gave better results with respect to the sex fixing activity of production of female offsprings.

Indications

1. Pharmaceutical aid
2. Acidifying agents
3. Buffering reagents in laboratory preparations and schedules
4. Female sex fixer

Dosage and Administration

For manipulation of sex ratio effect to produce female offsprings in cows and buffaloes, a dose of 200 to 250 ml can be given. In doses of less than 200 ml and up to 150 ml per animal the sex ratio manipulation activity was demonstrated but to a slightly lesser extent *i.e.* with 150 ml of Aulprofem the sex fixing activity was around 80 percent success rate. In trials with doses of 300 ml and above the unwanted effects of uneasiness probably resulting from acidity or nausea feeling and decreased appetite were observed. For moderate and effective use however, the dose regimen of 200 ml to 250 ml of Aulprofem is recommended.

Individualization of Dose

Available data do not suggest that a dosage adjustment is necessary with animals 15 to 20 years in age unless they have considerable hepatic and renal impairment. For animals with age of more than 20 years or more, the administration of Aulprofem is not recommended.

In animals with diseased conditions and especially with renal and hepatic impairments, the administration of Aulprofem is not recommended. For animals with liver cirrhosis the administration should not be done. In animals with starved or on empty stomach the administration should not take place. It is recommended that Aulprofem should be given to the animal which is hale and healthy and doing all his day to day activities in a perfectly normal and optimum way.

Contraindications

Aulprofem should not be administered to animals who previously have demonstrated hypersensitivity to Aulprofem, any of its other components, products or constituents. It is also contraindicated in cases of animals treated with alcoholic preparations, beverages, or acidifiers, centrally acting drugs, opioids or psychotropic drugs etc.

Warnings

Seizures

There has not been any history of animals reporting seizures with the administration of Aulprofem. Post marketing reports and clinical trials as well indicate that the animals are not prone to the occurrence of any seizure attack of any type with the administration of Aulprofem. However in patients taking selective serotonon reuptake inhibitors (SSRI antidepressants or anorectics), tricyclic antidepressants (TCAs), and other tricyclic compounds (cyclobenzaprine, promethazine etc), opioids, MAO inhibitors, neuroleptics etc the Aulprofem does not seem in any way to interfere in the otherwise usual incidences of seizures commonly seen during or in these types of administrations.

Other drugs that reduce the seizures threshold are not in any way reported to be interfered with the administration of Aulprofem. We at least have no such study or report to confirm such interference. However, even then the clinical correlation with this type of drugs being absent with the administration of Aulprofem; it is advised better not to go for the application of Aulprofem along with this type of drugs.

Use in GIT disorders

As the Aulprofem tends to increase the acidic contents in the rumen or the stomach; there are chances that the administration of Aulprofem being by the oral route, tends to compound the problem of acidity in rumen or stomach as the case may be. So, the clinical correlation of the acidic nature of Aulprofem is advised in this type of situation. But the amount of Aulprofem being very small *i.e.* 225 ml and the size of the rumen being very big; there are very less chances of this type of aggravation of acidity.

Loss of appetite

In post market reports there have been occurrences of the chances of animals resorting to less feeding for a day or two after the administration of Aulprofem but the animals normally start feeding after a day or two and there is nothing of much concern. If the problem persists for more than two days then treatment with mild anti acidity agents is recommended. The situation is not alarming in any way.

Mouth burning

In super sensitive animals to the application of acidic materials, the chances of mouth burns of animals are reported; almost in less than 2 percent of animals treated with Aulprofem. In some cases the mouth burning may cause the appearance of small to mild blisters to appear in the mouth or in the peri-buccal areas of nostrils, lips etc if the spillage of the material takes place on these areas due to mishandling of the drug while administering. So, it is advised to administer the drug by a gastric tube and should be practiced by an expert person trained in the veterinary or livestock disciplines.

Spillage on delicate organs, wounds, eyes etc

The spillage of the material of Aulprofem on the delicate organs of the animal like mucous membrane around the eyes, mouth parts etc or on the open wounds, burns etc may cause the aggravation in the generated or the associated problems. So, it is advised that in the first case the administration should be done by an expert in the veterinary or livestock fields so that such instances do not occur and secondly if anything like above happens; the affected parts should be washed thoroughly with normal water kept in temperature range of 15 to 30 degrees Celsius. If blisters rupture, then treatment with ordinary antiseptics or if need be with the ordinary antibiotics etc should be done.

Precautions

Selection of the animal

Only the healthy animal should be selected for the application of Aulprofem and care should be taken to avoid the use of this drug on weak, diseased or wounded animal. The Aulprofem as such does not act as a cause for any of such complications like skin burns on delicate or broken parts of skin, nausea, decreased appetite, uneasiness, acidity etc, but the prior occurrence of any of such conditions can lead to an embarrassment of the associated type in weak, starved or sick animals with these types of conditions. Care is advised to avoid selecting such types of animals and situations.

Storage conditions

The Aulprofem activity that is basically the sex fixing effect with respect to the offsprings, is seriously affected on elevated temperatures of storage near 40 degrees Celsius or so. So for maximum results the storage temperature should not increase above 30 degrees Celsius or so. That is the reason that optimum temperature for storage is advised in the range of 15 to 30 degrees Celsius. The success rate has been found to be heavily affected on temperatures above or below this range. The storage should also be in a shady and covered area and the direct contact of sunlight or sharp light of any nature should be avoided in any case.

AI or the natural service

The demarcation in the way this drug formula is administered is clear in the administration via natural and the artificial insemination routes. It is found that it is more effective if given after half an hour after the natural mating. The time is calculated when the male animal has ejaculated and not the time when it has dismounted. The time limit in case of artificial insemination is half an hour before this procedure. This is to avoid the complications in the field conditions due to the state of the sperm that has come out of the cryo sleep just a few moments back and it is in which condition of activity; is usually difficult to guess? So the application correlation of this type of administration of Aulprofem is advised.

Acute Abdominal Conditions

Care should be taken to avoid the application of Aulprofem in animals with acute abdominal disorders of any type. The GIT of the animals if subjected to any ruptures, blockage or diseased in any way should not lead to the administration of Aulprofem.

Use in Renal or Hepatic Disease

Animals with impaired renal or hepatic function should not be subjected to the administration of Aulprofem. Metabolism of Aulprofem is affected with animals with advanced or mild liver cirrhosis. In such animals therefore dosing is not recommended. So, the use of Aulprofem in animals with renal or hepatic malfunction should not be resorted to.

Drug Abuse and Dependence

Aulprofem does not show a tendency to induce drug dependence. To top to that it is a single dose medicament, so the chances of it to cause drug dependence are remote in the sense that it is intended to be used once only. This does not have the actions targeted in the area of CNS, spinal, sympathetic or parasympathetic nervous systems. It is not habit forming.

Drug Interactions

Aulprofem as such cannot enter into the body's metabolism or induce its own metabolism. For this its active constituents have to get converted into the acetyl group first and then it is metabolized via the acetyl-coA route that is central to almost all the major metabolic cycles of the body.

With alcohol, it can enter into a type of chemical interaction to yield an ester molecule. Hence its administration with alcohol or alcoholic contents is not recommended. The use of Aulprofem with preparations containing beverages or the foods that are constituted with beverages as essential parts should not be made a part with the administration of Aulprofem.

With acidic preparations like the ones containing citric acid, tartaric acid or citrates or tartarates or the like, the administration of Aulprofem should not be done in combination with these or similar compounds mentioned above. The acid radical from these drugs or agents tend to form a running equilibrium with the active acetyl radical of the Aulprofem and interfere in a way as to hinder the absorption of Aulprofem or the otherwise.

With salts of strong alkalies with weak acids like in the salts of sodium bicarbonate, potassium citrate or the ones like them; there is a strong chance of the acetic acid constituent of Aulprofem strongly to engage in the reaction with such preparations and any such administration mix up should be avoided.

With salts of sodium hydroxide; with the weak acids or even with strong acids, there are chances that sodium toxicity may develop with the conjugated administration with Aulprofem.

Carcinogenesis, Mutagenesis, Impairment of Fertility

Aulprofem was not found to be mutagenic in any of the reports and studies. Since the product is on a bigger use in the field and already has been tried on thousands of animals; I have not received a single such report or incidence of it causing the carcinogenic or mutagenic effect or even the effect of impairment of fertility in the animals, subjected to the use of it.

There was not a study in observation that showed any tumors or any unexpected growth of neoplasm. This finding is believed to suggest that Aulprofem is very safe in mammals, humans and dairy animals. No effects on fertility/sterility or impairment of fertility/sterility were observed for Aulprofem at oral dose levels up to 1 ml/kg body weight for male rats and 1.5 ml/kg body weight for female rats.

Pregnancy, Teratogenic Effects

Aulprofem is a drug that is basically intended to be given at the time of start of pregnancy, exactly timing a little before or after the mating or artificial insemination as the case may be. This is not something supposed to be given during the period of pregnancy. So the potential risk of loss of pregnancy should not rise in question as the intended use of this drug is not supposed to be happening during pregnancy.

The chances of non conception or the failure of the sperm to join with the ovum have not been seen in any of the studies. The usual rates of conception are observed with the application of Aulprofem as are the rates when no Aulprofem is given. So, the fetotoxic or embryotoxic studies were not conducted in case of Aulprofem.

No drug related teratogenic effects were observed in the progeny of rats, mice or rabbits which were treated with Aulprofem with oral route, in doses even higher than the ones that are used for dairy animals (1.5 ml/kg for rats 2.0 ml/kg for mice and 1.75 ml/kg for rabbits).

There was observed normal fetal formation, normal growth of nervous, skeletal, internal and external organs and the resultant progeny was of normal dimensions. No toxicity was observed in normal doses and doses slightly higher than the normal ones.

Labor and Delivery

Aulprofem is not supposed to be used during labor in pregnant animals. This drug formula is to be used before commencement of pregnancy or strictly speaking at the conception time of formation of a zygote. There is not supposed to be any use of this drug afterwards. Even then, the studies demonstrating the effect of Aulprofem on delivery and labor time and effects; on animals administered the dose at conception of the concerned pregnancy, were conducted. And the available data from such studies on animals treated and sex fixed for female offsprings with Aulprofem show that the animals treated with Aulprofem show normal delivery and labor as observed in other normal animals at the time of delivery.

The effect of Aulprofem or side effect if any, on the later growth, development and functional growth and maturation of offsprings delivered from animals which were treated with Aulprofem; are not reported in any of the studies or post market reports are unknown or non existent.

Lactating Mothers, Infant Use

The Aulprofem is not intended for mothers who have recently delivered. However the lactating animal which goes for the next mating or intercourse with the male animal when coming to heat after few months after delivery can be given the Aulprofem. There have not been any reports of the Aulprofem constituents or metabolites coming in the milk of the mother and there is no danger to infant feeding on mother's milk.

Use in Old Animals

In animals aged 15 years and more, the use of Aulprofem is not recommended. The aged animals are usually expected to vary to larger extents as regards to their ability to behave, exist, tolerate etc the variations in environment and subjecting to the use of drugs etc. So, the use of Aulprofem to animals 15 years or more is not recommended.

Overdosage

The cases of overdosage with Aulprofem are reported. Strictly speaking, any dose in excess to the normal dose which is 225 ml for a healthy animal is an overdose. But the minimum dose proving fatal for Aulprofem has been 1000 ml reported in a field trial. This was reported in an isolated case. There have been reports when the animals behaved hale and hearty when given doses of up to 1500 ml of the Aulprofem and they did not succumb to fatality. There have also been reports when the animals given normal dose, stopped feeding and they developed mouth burns and could only be normal before they were given a couple of days' treatment with antacids, antiseptics etc.

Special Precautions

Animals with known or reported alimentary tract diseases should not be given this drug. Weak, diseased or wounded animals should be abstained from this treatment. Proper temperature and storage norms should be adhered to. Spillage and leakage during storage and during application, should be taken care of. Caps should be airtight packed and not loose. Give this drug half an hour before in case of AI and half an hour after in case of mating and do not administer with other drug molecules containing sodium salts, weak or strong acids or alkalies etc. Avoid spillage on delicate organs, eyes, lips, wounds etc. during administration

Adverse Effects

Nausea, vomiting feeling, lack of appetite, decreased feeding for a day or two, mouth burns, estranged behavior, uneasiness, skin burns on delicate organs like mouth cavity, lips etc, compounding the affects of wounds, open skin parts if spillage takes place on these parts, etc.

Pharmacodynamics and Pharmacokinetics

These are discussed as under:

Pharmacodynamics

The acetyl group derived from acetic acid or sodium acetate is fundamental to the biochemistry of virtually all forms of life. When bound to coenzyme A, it is central to the metabolism of carbohydrates and fats **(Fig 28)**. Although the complete mode of action is still not understood, it appears that it performs the sex fixing activity due to selective binding of the ultimate binding sites on the body of the oocyte by a complicated mechanism. Bryant *et al.* (1993) has elucidated a method of binding of Y-sperms with Y-specific antibodies for immuno-separation of sperms. Zavos *et al.* (1995) preached non-protein diluents for such immuno-separation techniques. The monoclonal antibodies join with specific antigen binding sites on the bodies of Y-sperm, these sites being non existent on the bodies of X-sperm. This difference of binding sites on X and Y sperms are consistent with different and equal antibody receptor sites on the body of the ovum. Acetate induced metabolites block the Y sperms H-Y antigen binding specific antibody sites on the body of the ovum thereby ruling out the chances of Y sperms binding with the ovum and resulting in the only possible conjugation of X bearing sperms with the ovum leading to the formation of a female zygote (Fig 27). Apart form the sex fixing activity the acetyl group acts to speed recovery of liver and skeletal muscle glycogen. This has been proved in studies involving the intake of acetate in conditions of fatigue, after skipping meals, post exercise or as part of an athlete's breakfast on the day of competition (Takashi *et al.*, 2001). Intake of acetate also enhances the intestinal absorption of calcium (Kishi *et al.,* 1999). Acetic acid administered orally is immediately absorbed; uptake then occurs in liver and peripheral tissues (Cummings *et al.,* 1987; Pomare *et al.,* 1985). It is metabolized via acetyl-CoA in the tricarboxylic acid cycle in liver and skeletal muscle (Bellard, 1972; Crabtree *et al.,* 1990; Spydevold *et al.*, 1990).

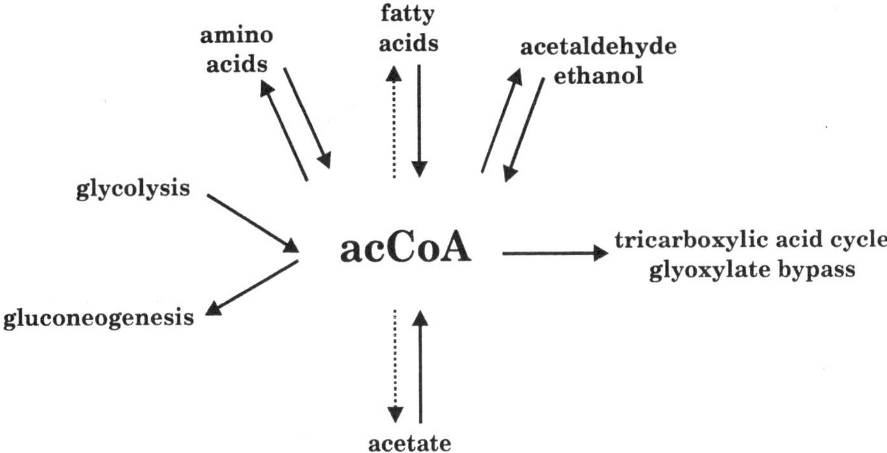

Fig 28. Acetyl-CoA (acCoA) sits at the crossroads of central metabolism

The relative contribution of acetic acid or sodium acetate in sex fixing depends on the concentration of these in plasma or body fluids or by an alternative mechanism to produce certain metabolites in plasma or body fluids that act as the effective blockers of binding sites for the Y-bearing sperm thereby clearing the way for the otherwise only possible combination with the X-bearing sperm resulting in the outcome of a female zygote (Fig 27). The plasma concentrations are reached very rapidly. It may take a quicker period for acetic acid to be absorbed and reach the metabolically effective phases in about an hour and half (Pomare *et al.,* 1985). However the metabolites produced by them in plasma or so may tend to remain in the living system for longer periods. The Aulprofem formula has been shown to remain effective for periods lasting from half an hour to about 12 hours after oral administration. The acetyl group as a sex fixer has found manifestation in many ways in the earlier sex fixing experiments. The fact can not be denied that the earlier success in sex fixing involving the *in vivo* vaginal techniques was primarily due to the use of acetic acid only in these types of experiments but the mistaken illusion with the role of –pH of the vaginal media in fixing the sex of the zygotes led to false interpretations and the use of other –pH manipulators and resulting in failures in effectively evolving a successful technique for this purpose. In therapeutic doses, Aulprofem doesn't have an effect on the heart rate, ventricular function or blood pressure. Some beneficial effects like improvement in appetite, enhancing of mineral absorption and speedy recovery from fatigue etc may take place (Takashi *et al.*, 2001).

Pharmacokinetics

The sex fixing property of the Aulprofem is due to presence of the acetyl group in the formula (Aulkh, 2008 Australian patent no. 2001235973). Aulprofem is present as a combination of acetic acid and sodium acetate both. The preparation being a combination of a weak acid and a salt of this very acid with a strong alkali furnishes a weakly dissociable equilibrium of acetate with the sodium ions. This weakly dissociable equilibrium is a good pharmacological environment for the rapid absorption of acetate in the body systems (Aulakh, 2008. U.S. Patant No. 7351581). Aulprofem is well absorbed orally with respect to both its constituents *i.e.* acetic acid and sodium acetate. Both the components follow a parallel time course in the body. The peak levels in blood are achieved within an hour. Maximum bioavailability of the constituents in blood plasma may be around 90% of the administered dose. A full therapeutic dose produces effective concentrations within half an hour or so. A therapeutic effective concentration in blood may be about one third of the total achievable concentration in blood. Concentration of components in blood start biochemical processes in the body to produce certain metabolites that act as blockers of specific H-Y antigen binding sites on the bodies of Y bearing sperms. The sex fixing activity is due to the binding of specific H-Y antigen binding sites on the bodies of Y bearing sperms; these sites being non-existent or different on the bodies of the X bearing sperms. This difference of H-Y antigen binding sites on the bodies of X and Y beaing sperms are consistent with the presence of different antibody receptor sites on the body of the oocyte. The result is that the chances of Y sperm to conjugate with the ovum are ruled out thereby leading to the binding of X bearing sperm with the ovum leading to the formation of a female zygote (Fig 27). The presence of certain metabolites YSBLM (Y sperm binding ligand mimics) generated by the Aulprofem reaction in the body lingers for comparatively longer periods as shown by hypothetical values **(Fig 29)** in practical studies conducted on animals in live trials (Aulakh, 2008. U.S. Patant No. 7351581). This effect was found to be effective by plus 90% efficacy rate even after 16 hours after oral administration (Fig 29). The effect seems to linger to plus 20 hours after administration to a decreased effectiveness. Aulprofem constituents *i.e.* acetic acid and sodium acetate are metabolized via various metabolic cycles in the body of the organism **(Figs 30-35)** and ultimately converted to carbon dioxide and water molecules and hence are excreted out of the body of the organism.

Absorption

Aulprofem is rapidly and almost completely absorbed in the blood with respect to both its constituents. The both constituents on absorption

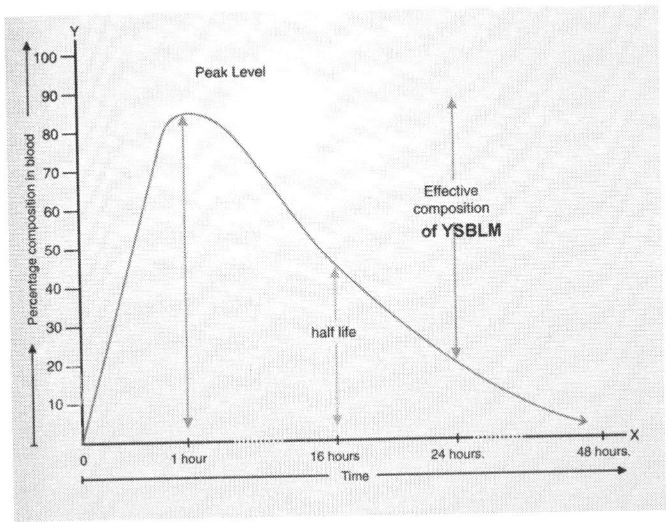

Fig. 29. Showing the hypothetical concentration of YSBLM (Y sperm binding ligand mimic) moeities in body fluids

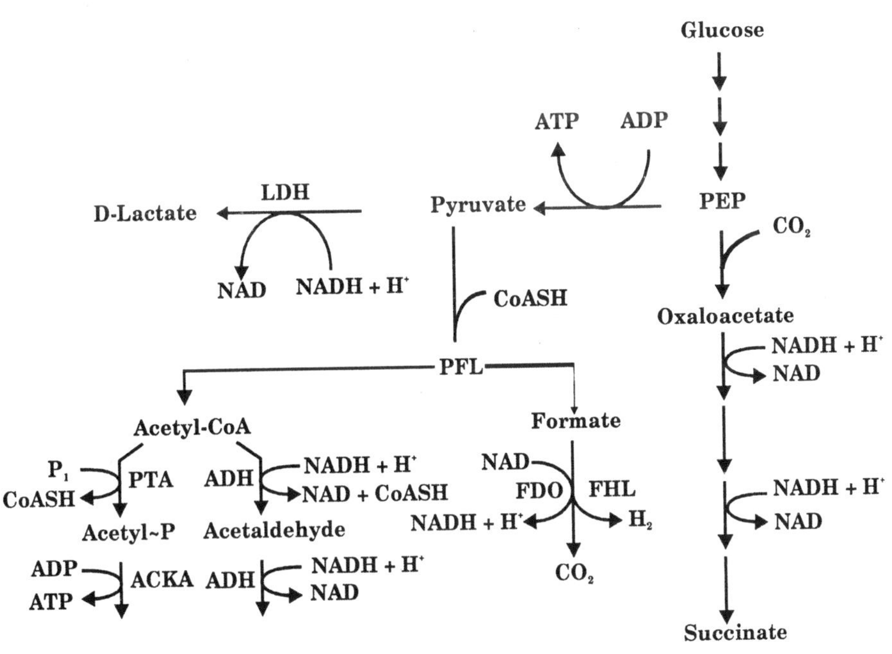

Fig 30. Pathways for the excretion of partially oxidized metabolites. Excreted metabolites are underlined. LDH, lactate dehydrogenase; PFL, pyruvate-formate lyase; PTA, phosphotransacetylase; ACKA, acetase, kinase; ADH, alcoho dehydrogenase; FDO, aerobic formate dehydrogenase; FHL, formate-hydrogen lyase

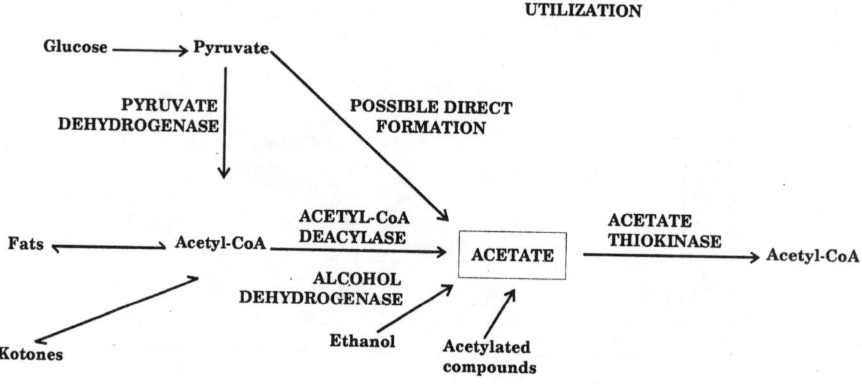

Fig 31. Metabolic pathways for acetate formation and conversion to acetyl-CoA

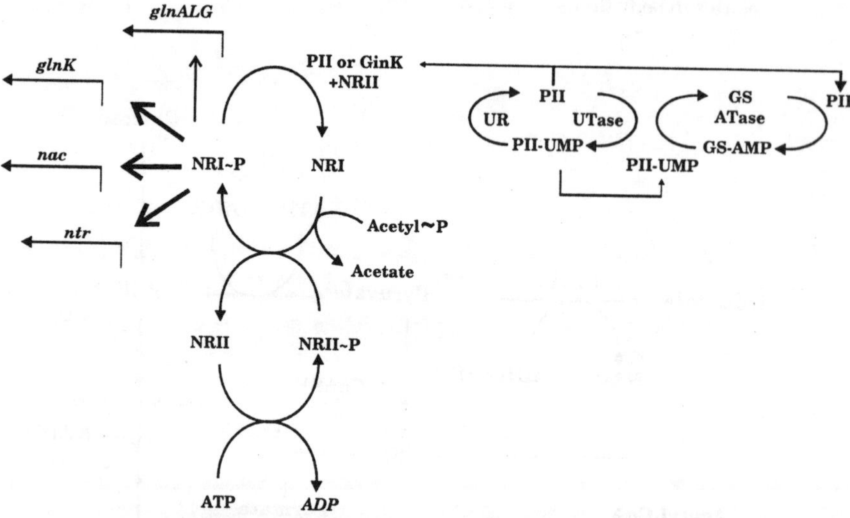

Fig 32. Regulation of nitrogen assimilation. Uridylytransferase (UTase) and uridylyl-removing enzyme (UR) sense the relative amount of nitrogen. Under limiting conditions, UTase uridylylates PII to PII-UMP. In excess nitrogen, the UR deuridylylates PII-UMP to PII. PII favors adenylytransferase (ATase) activity. ATase inactivates GS by adenylylating it to GS-AMP. PII (or its ortholog GinK) also enhances the phosphatase activity of the histidine kinase/phosphatase NRII, which then dephosphorylates NRI~P. This diminishes transcription from the nitrogen-regulated promoters such as *ntr, nac, glnK*, and *glnALG*. In contrast, PII-UMP enhances the deadenylylation of GS-AMP, hence activating the enzyme. PII-UMP exerts no direct effect on NRII; however, the lack of PII favors the kinase activity of NRII, which donates its phosphorylgroup to NRL. Acetyl~P also donates its phosphoryl group to NRL. Transcription from the *glnALG* promoter requires low amounts of NRI~P (thin arrow), whereas transcription from the other promoters requires high amounts of NRI~P (thick arrow)

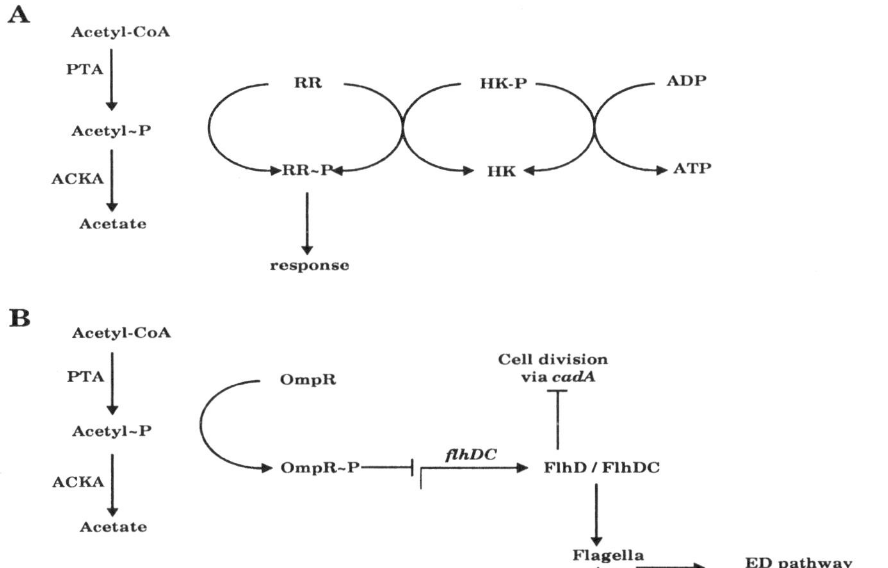

Fig 33. Acetyl~P can donate its phosphoryl group to two component RRs. HK-P, histidine kinase/phosphatase. (B) Model depicting how acetyl~P can influence diverse cellular processes through the, RR OmpR and its ability to repress transcription of *flhDC*. In contrast to acetyl~P-dependent control of flagellar biosynthesis, its control of cell division does not require FlhC. ED, Entner-Doudoroff

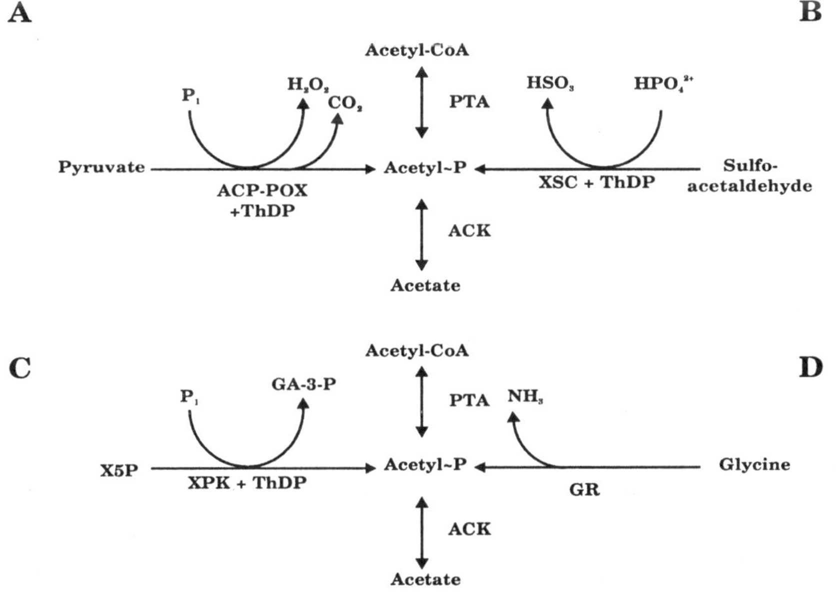

Fig 34. Other acetyl~P-forming enzymes. (A) ACP-POX, acetyl~P-forming pyruvate oxidase; ThDP, thiamine diphosphate. (B) XSC, sulfoacetaldehyde acetyltransferase. (C) XPK, xylulose 5-phosphate phosphoketolase; GA-3-P, glyceraldehyde-3-phosphate. (D) GR, glycine reductase

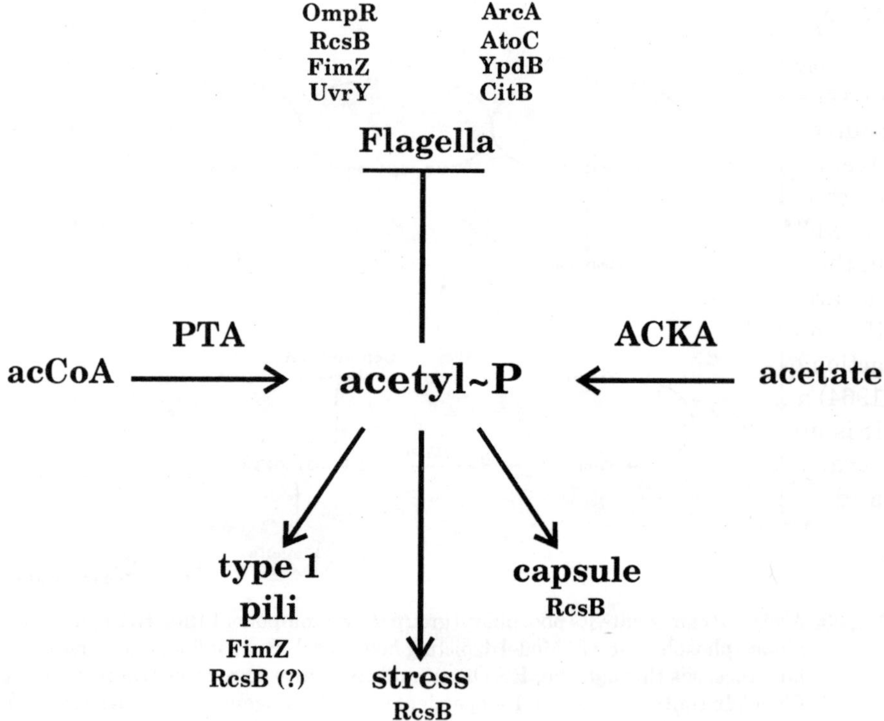

Fig 35. Acetyl~P-responsive processes and the RRs known to regulate them

cause the blood concentrations to rise very rapidly and peak levels are observed within an hour or so (Fig 29). The mean absolute bioavailability of the 225 ml dose is approximately 90% of the active constituents constituting the therapeutic application.

Food Effects

Oral administration of Aulprofem with food does not significantly affect its rate of absorption or extent of absorption; therefore it can be administered without regard to food.

Distribution

The volume of distribution of Aulprofem with regard to both its constituents is following a 225 ml oral dose. The binding of Aulprofem to plasma proteins with respect to both its components is directed to keep the concentration of free acetic acid to low levels to stop disrupting the control of –pH of cell contents and this also appears to be dependent of concentration coefficient up to 1 ml/kg body weight. However, saturation of plasma protein binding occurs only at concentrations outside the clinically relevant range.

Metabolism

Acetate is an important biochemical component naturally found in body in several activated forms. It mainly functions in the body as energy source via TCA cycle or synthesis of lipids, carbohydrates and proteins. Acetate is generated in the body by both exogenous and endogenous mechanisms from fats, carbohydrates, ketone bodies, several amino acids etc which are degraded to yield acetyl-co-enzyme-A but acetyl Co-A formed in this way would normally be used for synthetic reactions or metabolized to carbon dioxide and water via the tricarboxylic acid cycle. It is possible that a portion the acetyl-co-A could be converted to free acetate by the action of acetyl-coA deacetylase (Fig 28). Hochheuser *et al.* (HocimEusEit, 1964) have shown that acetate can be formed from pyruvate in erythrocytes. It is unlikely that acetyl-co-A would be an intermediate in this pathway, because the pyruvate dehydrogenase needed is a mitochondrial enzyme and thus is absent from mature erythrocytes. Hocceuser *et al.*

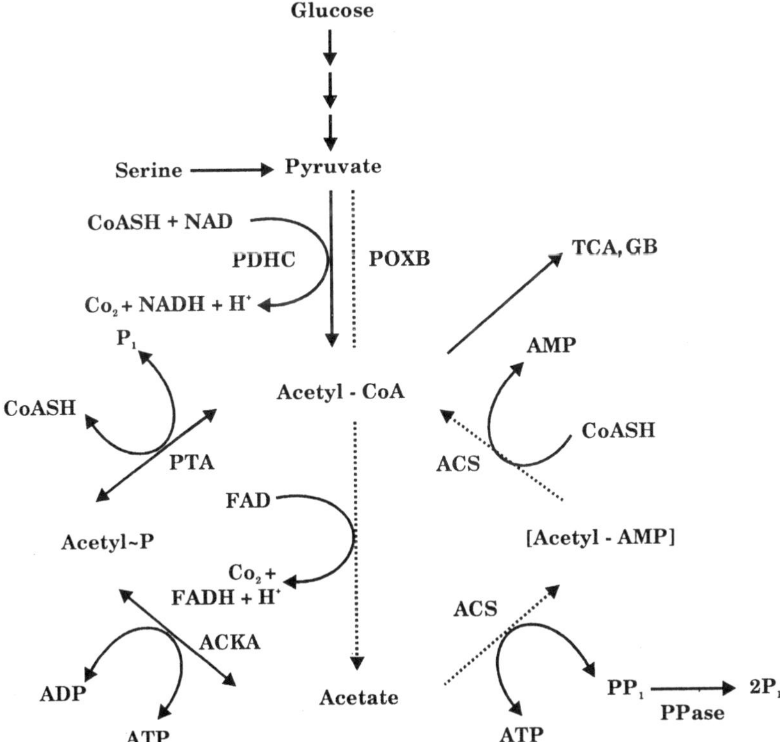

Fig 36. Acetate activation pathways. PDHC, pyruvate dehydrogenase complex; POXB, pyruvate oxidase; PTA, phosphotransacetylase; ACKA, acetate kinase; ACS, AMP-forming acetyl-CoA synthetase; PPase pyrophosphatase; TCA, Tricarboxylic acid cycle; GB, glyoxylate bypass. The dotted arrows denote the proposed PDHC bypass formed by POXB and AMP-ACS

(HocimEusEit, 1964) has suggested that an oxidative dismutation reaction could explain the production of acetate cycle, is suppressed either by malonate or by omitting phosphate or adenosine diphosphate **(Fig 36)** from the medium (VonKroff, 1967; Davis, 1968).

For metabolism the acetate must be converted to acetyl-coA by acetate thiokinase (Beinert, 1953). Acetyl phosphate (acetyl_P), the high energy intermediate of the dissimilation pathway, might function as a global signal (McCleary, 1993; Wanner, 1993). Today mounting evidence suggests that, indeed acetyl_P plays such a role, regulating cellular processes as diverse as nitrogen assimilation (Fig 32), osmo-regulation, flagellar biosynthesis (Fig 35), pilus assembly, capsule biosynthesis, bio-film development, and pathogenicity (Figs 33, 34). Acetyl-coA sits at the crossroads of the central metabolism (Fig 28). For several reasons, there also exists renewed interest in the acetate assimilation enzyme AMP-ACS is a prototype for enzymes involved in the synthesis of fatty acids (Fig 36), some antibiotics and certain anti cancer drugs as well; as well as the degradation of pollutants (Starai, 2004) second the AMP-ACS activity is regulated by an acetylation-deacetylation system homologous to that used by eukaryotes to control chromatin structure, silencing mitochondrial signaling and ageing.

Elimination

Aulprofem is primarily eliminated by the metabolism involving the basic metabolic pathways in the body. The mean terminal half-life of the Aulprofem YSBLM constituents calculated hypothetically **(Fig 37)** is around 12-16 hours with fluctuation of ± 2 hours. The plasma elimination half life of Aulprofem constituents can be increased with the administration of multiple dosing. The total elimination can take place from the body in a span extending from 24 to 48 hours after administration.

References

Aulakh, B.S. 2008. U.S. patent no. 7351581.

Aulakh, B.S. Australian patent no. 2001235973.

Beinert, H.D. *et al.*, 1953. The acetate activating enzyme system of heart muscle. *I. Biol. Chem.* **203**: 35.

Bellard, F.J. 1972. Supply and utilization of acetate in mammals. *Am. J. Clin. Nutr.* **25**: 773-79.

Bryant *et al.*, 1993. U.S. patent no. 4,448,767.

Crabtree, B. *et al.*, 1990. Measurement of the rates of acetyl-CoA hydrolysis in synthesis from acetate in rat hepatocytes and the role of these fluxes in substrate cycling. *Biochem. J.* **270**: 219-25.

Cummings, J.H. *et al.*, 1987. Short chain fatty acids in human large intestine, portal, hepatic and venous blood. *Gut.* **28**: 1221-27.

Davis, E.J. 1968. On the nature of malonate-insensitive oxidation of pyruvate and glutamate by heart sarcosomes. *Biochem. Biophys. Acta* **162**: 1.

HocimEusEit, W.H. *et al.*, 1964. Uber den Stoffwechsel der freien Essigsaure im Tierk {246}rper und seine Beziehung zum Atherogenesproblem. *Z. Kim. Chem.* **6**: 75.

Kishi, M. *et al.*, 1999. Enhancing effect of dietary vinegar on the intestinal absorption of calcium in ovariectomised rats. *Biosci. Biotechnol. Biochem.* **63**: 905-10.

McCleary, W.R. 1993. Is acetyl phosphate a global signal in *Escherichia coli*? *J. Bacteriol.* **175**: 2793-98.

Pomare, E.W. *et al.*, Carbohydrate fermentation in the human colon and its relation to acetate concentrations in venous blood. *J. Clin. Investig.* **75**: 1448-54.

Spydevold, O. *et al.*, 1976. Replenishment and depletion of citric acid cycle intermediates in skeletal muscle. *Eur. J. Biochem.* **71**: 155-65.

Starai, V.J. and Escalante-Semerena, J.C. 2004. Acetyl-coenzymeA synthetase (AMP forming). *Cell. Mol. Life Sci.* **61**: 2020-30.

Takashi, F. *et al.*, 2001. Acetic acid feeding enhances glycogen repletion in liver and skeletal muscle of rats. *J. Nutr.* pp. 1973-1977.

Von Kroff, R.W. 1967. Changes in metabolic control sites of rabbit heart mitochondria. *Nature* **214**: 23.

Wanner, B.L. 1993. Gene regulation by phosphate in enteric bacteria. *J. Cell. Biochem.* **51**: 47-54.

Zavos *et al.*, 1995. U.S. patent no. 4,999,283.

PART - G

Clinical trials and Pharmacology and Toxicological Studies

- Clinical Trials
- Studies for Toxicology and Pharmacology
- Some More Studies

Clinical Trials

A brief account of the clinical trials on aulprofem is as under:

Detailed Summary of the Clinical Trials on Aulprofem

Clinical trials with female sex fixer drug Aulprofem were conducted on dairy animals on the research farm of Gregor Mendel Institute for Research in Genetics located at village Alechak, district Gurdaspur of Punjab province of India and various procedures and results were as given below. The aim of the trials was to ascertain its efficacy to produce female progenies in dairy animals. As per records of this institute, trials and test data are as noted below:

1. **Clinical Trials:** Clinical trials were conducted as a study to evaluate the sex fixing effect of sodium ethanoate solution in ethanoic acid in dairy animals, cows and buffaloes. Apart from this studies were conducted on other animals and even other mammals to know the metabolism and toxicological studies. This information can be divided into two categories as under:

    A. Clinical testing for efficiency of sex fixing activity

    B. Studies for toxicology and pharmacology

    Studies for toxicology and pharmacology are covered in the next chapter. Where as the clinical testing for effeciency trials is as given below:

Clinical testing for efficiency of sex fixing activity

The various details regarding abstract, introduction, materials and methods, results and discussion etc are as given below:

Abstract

Sex fixing is a fascinating new concept in applied clinical pharmacology. Until now it was reported in certain ratites and lower reptiles. Of course, there have been numerous attempts to manipulate sex ratios in mammals but by and large, success has eluded them. Mammals are an important family and any success in producing desired sexed babies in them has considerable economic and social benefits. The present research is of a drug formula called 'aulprofem' which is a combination of sodium ethanoate and ethanoic acid in solution that when given orally to females of dairy animals, cows and buffaloes; yields female off-springs with near hundred percent success rate. The active constituents are therapeutically safe and pharmacologically compatible and do not exhibit any toxic or side effects. The various outcomes of this discovery may be a rapid increase in the numbers of females of high yielding milch varieties of dairy animals and thereby increasing milk production to unprecedented heights across the globe. Similarly, meat and wool productions can also be increased by increasing the overall numbers of meat and wool etc producing animals by producing more female off-springs of these animals with the help of this discovery.

Introduction

Mammals are the most dominant animals on earth today. So the production of sexed offsprings in them has marked economic and social implications. Many scientists have worked on this since the first half of the last century. There were mainly three lines on which such research was conducted. First were the attempts to alter the –pH of the female genital tract, second were the attempts to produce sexed semen (a sexed semen is one containing the accentuated proportions of either X or Y bearing spermatozoa) and the third were the use of certain chemicals, harmones, sera etc to produce a shift in the sex ratio. Recently much emphasis is laid on the attempts to understand the exact mechanism of sperm-egg fusion. Candidate molecules have been identified that act as receptors on the ZP (zona pellucida) or the ovum membrane and bind with possible sperm ligands. Along with this the differential *in vitro* binding of X and Y containing sperms with different antigen-antibody media based on the assumption of differential binding sites on them and their successive elutriation in accentuated proportions has gained attraction. The present research is an *in vivo* attempt to produce female offsprings where sex ratios are manipulated probably by binding of receptor sites for Y sperm on the ovum by the action of certain moieties that are generated in the living system by the action of the oral administration of a solution of sodium ethanoate in ethanoic acid given to the female animals at the time of mating. The important consequences of this may be rapid increase in milk, meat and wool etc production in India and the world over to unfathomable heights by preferential production of females of dairy and related fields and the great social and economic benefits and changes associated with this. In a projection when only 10 percent of the total dairy animal wealth of India is brought under the application of this technique for a smaller time period of only 10 years; considering the normal national levels of conception failure, calf mortality, adult mortality etc, nearly 160 million tonnes of increase in national milk production can be achieved. This is no short of a revolution.

Materials and Methods

The present technology is an *in vivo* technique. It is based on the assumption of the production of certain moieties in the living system by the action of administration of certain chemicals that can block the binding or fusion of a particular X or Y sperm receptor on the ovum. The study was conducted in the form of *in vivo* trials on healthy dairy cows in the age group between 4 to 8 years in age in the dairy farms selected for this purpose. All the animals formed the test group. The other materials included the formulation that comprised a solution of monosodium ethanoate 12.5 mg dissolved in ethanoic acid 1.45 M q.s. to make the total solution to 225 ml. The solution was administered to the animal 30 minutes before artificial insemination. The administration was by a single dose via the oral route to the female animal. The results were recorded and calculated in simple percentage methods. The percentage of male or female births is assumed as the indicator of the extent of binding of Y sperm receptor sites on the ZP or oocyte membrane.

Table 1. The results of cows subjected to manipulation of sex ratios

Sr. No	No. of Inseminations	No. of pregnancies	No. of Miscarriages	Male Offsprings	Female Offsprings
1.	94	48	01	04	43

Results

Study was conducted on 94 female cows. Animals were given the material by the oral route 30 minutes before the artificial insemination. There were 46 repeat inseminations as many animals could not conceive. This is a natural practice in dairy and livestock circles with the national conception rate in India standing around 40 percent. 48 animals conceived and one underwent miscarriage and remaining 47 delivered healthy and viable offsprings; 4 of them males and 43 females. The sex of the aborted calf was not noted. The percentage of the male calves was nearly 8.5 percent and the females 91.5 percent in the present study. So, it seems to turn out that plus ninety percent receptors on the ZP or the oocyte membrane meant for the binding and subsequent fusion of Y bearing sperms are blocked by the moiety 'YSBLM' generated by the reaction of the material of the present research in the living system of the animal under treatment of this chemical.

Discussion

Most of the animals that received the treatment delivered female offsprings. The failures can be attributed to manual error, spillage and faulty dose adjustment in case of heavier animals. Sex fixing in mammals has a long history. Many scientists have worked on this including Schroeder (1941), Bhattacharya *et al.* (1966), Ericsson *et al.* (1973), Gordon (1958), Lindahl (1956), Gledhill (1983), Corson *et al.*(1984), Beernik and Ericsson (1982), Barrat and Leger (1979), Sampson *et al.* (1983) etc. The major drawbacks of these techniques remained the lack of laboratory tests to evaluate the degree of sperm separation (Hafez, 1982) and the inability to know the precise mechanism of binding and fusion of mammalian sperm with ova. Overall, as opined by Prof. R.V. Short (1982), the manipulation of mammalian sex ratio has still remained a mirage on the horizon (Hunter, 1982). Recently there have been attempts to study the effects of certain compounds like tetraspanins like CD_9 (Hemler, 2003), CD_{81} (Cormier, 2004), glycosylphosphatidylinositol anchored proteins like CD_{55} (Coonrad, 1999), integrins (He *et al.*, 2003), disintegrins (Primakoff *et al.*, 2000), fertilins, cyritestin (Cho *et al.*, 1998) etc on the process of gamete fusion. These researches have clearly established the receptor sites on oocyte and their respective ligands on the surface of sperms. But the bigger questions are far from being answered. The 'tetraspanin-web' is still a big mystery (Boucheix *et al.*, 2001). The role of other candidate molecules the likes of which are mentioned above, needs a lot more investigating. Recently there have been attempts to differentially bind X and Y bearing spermatozoa using H-Y antigen antibody interactions using non protein substrata such as agarose beads and U.S. patents granted for them (Zavos *et al.*, 1991; Bryant *et al.*, 1984). These developments clearly demonstrate the existence of differential binding sites for X and Y sperms and this difference can be exploited in developing a viable technology for sex fixing in mammals, either *in vivo* or *in vitro*. The present technology is an *in vivo* technique and it can be hypothesized that there are differential receptors sites on the body of the ovum and their corresponding ligands on the bodies of the X and Y bearing sperms. A receptor site for X sperm does not receive the binding ligand of Y sperm as it does not correspond to its binding morphology and physiology and *vice versa*. The material of present research produces certain Y sperm binding ligand mimics (YSBLM) moieties in the living system and also in the genital system of the female animals undergoing this treatment in such a way that these YSBLM moieties differentially bind with receptor sites on the ovum thus occupying these sites and inhibit their subsequent binding and/or fusion with true binding ligands on the bodies of the Y sperms and this way over rule the chances of an XY zygote being formed. This leaves all the chances of an X sperm to come and combine freely with the X sperm specific receptor sites on the ovum and form an XX zygote. So, it can be inferred that the most appropriate candidate molecules that act as receptors on the oocyte like CD_9 and binding ligands on the sperm like fertilin beta are also X and Y sperm specific. So we can have tetraspanins like CD_9x and CD_9y and fertilins like fertilin beta-x and fertilin beta-y. The importance of this technology becomes even more marked with the view that it will not only provide an insight into the differential behaving and working of X and Y bearing sperms in the process of fertilization but will also definitely provide more valuable data and knowledge on the role played by the candidate molecules in understanding fully well the exact mechanism of gamete fusion in further research on this subject.

Summary

The drug formula or material involved in above study is effective in altering the sex ratios in dairy animals as is clear from the results obtained in this experiment. The rate of success in modification of sex ratios is tremendous. This only leads to the derivation of one and only one conclusion that the first successful mammalian sex fixer of the world has arrived on the scene and is very much here to stay. The manipulation of mammalian sex ratio is not a dream or a mirage on the horizon (Short, 1982) now but a living reality and it cannot be ruled out that more and more of sex fixers meant for mammals will be discovered in the future to come.

References

Barra , J. and Leger, D. 1979. Aveir des grossesses obstenues après stimulation de l' ovulation. A propos de 519 grossesses. *J. Gynecol. Obstet. Biol. Reprod. (Paris)* **8**: 332.

Beernik, F.J. and Ericsson, R.J. 1982. Male sex preselection through sperm isolation. *Fertil. Steril.* **38**: 493.

Bhattacharya, B.C. *et al.*, 1966. An attempt to determine the sex ratio of calves by artificial insemination with spermatozoa separated by sedimentation. *Nature* **211**: 863.

Boucheix, C. *et al.*, 2001. Tetraspanins. *Cell Mol. Life Sci.* **58**: 1189-1202.

Bryant, F. 1984. U.S. patent no.4, 448, 767.

Cho, C. *et al.*, 1998. Fertilization defects in sperm from mice lacking fertilins beta. *Science* **281**: 1857-59.

Coonrad, S.A. *et al.*, 1999. Treatment of mouse oocytes with PI-PPLC releases 70kDa (pl5) and 35 to 45 kDa (pl5.5) protein clusters from the egg surface and inhibits sperm oolema binding and fusion. *Dev. Biol.* **207**: 334-49.

Cormier, E.G. *et al.*, 2004. CD81is an entry coreceptor for hepatitis C virus. *Proc. Natl. Acad. USA.* **101**: 7270-74.

Corson, S.L. *et al.*, 1984. Sex selection by sperm separation and insemination. *Fertil. Steril.* **42**: 756.

Ericsson, R.J. *et al.*, 1973. Isolation of fractions rich in human Y sperm. *Nature* **246**: 421.

Gledhil, B.L. 1983. Control of mammalian sex ratio by sexing sperm. *Fertil. Steril.* **40(5)**: 572.

Gordon, M.J. 1958. The control of sex. *Scientific Am.* **199**: 87.

Hafez, E.S.E. 1982. Reproduction in farm animals. Reprint 5th edn. Lea and Fabiger, p. 499.

He, Z.Y. 2003. None of the integrins known to be present on the mouse egg orto be ADAM rreceptors are essential for sperm egg binding and fusion. *Dev. Biol.* **254**: 226-37.

Hemler, M.E. 2003. Tetraspanin proteins mediate cellular penetration, invasion and fusion events and define a novel type of membrane micro domain. Annu. Rev. Cell Dev. Biol. 19: 397-422.

Hunter, R.H.F. 1982. Reproduction of farm animals. Longman. London, pp. 138-39.

Lindahl, P.E. 1956. Counterstreaming centrifufation of bull spermatozoa. *Nature* **178**: 491.

Primakoff, P. *et al.*, 2000. The ADAM gene family: surface proteins with adhesion and protease activity. *Trends. Genet.* **16**: 83-7.

Sampson, J.H. *et al.*, 1983. Gender after induction of ovulation and artificial insemination. *Fertil. Steril.* **40**: 481.

Schroeder, V. 1941. Physica-chemical methods of sex regulation of progeny of mammals. Russian contributions. 1939 Genetics Congr. Am. Documentation Inst. Doc. 1565. *Abstr. In J. Hered.* **32**: 248.

Short, R.V. 1982. As quoted in reproduction of form animals by RHF Huntes. London. pp. 138-39.

Zavos, M and Dawson, K.A. 1991. Method for X and Y spermetozoa separation. U.S. Patent no. 4,999,283.

Studies for Toxicology and Pharmacology

The various studies to evaluate the pharmacology and toxicology criteria of the aulprofem formula are given as follows:

Study no.1 : Studies to evaluate the effect of Aulprofem administration on liver toxicology

Title : **A study to know the effects of Aulprofem Administration on the liver function and toxicology**

Introduction

The test to estimate the alkaline phosphatase content is one of the most important and common tests to evaluate the normal liver function. So a study was undertaken to measure this for animals under the treatment of Aulprofem. This included the study into effect of Aulprofem treatment on the secretion of alkaline phosphatase and the effect of Aulprofem administration on the secretion of this. This is discussed as below:

Materials and Methods

10 healthy female animals without any history of disease and ailment in the age group 4 to 8 years in age were selected for this study. They were given the Aulprofem solution 225 ml which is a standard dose for the sex fixing act. The other materials and methods in this study included the standard Autozyme alkaline phosphatase apparatus, spectrophotometer set at 405 nm and procedure for the estimation of alkaline phosphatase activity. The various reagents used for working solution included diethanolamine buffer, 1 Mol/L, at –pH 9.8, p-nitro phenyl phosphate, 10 m Mol/L, magnesium chloride 0.5 m Mol/L.

Observations

Blood of the animals after 5 hours from the time of administration of Aulprofem was taken in a clean dry container. For plasma separation heparin 200 IU/ml blood was used. Sodium fluoride 10 mg/ml blood could also be used. Blood plasma was immediately separated by centrifugation at 2000 rpm for 5-10 minutes. Alkaline phosphatase is stable at 2-8 degree Celsius for 4 days. Took 0.02 ml plasma and transferred to 1.0 ml working solution at 37 degree Celsius. Mix thoroughly and transfer the assay mixture immediately to the thermostatic cuvette and start the stop watch simultaneously. Record the first reading at 60 seconds and subsequently three more readings at 30 seconds interval at 405nm.

Table S1. Table showing the alkaline phosphates activity of the adult female animals which were given a 225 ml dose of Aulprofem

S. no.	The age of animals in months	Alkaline phosphatase activity
1.	74	111
2.	82	115
3.	65	108
4.	68	131
5.	78	126
6.	94	109
7.	77	137
8.	69	109
9.	78	128
10.	64	114

Results

The average change in absorbance per minute was calculated. The alkaline phosphatase (IU/L) = ΔAbs./min × 2720. The various results of the alkaline phosphatase activity were calculated as given below in table.

Discussion

The various animals which were given the standard dose of Aulprofem and later subjected to observation for 5 long hours which are sufficient for Aulprofem to give any unwanted effects on liver and later testing for the alkaline phosphatase activity testing showed normal values of the alkaline phosphatase activity which is an indicative of the normal liver function. None of the animals showed any abnormal or dangerous values. Hence it can be concluded that the administration of Aulprofem is not toxic to liver at all.

Study no. 2: The study into the effects of Aulprofem on renal (kidney) function

Title : **A study to know the effects of Aulprofem on renal function**

Introduction

The most important and common test to evaluate the normal renal function for animals under the treatment of Aulprofem was under taken. These included the effect of Aulprofem treatment on the secretion of creatinine activity. Creatinine plasma level is a better indicator of the health of kidneys and renal function than the measurement of urea content. So this test was preferred. This is discussed as below:

Materials and Methods

10 healthy female animals without any history of disease and ailment in the age group 4 to 8 years in age were selected for this study. They were given the Aulprofem solution 225 ml which is a standard dose for the sex fixing act. The other materials and methods in this study included the standard apparatus, spectrophotometer set at 505nm and procedure for the estimation of plasma creatinine level. The various reagents used for working solution included unhaemolyzed serum, picric acid, sodium hydroxide, creatinine standard, clean and dry glass ware, laboratory glass pipettes or micropipettes, bio-chemistry analyzer etc.

Observations

Blood of the animals after 5 hours from the time of administration of Aulprofem was taken in a clean dry container. For plasma separation heparin 200 IU/ml blood was used. Sodium fluoride 10mg/ml blood could also be used. Blood plasma was immediately separated by centrifugation at 2000 rpm for 5-10 minutes. Working reagent was prepared by combining equal volumes of picric acid and sodium hydroxide and mixed by gentle swirling and allowed this reagent mixture to stand for 5 minutes. Took 0.05 ml plasma and transferred to 1.0 ml working solution at 37 degree Celsius. Mix thoroughly and transfer the assay mixture immediately to the thermostatic cuvette and start the stop watch simultaneously. Record the first reading (A_1) at 30 seconds and second reading (A_2) after 120 seconds interval at 505 nm. Calculated $\Delta A = A_2 - A_1$. Repeat this with the standard creatinine solution containing 2.0 mg/dl as well.

Results

The creatinine concentration in mg/dl is calculated as the following equation:

$$\text{Creatinineconc.mg/dl} = \frac{\Delta A \text{ Sample}}{\Delta A \text{ Standard}} \times 2$$

Table S2. Table showing the plasma creatinine concentration of the adult female animals which were given a 225 ml dose of Aulprofem.

S. no.	The age of animals in months	Creatinine concentration mg/dl
1.	67	0.7
2.	85	0.8
3.	62	1.1
4.	68	1.1
5.	75	1.2
6.	88	1.0
7.	77	1.2
8.	92	0.7
9.	78	0.8
10.	66	0.9

The various results of the creatinine concentration in plasma are pasted in the following table.

Discussion

The various animals which were given the standard dose of Aulprofem and later subjected to observation for 5 long hours which are sufficient for Aulprofem to give any unwanted effects on renal function and later testing for the creatinine concentration testing showed normal values which is an indicative of the normal renal function. None of the animals showed any abnormal or dangerous values. Hence it can be concluded that the administration of Aulprofem is not toxic to Kidneys at all.

Study no. 3: The study into the effects of Aulprofem on cardiovascular health

Title : **A study to evaluate the effect of Aulprofem on cholesterol levels in blood**

Introduction

The most important and common test to evaluate the normal cardiovascular function is to measure the cholesterol levels in blood. Therefore this test for animals under the treatment of Aulprofem was under taken. These included the effect of Aulprofem treatment on the cholesterol levels in blood of the animals. This is discussed as below:

Materials and Methods

10 healthy female animals without any history of disease and ailment in the age group 4 to 8 years in age were selected for this study. They all had cholesterol levels of less than 200 mg/dl. They were given the Aulprofem solution 225 ml which is a standard dose for the sex fixing act. The other materials and methods in this study included the standard apparatus, spectrophotometer set at 505nm and procedure for the estimation of cholesterol levels in plasma. The various reagents used for working solution included unhaemolyzed serum, enzyme reagent containing standard cholesterol sterase and cholesterol oxidase enzymes, cholesterol standard reagent 200 mg/dl, clean and dry glass ware, laboratory glass pipettes or micropipettes, bio chemistry analyzer etc.

Observations

Blood of the animals after 5 hours from the time of administration of Aulprofem was taken in a clean dry container. For plasma separation heparin 200 IU/ml blood was used. Sodium fluoride 10mg/ml blood could also be used. Blood plasma was immediately separated by centrifugation at 2000 rpm for 5-10 minutes. Working reagent was prepared by taking enzyme reagent ready to use and by bringing it to room temperature and gently swirling and allowed this reagent mixture to stand for 5 minutes. Took 0.01 ml plasma and transferred to 1.0 ml working solution at 37 degree Celsius. Mix thoroughly and transfer the assay mixture immediately to the thermostatic cuvette and incubated for 10 minutes.

Table S3. Table showing the plasma creatinine concentration of the adult female animals which were given a 225 ml dose of Aulprofem.

S. no.	The age of animals in months	Cholesterol concentration mg/dl before aulprofem administration	Cholesterol concentration mg/dl after aulprofem administration
1.	74	156	156
2.	83	143	143
3.	76	178	175
4.	68	168	168
5.	78	156	155
6.	84	147	147
7.	77	185	185
8.	76	165	166
9.	68	156	154
10.	64	144	144

Record the first reading after 10 minutes for blank containing only reagent (D_1) at 505 nm, the other for reagent with standard (D_2), and the third with reagent and sample (D_3). Caculate Δ Dst = D_2 - D_1 and Δ Dsm = D_3 - D_1. Repeat this procedure with 3 ml of enzyme reagent as well for more consistancy and accuracy. This set of experiments was reapeated with the animals 5 hours before and 5 hours after the administration of Aulprofem standard dose of 225 ml.

Results

The cholesterol concentration in mg/dl is calculated as the following equation:

$$\text{Creatinineconc.mg/dl} = \frac{\Delta \text{Dsm}}{\Delta \text{Dst}} \times 200$$

The various results of the creatinine concentration in plasma are pasted in the following table.

Discussion

The various animals which were tested for cholesterol levels and then given the standard dose of Aulprofem and later subjected to observation for 5 long hours which are sufficient for Aulprofem to give any unwanted effects on cardiovascular function and later testing for the cholesterol concentration testing which showed normal values again which is an indicative of the normal cardiovascular function. Very slight fluctuations in some cases from the previous values may be due to experimental error factor or slight metabolic changes in the bodies of the animals under test in 10 long hours and these fluctuations if any have negligible or no clinical value. None of the animals showed any abnormal or dangerously increased values after the Aulprofem administration. Hence it can be concluded that the administration of Aulprofem is not toxic to cardiovascular system at all.

Study no. 4: The study into the effects of Aulprofem on CNS

Title : **A study to evaluate the effects of Aulprofem administration on CNS**

Introduction

CRP (C. reactive protein) latex slide is an important test to evaluate the normal CNS function. C. reactive proteins are secreted in the body when there is acute injury, or onset of inflammation and their concentration may reach as high as 20 times the normal value. This test is used for detection of meningitis (inflammation in central nervous system) and neonatal septicemia. So this test was done. This is discussed as below:

Materials and Methods

5 healthy female animals without any history of disease and ailment in the age group 4 to 8 years in age were selected for this study. They were given the Aulprofem solution 225 ml which is a standard dose for the sex fixing act. The other materials and methods in this study included the standard apparatus, CRP latex antigen, positive control, negative control, disposable plastic slides, rubber teats, disposable applicator sticks, disposable plastic droppers.

Observations

Blood of the animals after 5 hours from the time of administration of Aulprofem was taken in a clean dry container. For plasma separation heparin 200 IU/ml blood was used. Sodium fluoride 10 mg/ml blood could also be used. Blood plasma was immediately separated by centrifugation at 2000 rpm for 5-10 minutes. Placed one drop of the plasma on the slide using disposable plastic dropper in circled area of the slide, provided for this test. Added one drop of the CRP latex reagent to the above drop and mix with the disposable applicator stick. Then rocked the slide to and fro, observed for agglutination macroscopically after two minutes. Then diluted the plasma to ratios 1:2, 1:4, 1:8, 1:16, 1:32 and 1:64 using normal saline. Then again, performed the above test as illustrated before. Then observed for agglutination.

Table S4. Table showing the CRP agglutination in the adult female animals which were given a 225 ml dose of Aulprofem

S. no.	The age of animals in months	CRP agglutination
1.	69	Nil
2.	85	Nil
3.	68	Nil
4.	62	Nil
5.	77	Nil

Results and Discussion

The various animals which were given the standard dose of Aulprofem and later subjected to observation for 5 long hours which are sufficient for Aulrofem to give any unwanted effects on CNS. The CRP test was done. No agglutination appeared. This showed that no antibodies that are a potential threat to the central nervous system and can cause damage to the central nervous system were produced with the administration of Aulprofem in the animals under the administration of Aulprofem. So, it can be inferred that the administration of Aulprofem is not toxic to CNS in any way. Also it shows that the administration of Aulprofem is not associated with the development of neonatal septicemia (the growth of cancerous tissues) of any type in the bodies of the animals under its treatment.

Study no 5: The effect of Aulprofem on Genital system functioning in the female animals

Title : **A study to evaluate the small and long term effects of Aulprofem administration on Genital system and production efficiency of the cows**

Table S5A. Table showing the number of deliveries in six years by the animals in test group of animals under Aulprofem administration

S. no.	The age of animals in months at the start of trials	The age of animals at the end of trials	No. of deliveries
1.	42	102	5
2.	38	98	5
3.	39	99	4
4.	43	103	5
5.	41	101	5

Table S5A. Table showing the number of deliveries in six years by the animals in standard group of animals not under Aulprofem administration

S. no.	The age of animals in months at the start of trials	The age of animals at the end of trials	No. of deliveries
1.	44	104	5
2.	39	99	5
3.	41	101	5
4.	43	103	4
5.	42	102	

Introduction

Genital system is one of the main organ systems of the living body. This is designed in female animals basically by nature to produce the progenies. So, production efficiency is the most important criteria to judge the effect of Aulprofem on the genital system. Since Aulprofem is intended for administration to the female animals. So, a study was designed to evaluate the short and long term effects of the administration of Aulprofem on the bodies of the female animals on which it is administered. So this test was done. This is discussed as below:

Materials and Methods

5 healthy female animals without any history of disease and ailment in the age group 4 to 8 years in age were selected for this study. They were given the Aulprofem solution 225 ml which is a standard dose for the sex fixing act at the time of mating. They were subjected to a long term observation for a period of 6 years and every time they mated, they were given Aulprofem 225 ml for sex fixing. The various observations are as given below.

Observations

3 of the 5 cows that formed the group of the animals under test, conceived in the first attempt. The remaining two conceived in the next attempt after about 3 weeks later. All animals delivered female calves on completion of their pregnancies. All these five animals conceived in the next 3 to 5 months and were administered Aulprofem again to deliver female calves again on the completion of their terms of pregnancy. This exercise was kept on repeating up to six long years. Same exercise was done with 5 other animals in the control group which were not given Aulprofem. They delivered mixed offsprings; males and females. The various results are as under.

Results and Discussion

The various animals which were given the standard dose of Aulprofem and later subjected to observation for sex fixing produced female offsprings. These animals in six years produced total female offsprings; 24 in number. The animals which were in the standard group without any administration of Aulprofem produced mixed offsprings but the total number of progenies in their case turned out also to be 24. This shows that by and large the reproductive efficiency and the health and state of the genital systems in all these animals was the same or identical. This infers that whether the use of Aulprofem or not; the genital system functions in the normal way with and also without it and there is no

Table S6: Table showing the effect of Aulprofem administration on eyesight and Optic system in human volunteers. (Legend: M = Male, F = Female, N = Normal)

S. no.	Age of the subjects	Sex of Subjects	Eyesight before trials		Eyesight 6 hours after trials		Eyesight 5 days after trals		Eysight 20 days after trials		Eyesight 100 days after trials	
			Left eye	Right eye	Left eye	Right eye	Left eye	Right eye	Left eye	Right eye	Left eye	Right eye
1.	24	M	N	N	N	N	N	N	N	N	N	N
2.	27	M	N	N	N	N	N	N	N	N	N	N
3.	26	F	+0.5	N	+0.5	N	+0.5	N	+0.5	N	+0.5	N
4.	29	M	N	N	N	N	N	N	N	N	N	N
5.	22	F	N	N	N	N	N	N	N	N	N	N
6.	30	F	N	N	N	N	N	N	N	N	N	N
7.	26	F	-0.5	-0.5	-0.5	-0.5	-0.5	-0.5	-0.5	-0.5	-0.5	-0.5
8.	28	M	N	N	N	N	N	N	N	N	N	N
9.	25	F	N	N	N	N	N	N	N	N	N	N
10.	25	F	N	N	N	N	N	N	N	N	N	N

difference between the two. So, Aulprofem administration has no side or bad effects on the genital system of the female animals for which it is intended.

Study no 6: The effect of Aulprofem on Optical system functioning in the human volunteers

Title : **A study to evaluate the effects of Aulprofem administration on Optical system functioning in human volunteers**

Introduction

Optical system is one of the main organ systems of the living body and eyesight is a test for the state or health of the optical system. This was designed in human volunteers who willingly opted for the testing and the various results and observations were noted and recorded as given below.

Materials and Methods

10 healthy human volunteers; both male and female, in the age group 22 to 30 years in age were selected from the many people who could come forward for the testing of Aulprofem effects on eye sight or otherwise, were selected. They were given Aulprofem 30 ml each (depending upon the average body weight of 60 to 75 kgs of the human volunteers as compared to animals weighing 400 to 500 kgs in body weight). These were the people without any recent history of disease and ailment all had normal eyesight except two people who were little near the normal eyesight. The various observations are as given below.

Observations

The volunteers had a normal diet of lunch one hour before the administration of Aulprofem. They were given Aulprofem 30mls each. Then they were made to retire in a common rest hall with a 23 inch color television, two table top and two laptop computers and lot of cine, news and gossip magazines and newspapers. They rested, watched television, ate snacks, tea, cold drinks and milk and behaved in the normal way as they felt good but remained in the designated place for six hours. After six hours they were tested for the eyesight and asked to leave. They reported for eyesight testing on day 5, day 20 and day 100. The various data were recorded and kept. The various results are as under.

Results and Discussion

The different subject volunteers who came for the testing of Aulprofem had tested for their eyesight. They were given the treatment and took normal living schedules like watching of television, reading of newspapers, magazines, operating of laptops, table tops etc. they tested for the eyesight after the trial of Aulprofem. The various eyesight figures that they had before the trials were reported on their testing after 6hours of the Aulprofem administration. The same results were reported on day 5, day 20 and day 100. The results were unaltered. This clearly shows that the Aulprofem does not produce any short term and long term adverse effects on eyesight. Hence, it can be concluded that the administration of Aulprofem does not produce any side effects or harmful effects on the Optic System.

Study no. 7: The study into the effects of Aulprofem on Immune system

Title : **A study to evaluate the effects of Aulprofem administration on Immune System**

Introduction

The human body produces auto antibodies against the host antigen. The role this plays is well known. The immunoglobulins of the class IgG, IgM, IgA and IgE auto antibodies are also produced as a result of the immune response generated by the living system along with other antibodies. This sometimes may lead to diseases like rheumatoid arthritis. So, rheumatoid factor test is conducted to evaluate ay incidence of induction of such immunity on the administration of Aulprofem to animals. to know the long term effects of decreased immunity a very large data has been generated to show that the

Table S7: Table showing the RF agglutination in the adult female animals which were given a 225 ml dose of Aulprofem

S. no.	The age of animals in months	RF agglutination
1.	62	Nil
2.	69	Nil
3.	92	Nil
4.	62	Nil
5.	74	Nil

Aulprofem administration does not be a cause to the occurrence of any type of occasional infections like common cold, GIT disturbances, and common fevers etc. this has been proved in countless studies involving the administration of Aulprofem to Animals and human subjects. In none of the cases there have been reported the instances of reduced immunity in animals and human subjects undergoing Aulprofem Administration *e.g.* Study no 5 and 6 in the present case. This is discussed as below:

Materials and Methods

5 healthy female animals without any history of disease and ailment in the age group 4 to 8 years in age were selected for this study. They were given the Aulprofem solution 225 ml which is a standard dose for the sex fixing act. The other materials and methods in this study included the standard apparatus, RF latex antigen (Gamma globulin), positive control, negative control, disposable plastic slides, rubber teats, disposable applicator sticks and disposable plastic droppers.

Observations

Blood of the animals after 5 hours from the time of administration of Aulprofem was taken in a clean dry container. For plasma separation heparin 200 IU/ml blood was used. Sodium fluoride 10mg/ml blood could also be used. Blood plasma was immediately separated by centrifugation at 2000 rpm for 5-10 minutes. Placed one drop of the plasma on the slide using disposable plastic dropper in circled area of the slide, provided for this test. Added one drop of the RF latex reagent to the above drop and mix with the disposable applicator stick. Then rocked the slide to and fro, observed for agglutination macroscopically after two minutes. Then diluted the plasma to ratios 1:2, 1:4, 1:8, 1:16, 1:32 and 1:64 using normal saline. Then again, performed the above test as illustrated before. Then observed for agglutination.

Results and Discussion

The various animals which were given the standard dose of Aulprofem and later subjected to observation for 5 long hours which are sufficient for Aulprofem to give any unwanted effects on immune system. The RF test was done. No agglutination appeared. This showed that no antibodies that are a undue activation of immune system and can cause damage to the central nervous system and other body systems were produced with the administration of Aulprofem in the animals under the administration of Aulprofem. So, it can be inferred that the

Table S8. Table showing the effect of Aulprofem administration on Pulmonary system in human volunteers (Legend: M = Male, F = Female, N = Normal lungs).

S. no.	Age of the subjects	Sex of Subjects	Lungs x-R-ray status report before trials	Lungs x-R-ray status report 6 hours after trials	Lungs x-R-status report 5 days after trials	Lungs x-R-ray starus report 20 days after trials	Lungs x-R-ray status report 100 days after trials
1.	22	M	N	N	N	N	N
2.	25	M	N	N	N	N	N
3.	29	F	N	N	N	N	N
4.	30	M	N	N	N	N	N
5.	27	F	N	N	N	N	N
6.	30	M	N	N	N	N	N
7.	26	F	N	N	N	N	N
8.	28	M	N	N	N	N	N
9.	26	F	N	N	N	N	N
10.	24	F	N	N	N	N	N

administration of Aulprofem is not toxic to induce unwanted immune response in any way. The chances of reduced immunities if any could have been easily demonstrated in many of the other studies conducted so far with the Aulprofem on animals and human subjects like study nos. 5 and 6 in the present case and lot more studies like U.S. patent no 7351581 and many more etc.

Study no 8: The effect of Aulprofem on Pulmonary system (lungs) functioning in the human volunteers

Title : **A study to evaluate the effects of Aulprofem administration on pulmonary system (lungs) functioning in human volunteers**

Introduction

Pulmonary system is one of the main organ systems of the living body and the occurrence of common pulmonary diseases like cough, cold, pneumonia etc are a usual instances of lungs not working properly. Also the occurrences of lesions or scars in the lungs that are caused by diseases like tuberculosis etc are a test to show that lungs function properly or not. This can be tested by the radiological examination of the human subjects at regular intervals after administration of Aulprofem say after 6 hours, after day1, day 5, day 20 and day 100. The subjects can be observed for such periods and the findings recorded for such purposes as given below.

Materials and Methods

10 healthy human volunteers; both male and female, in the age group 22 to 30 years in age were selected from the many people who could come forward for the testing of Aulprofem effects on eye sight or otherwise, were selected. They were given Aulprofem 30 ml each (depending upon the average body weight of 60 to 75 kgs of the human volunteers as compared to animals weighing 400 to 500 kgs in body weight). These were the people without any recent history of disease and ailment all had normal lung functioning which wasdetermined with their OPD testing and radiological examination of the chest X-ray photographs taken 1 day before the start of such an exercise. The various observations are as given below.

Observations

The volunteers had a normal diet of lunch one hour before the administration of Aulprofem. They were given Aulprofem 30 mls each.

Then they were made to retire in a common rest hall with a 23 inch color television, two table top and two laptop computers and lot of cine, news and gossip magazines and newspapers. They rested, watched television, ate snacks, tea, cold drinks, milk and behaved in the normal way as they felt good but remained in the designated place for six hours. After six hours they were tested for the occurrence of common pulmonary infections like cold, cough etc and subjected to X-ray radiographic study and asked to leave. They reported for testing on day 5, day 20 and day 100. The various data was recorded and kept. The various results are as under.

Results and Discussion

The different subject volunteers who came for the testing of Aulprofem had tested for the occurrence of common pulmonary infections like cough, cold etc and screened with radiological X-ray filming to know of the normal lung functioning. They were given the administration of Aulprofem 30mls and took normal living schedules like watching of television, reading of newspapers, magazines, operating of laptops, table tops etc. they tested for the occurrence of pulmonary infections like cough, cold etc after the trial of Aulprofem on them. The various radiological findings that they had before the trials, similar findings were reported on their testing after 6hours of the Aulprofem administration. The same results were reported on day 5, day 20 and day 100. The results were unaltered. This clearly shows that the Aulprofem does not produce any short term and long term adverse effects on the pulmonary system. Hence, it can be concluded that the administration of Aulprofem does not produce any side effects or harmful effects on the Pulmonary System.

Study no. 9: The study the absorption of Aulprofem in the living system

Title : **A study to know the absorption of Aulprofem in the living system.**

Introduction

The sodium acetate and acetic acid is converted in the body of the living organisms into Acetyl-coA which in turn acts to convert oxaloacetate to citrate. The oxaloacetate required in this reaction is formed from L-malate and nicotinamide-adenine-dinucleotide (NAD+). The NAD+ is reduced to NADH. The amount of NADH is measured by the increase in absorbance at 340nm.

Materials and Methods

10 healthy female animals without any history of disease and ailment in the age group 4 to 8 years in age were selected for this study. They were given the Aulprofem solution 225 ml which is a standard dose for the sex fixing act. The other materials and methods in this study included the standard apparatus, spectrophotometer set at 340nm and procedure for the estimation of acetate level in serum. The various reagents used for working solution included unhaemolyzed serum, triethanolamine buffer at –Ph, 8.4, 0.4M, NAD+, ATP freeze dried powder, L malate dehydrogenase/citrate suspension, Acetyl-coA synthetase suspension, acetic acid standard solution (5 ml, 1.45N, 0.7725N, blank sample etc).

Observations

Blood of the animals after 5 hours from the time of administration of Aulprofem was taken in a clean dry container. For plasma separation heparin 200 IU/ml blood was used. Sodium fluoride 10mg/ml blood could also be used. Blood plasma was immediately separated by centrifugation at 2000 rpm for 5-10 minutes. Working reagent was prepared by taking triethanolamine buffer, then preparing a solution of NAD+/ATP dry powder, and stored at -20 degree Celsius, then took suspensions of L-malate dehydrogenase/citrate suspension, Acetyl-coA, synthetase suspension, and acetic acid standard solution separately and kept. Took 1.0 ml of Triethanolamine and pipetted into cuvette, added 0.2 ml NAD+/ATP solution and added 0.1 ml sample standard acetic acid and distilled water 1.9 ml. Mix the reagents and measured the absorbance (A_0) then added the L-malate dehydrogenase/citrate suspension, 0.025 ml mix and read the absorbance (A_1) after 4 minutes. Then added acetyl-coA synthetase suspension, 0.02 ml and waited for 10-15 minutes and noted the absorbance (A_2). Repeated these readings with the blank sample and the standard acetic acid solution.

Results

The Δ A is calculated as follows:

$$\Delta \text{ A acetic acid} = [(A_2 - A_0)\text{ Sample} - \frac{(A_1 - A_0)_2 \text{ blank}}{(A_1 - A_0)\text{ blank}}] - \times [(A_2 - A_0)\text{ blank} - \frac{(A_1 - A_0)_2 \text{ blank}}{(A_1 - A_0)\text{ blank}}$$

The concentration is calculated as follows:

$$C = \frac{V + MW}{6.3 \times 1 \times 0.1 \times 1000} \times \Delta A$$

Table S9A. Table showing the active constituents concentration w.r.to sodium acetate in plasma levels of the animals under administration of Aulprofem

Age of the animals in months	Plasma concentration levels of sodium acetate in mg/L after						
	15 min	30 min	60 min	2 hrs	10 hrs	24 hrs	48 hrs
76	0.012	0.016	0.042	0.035	0.026	0.020	0.00
62	0.013	0.016	0.041	0.037	0.025	0.019	0.00
84	0.011	0.017	0.040	0.034	0.024	0.018	0.00
65	0.011	0.015	0.042	0.033	0.025	0.019	0.00
76	0.010	0.016	0.043	0.035	0.025	0.019	0.00
87	0.012	0.015	0.040	0.036	0.025	0.020	0.00
63	0.011	0.014	0.042	0.033	0.024	0.018	0.00
72	0.012	0.016	0.042	0.034	0.026	0.019	0.00
75	0.010	0.015	0.040	0.035	0.024	0.020	0.00
66	0.011	0.015	0.042	0.036	0.024	0.018	0.00

Table S9B. Table showing the active constituents concentration w.r.to Acetic acid in plasma levels of the animals under administration of Aulprofem

Age of the animals In months	Plasma concentration levels of sodium acetate in gm/L after						
	15 min	30 min	60 min	2 hrs	10 hrs	24 hrs	48 hrs
76	0.031	0.060	0.16	0.120	0.105	0.080	0.00
62	0.030	0.073	0.17	0.125	0.100	0.075	0.00
84	0.035	0.068	0.15	0.124	0.108	0.080	0.00
65	0.036	0.060	0.16	0.120	0.102	0.074	0.00
76	0.037	0.066	0.17	0.122	0.098	0.085	0.00
87	0.035	0.072	0.17	0.123	0.095	0.082	0.00
63	0.038	0.075	0.16	0.123	0.105	0.080	0.00
72	0.037	0.060	0.15	0.120	0.098	0.084	0.00
75	0.036	0.067	0.16	0.120	0.104	0.075	0.00
66	0.038	0.068	0.16	0.124	0.098	0.082	0.00

Where as V = final volume = 3.245 ml and

MW = Molecular weight of acetic acid = 60.05.

So, the equation becomes:

$C = 0.309 \times \Delta A$

Standard correlation of the study: the above findings are compared for the ingredients under study in the animal with the mean of the concentrations thus found in ten consecutive prior days without the application of Aulprofem.

Discussion

The absorption of the acetic acid and sodium acetate is the total amount of these in blood minus the amount of these chemicals in the blood at the given time. The usual method is to calculate the amount if any of these compounds in blood then go for the administration of the Aulprofem and then at fixed intervals of 15 minutes, 30 minutes, 1 hour, 2, hours, 5 hours, 10 hours, 24 hours, 48 hours is calculated and extrapolated as a concentration curve. The amount /ml of these chemicals multiplied by the total body weight in grams is the total amount of acetate content in the body of the animal. The amount of the sodium acetate is calculated as the proportional weight of this chemical against the total Acetate concentration in the body of the animal. The amount of acetic acid alone is the amount of the total acetate minus the amount of sodium acetate in the body of the animal at a given time.

Study no. 10: The study in the elimination of Aulprofem from the living system

Title : **A study to know the elimination of Aulprofem in the living system in cows**

Introduction

The basic principle is same for study of elimination of Aulprofem from the body of the organism except in that it is reversed w.r.to decreasing concentrations of active constituents. The sodium acetate and acetic acid is converted in the body of the living organisms into Acetyl-coA which in turn acts to convert oxaloacetate to citrate. The oxaloacetate required in this reaction is formed from L-malate and nicotinamide-adenine-dinucleotide (NAD+). The NAD+ is reduced to NADH. The amount of NADH is measured by the increase in absorbance at 340 nm. Acetate which is present in the body is simultaneously metabolized and excreted through kidneys via urine and can be tested as told above. A point here to note is that the elimination is near total when concentration levels in blood are near zero.

Materials and Methods

10 healthy female animals without any history of disease and ailment in the age group 4 to 8 years in age were selected for this study. They were given the Aulprofem solution 225 ml which is a standard dose for the sex fixing act. The other materials and methods in this study included the standard apparatus, spectrophotometer set at 340nm and procedure for the estimation of acetate level in urine. The various reagents used for working solution included unhaemolyzed serum, triethanolamine buffer at –Ph, 8.4, 0.4M, NAD+, ATP freeze dried powder, L malate dehydrogenase/citrate suspension, Acetyl-coA synthetase suspension, acetic acid standard solution (5 ml, 1.45N, 0.7725N, blank sample etc.).

Observations

Urine of the animals after fixed intervals of 15 minutes, 30 minutes, 60 minutes, 2 hours, 10 hours, 24 hours, 48 hours from the time of administration of Aulprofem was taken in a clean dry container. Working reagent was prepared by taking triethanolamine buffer, then preparing a solution of NAD+/ATP dry powder, and stored at -20 degree Celsius, then took suspensions of L-malate dehydrogenase/citrate suspension, Acetyl-coA, synthetase suspension, and acetic acid standard solution separately and kept. Took 1.0 ml of Triethanolamine and pipetted into cuvette, added 0.2 ml NAD+/ATP solution and added 0.1 ml sample standard acetic acid and distilled water 1.9 ml. Mix the reagents and measured the absorbance (A_0) then added the L-malate dehydrogenase/citrate suspension, 0.025 ml mix and read the absorbance (A_1) after 4 minutes. Then added acetyl-coA synthetase suspension, 0.02 ml and waited for 10-15 minutes and noted the absorbance (A_2). Repeated these readings with the blank sample and the standard acetic acid solution. Also repeat these readings at fixed intervals of 15 minutes, 30 minutes, 60 minutes, 2 hours, 10 hours, 24 hours, and 48 hours after the administration of Aulprofem to the animal.

Results

The Δ A is calculated as follows:

$$\Delta \text{ A acetic acid} = [(A_2 - A_0)\text{Sample} - \frac{(A_1 - A_0)_2 \text{ blank}}{(A_1 - A_0) \text{ blank}}] - \times [(A_2 - A_0) \text{ blank} - \frac{(A_1 - A_0)_2 \text{ blank}}{(A_1 - A_0) \text{ blank}}$$

Table S10A. Table showing the active constituents concentration w.r.to sodium acetate in urine levels of the animals under administration of Aulprofem

Age of the animals in months	Plasma concentration levels of sodium acetate in mg/L after						
	15 min	30 min	60 min	2 hrs	10 hrs	24 hrs	48 hrs
76	0.0013	0.012	0.022	0.030	0.026	0.028	0.00
62	0.0011	0.014	0.021	0.031	0.025	0.030	0.00
84	0.0010	0.012	0.019	0.029	0.024	0.028	0.00
65	0.0010	0.011	0.022	0.027	0.025	0.029	0.00
76	0.0011	0.012	0.023	0.025	0.025	0.026	0.00
87	0.0010	0.013	0.019	0.028	0.025	0.032	0.00
63	0.0013	0.010	0.018	0.028	0.024	0.031	0.00
72	0.0012	0.011	0.022	0.028	0.026	0.030	0.00
75	0.0010	0.011	0.020	0.030	0.024	0.032	0.00
66	0.0012	0.010	0.019	0.031	0.024	0.029	0.00

Table S10B. Table showing the active constituents concentration w.r.to Acetic acid in urine levels of the animals under administration of Aulprofem

Age of the animals in months	Plasma concentration levels of sodium acetate in mg/L after						
	15 min	30 min	60 min	2 hrs	10 hrs	24 hrs	48 hrs
76	0.0030	0.045	0.06	0.104	0.125	0.108	0.00
62	0.0027	0.042	0.08	0.103	0.120	0.107	0.00
84	0.0028	0.039	0.07	0.102	0.119	0.108	0.00
65	0.0028	0.038	0.08	0.098	0.118	0.106	0.00
76	0.0029	0.040	0.07	0.097	0.109	0.105	0.00
87	0.0030	0.039	0.07	0.097	0.120	0.106	0.00
63	0.0029	0.038	0.06	0.102	0.118	0.108	0.00
72	0.0028	0.042	0.09	0.104	0.119	0.104	0.00
75	0.0027	0.045	0.08	0.100	0.117	0.105	0.00
66	0.0028	0.048	0.07	0.104	0.118	0.104	0.00

The concentration is calculated as follows:

$$C = \frac{V + MW}{6.3 \times 1 \times 0.1 \times 1000} \times \Delta A$$

Where as V = final volume = 3.245 ml and

MW = Molecular weight of acetic acid = 60.05.

So, the equation becomes:

$C = 0.309 \times \Delta A$

Standard correlation of the study: the above findings are compared for the ingredients under study in the animal with the mean of the concentrations thus found in ten consecutive prior days without the application of Aulprofem.

Discussion

The elimination of the acetic acid and sodium acetate is the total amount of these in urine minus the amount of these chemicals in the urine at the given time. The usual method is to calculate the amount if any of these compounds in urine before the administration starts, then go for the administration of the Aulprofem and then at fixed intervals of 15 minutes, 30 minutes, 1 hour, 2, hours, 5 hours, 10 hours, 24 hours, 48 hours is calculated and extrapolated as a concentration curve. The amount /ml of these chemicals excreted in total urine is the total amount of acetate content excreted through the body of the animal. The amount of the sodium acetate is calculated as the proportional weight w.r.to the original formulation and excreted in urine against the total Acetate concentration excreted from the body of the animal. The amount of acetic acid eliminated alone is the amount of the total acetate excreted minus the amount of sodium acetate excreted from the body of the animal at a given time.

Study no. 11: A study in the effects of Aulprofem on Digestive system function

Title : **A study to evaluate the effects of administration of Aulprofem on the functioning of digestive system in cows**

Introduction

The most important and common test to evaluate the normal digestive function for animals under the treatment of Aulprofem was under taken.

This included the effect of Aulprofem treatment on the secretion of amylase enzyme activity. Amylase plasma level is a better indicator of the health of digestive system function than the measurement of other criteria. So this test was preferred. This is discussed as below:

Materials and Methods

10 healthy female animals without any history of disease and ailment in the age group 4 to 8 years in age were selected for this study. They were given the Aulprofem solution 225 ml which is a standard dose for the sex fixing act. The other materials and methods in this study included the standard apparatus, a dry state analyzer manufactured by Johnson & Johnson. These fully automatic dry state analyzers are available in three configurations; model DTSC for enzymes studies, model DT60 for glucose, proteins etc level studies and model DTE for studying electrolyte studies such as sodium chloride etc. In the present study model DTSC was used for obvious reasons. This reads amylase levels in the range of 5-900 U/L. Other materials used include clean and dry glass ware, laboratory glass pipettes or micropipettes, isotonic saline etc.

Observation

Blood of the animals after 5 hours from the time of administration of Aulprofem was taken in a clean dry container. For plasma separation heparin 200 IU/ml blood was used. Sodium fluoride 10mg/ml blood could also be used. Blood plasma was immediately separated by centrifugation at 2000 rpm for 5-10 minutes. 0.10 ml plasma sample was transferred to the fully automatic analyzer. It read the readings using its software

Table S2. Table showing the plasma amylase concentration of the adult female animals which were given a 225 ml dose of Aulprofem

S. no.	The age of animals in months	Amylase concentration U/L
1.	62	54
2.	78	78
3.	67	94
4.	66	85
5.	87	65
6.	88	72
7.	72	105
8.	93	88
9.	78	92
10.	68	68

installed in it for this purpose. Repeated this with all the 10 animals and noted the results.

Results

The results of amylase concentrations are read automatically by the dry state fully automatic analyzer.

The various results of the amylase concentrations in plasma are pasted in the following table.

Discussion

The various animals which were given the standard dose of Aulprofem and later subjected to observation for 5 long hours which are sufficient for Aulprofem to give any unwanted effects on the digestive function and later testing for the amylase concentration testing showed normal values which is an indicative of the normal digestive function. None of the animals showed any abnormal or dangerous values. Hence it can be concluded that the administration of Aulprofem is not toxic to digestive system at all.

❋❋

Some More Details of Sex Fixing Trials

Some mere studies for evaluating the sex fixing activity of Aulprofem, were conducted on cows and buffaloes. We are giving those details as per the institute records as given below:

1. Number of Animals

The research farm had a population of nearly 140 animals. Out of these 96 healthy and mature animals were put to trials.

2. Breed

The animals included cattle which were a mixture of zebu (Bos indicus), and cross bred of zebu cattle with Jersey and Holstein Friesian. The buffaloes included animals from the Murrah breed of buffaloes, commonly found in north India.

3. Age of Each Animal

The animals were tagged 1 to 96. The age of each animal at the time of mating and administration of drug material is as given below (Abbreviations used: C for cow, B for buffalo, I for Native Indian cattle, IJ for cross bred of Indian native cow with Jersey cattle, IH for cross bred of Indian native cow with Holstein Friesian cattle, BM for Murrah buffalo,

S. no. / tag no.	Age of animal at mating (months)	Animal / Breed	Pregnancy / Delivery
1.	37	IJ	D
2.	45	IH	D
3.	58	BM	D
4.	49	BM	D
5.	53	BM	NC
6.	47	BM	D
7.	42	IJ	D
8.	41	IJ	D
9.	41	IJ	NC
10.	56	BM	D
11.	43	IJ	D
12.	74	BM	NC
13.	67	BN	D
14.	78	BO	D
15.	38	IJ	D
16.	49	BM	M
17.	57	BM	D
18.	41	IJ	NC
19.	65	BM	D
20.	53	BN	D
21.	56	BN	NC
22.	48	I	D
23.	70	BO	D
24.	39	IJ	NC
25.	65	BO	D
26.	62	BM	D
27.	48	BM	D
28.	46	IH	D
29.	57	BN	D
30.	57	BM	NC
31.	48	BO	D
32.	45	BM	D
33.	52	IH	D
34.	38	IJ	NC
35.	39	IJ	D
36.	45	BM	D
37.	46	BN	D
38.	57	BO	D
39.	65	I	D
40.	62	BN	D
41.	39	IJ	D
42.	47	BN	D
43.	43	IJ	D
44.	64	BM	NC
45.	65	BM	D
46.	41	IJ	NC
47.	44	IJ	D

S. no. / tag no.	Age of animal at mating(months)	Animal / Breed	Pregnancy / Delivery
48.	56	BO	D
49.	67	BN	D
50.	61	BM	M
51.	48	BM	D
52.	42	IJ	D
53.	39	IJ	NC
54.	40	IJ	D
55.	58	BN	NC
56.	59	BO	D
57.	57	BO	D
58.	57	IH	D
59.	65	BM	NC
60.	69	BM	D
61.	76	BN	NC
62.	49	BN	D
63.	48	IH	D
64.	48	BO	D
65.	48	IH	D
66.	47	IH	D
67.	42	BM	D
68.	75	BN	D
69.	74	BM	NC
70.	43	BM	D
71.	39	IJ	D
72.	65	BM	NC
73.	63	BM	D
74.	40	IJ	D
75.	54	BN	D
76.	54	BO	D
77.	68	BO	D
78.	63	BO	D
79.	41	IJ	M
80.	49	BM	D
81.	59	BM	NC
82.	63	BO	D
83.	83	BO	D
84.	67	BM	D
85.	39	IJ	NC
86.	56	BN	D
87.	54	BN	D
88.	78	BO	NC
89.	49	BO	D
90.	49	IH	NC
91.	53	BM	D
92.	56	BM	NC
93.	67	BN	D
94.	69	BM	D
95.	48	IH	D
96.	46	BM	D

BN for Nili Ravi buffalo, BO for buffalo of other breeds, D for delivered, M for miscarriage, NC for not conceived).

4. Owner Authorization

The ownership of the research farm belongs to the Gregor Mendel Institute for Research in Genetics and it exists as a separate Clinical Testing and Trials Unit of the above institute. The proper and prior approval of institute authorities was taken before the start of these trials. The ethical committee of the institute had also approved the trials. The in charge of the Clinical Testing and Trials Unit, Mr. John Sahotra has also authorized and signed in this document in a separate column in a declaration given below.

5. Statistical Analysis

The present trials involved 96 animals, of which 32 (33.33 %) were cows and 64 (66.66 %) were buffaloes which were administered the Aulprofem. Of these in cows, there were 2 cows of pure Indian native origin, 21 were crossbreds of Jersey with Indian native cows and 9 were crossbreds of Holstein Friesian with Indian native cows which respectively formed 6.25 %, 28.125 % and 65.625 % of the total cows involved in trials. In buffaloes there were 32 buffaloes from the Murrah breed, 16 buffaloes from the Nili Ravi breed and remaining 16 were a mixture of other local breeds commonly found in northern India. These respectively formed 50 %, 25 % and 25 % of the total buffalo populations involved. The cattle reported 1 miscarriage. This is 3.125 % of the total cattle involved. Again the miscarriage reported was in the category of crossbred of Indian native with the jersey cattle. Of the total 21 cows from this category this formed 4.76 % of the total cattle involved of this crossbred of Jersey cattle with the native Indian cows. In buffaloes there were reported 2 miscarriages from a population of 64 buffaloes. This formed 3.125 % of the total buffalo populations involved. Again in buffaloes, these miscarriages wee from the Murrah breed which meant a miscarriage ratio of 6.25 % in Murrah buffaloes. There were 8 animals not conceived in cows which meant a percentage of 25 %. Further, these non-conceptions were 7 (*i.e.* 33.33 %) in crossbreds of Indian native cows with Jersey cattle and 1 (*i.e.* 11.11 %) in crossbreds of Indian native with Holstein Friesian cattle. There were 13 (20.31 %) cases of no conceivement with the buffaloes. Of these there were 9 (28.125 %) in Murrah buffaloes, 3 (18.75 %) in Nili Ravi buffaloes and 1 (06.25 %) in other buffaloes. There were 72 (*i.e.* 75% of the total no of animals involved in the trials) successful deliveries. Of them 72 (*i.e.* 100 %) were of the offsprings of the female sex. This is the most important outcome of the total trials exercise *i.e.* 100% efficacy of the Aulprofem to produce babies of female sex in dairy bovines *i.e.* cattle and buffaloes.

6. Time Table of the Clinical Trial

Time period of testing ranged from March, 1998 to September, 1999.

Brief summary of the Trials: Major points that can sum up above discussion are as follows:

Selection of the animals: 96 healthy females were selected for the testing, 32 cows and 64 buffaloes. All animals selected had recorded history as good conceivers.

Procedures/Materials/Methods: The Aulprofem preparation as prepared with 12.5 mg of sodium ethanoate dissolved into 1.45 M ethanoic acid to make the total volume to 225 ml was used as the test material to be given by the oral route to the female animal as a single dose 30 minutes after natural service.

Results: There were 72 successful deliveries, 23 in cows and 49 in buffaloes. All animals delivered female calves. There were no male births reported.

Side effects: Mild signs of acidity and decreased feeding for a day or two in less than 5% of cases treated with Aulprofem. May cause itching or mild skin reactions on delicate parts like nostrils, buccal cavity, open wounds etc. Wash with water if necessary.

Special observations: Best suited for animals aged below 10 years.

Inference: The drug formula Aulprofem *i.e.* 12.5 sodium ethanoate in 1.45 M ethanoic acid to produce the final volume to 225 mls, given as a single, oral dose; definitely has a successful role in manipulation of sex ratios in dairy animals, cows and buffaloes and this treatment is pharmacologically safe and pharmaceutically compatible and is a preparation free from any unwanted and toxic side effects.

Some More Studies into Sex Fixing with Monosodium Ethanoate in Ethanoic Acid

A brief account of some further trials with Aulprofem on other animals is as given as follows:

A. Sex assignment and fixing in horses with the help of monosodium ethanoate solution in ethanoic acid

Objectives

The objective of the study was to ascertain the sex fixing effect of monosodium ethanoate in ethanoic acid given by oral route in mammals particularly horses. The horses are an important member of family mammalian and they have a vital role in the economy of many countries of the world particularly Asian and Arab countries, Russia, China, India, and many European countries. Almost in all the nations of the world the horses are an important source of transport, ride and games such as racecourses. They are a valuable asset in numerous stud farms of the world. They are also used as a source of meat and they have a global role in the economy of the world. So, an attempt to alter the sex ratios in horses has a lot to do with the world economy.

Prior art

Mammals are an important family. Hence, prior art is replete with attempts to alter sex ratios in mammals and it has been one of the most favorite subjects of reproductive physiologists. Way back in 1939, Russian scientist Schroder, V. discussed the efficacy of physicochemical methods to alter sex ratios in mammals. After that many workers have reported on the line. These scientists have worked mainly on two lines. One was the production of sexed semen. Sexed semen is the one which contains either X- or Y-chromosome bearing sperms in greater or accentuated proportions. A list of scientists who based their attempts to produce sexed semen is long (Gordon, 1958; Lindahl, 1956; Bhattacharya *et al.*, 1966; Meistrich, 1982; Ericsson, 1973; Gledhil, 1983).

Some other workers tried a different line of using certain chemicals, enzymes, sera, anti-sera etc. They include many scientists (Corson *et al.*, 1984; Beernik *et al.*, 1984. Barrat and Leger. 1979; Sampson *et al.*, 1983; Mitra and Chowdhury. 1990).

The present research is in the line of second type of sex pre-assignment and determination where the sex of the off-spring is fixed with the help of an oral liquid dosage form. The exact chemical nature of this material has been discussed in the next heading of materials and methods as given below.

Materials and Methods

The study was conducted on the research farms controlled by Gregor Mendel Institute or Research in Genetics, Ludhiana, Punjab, India. The animals were divided into control and test groups and the material was administered to the animals in the test group. The material comprised a solution of monosodium ethanoate, 12.5 mg dissolved in a solution of ethanoic acid, 1.75 M to make the total solution to 250 ml. The solution was administered to the animal at the time of mating 30 minutes after mating. The administration was by a single dose by the oral route to the female animal. The results were recorded and calculated in simple percentage methods.

Observations

Study was conducted on 58 female animals, divided into two groups of control and test animals. The animals in the control group were not given any treatment of the test material and the animals in the test group were given the material by the oral route, 30 minutes after mating. The control group animals reported 13 animals as not conceived and none of the animals underwent miscarriage. The remaining 16 animals produced 16 off-springs, 09 males and 07 females. The animals in the test group reported 12 animals not conceived and none of the animals underwent miscarriage and remaining 17 animals produced 17 off-springs, 03 males and 14 females.

	No. of animals	Animals not conceived	No. of miscarriages	Offspring	
				male	female
Test	29	12	0	3	14
Control	29	13	0	09	07

Results

The animals in the control group yielded mixed offspring, 09 males *i.e.* 56.25% and 07 females *i.e.* 43.75%. The animals in the test group yielded 17 off-springs; 03 males *i.e.* 17.65 % and 14 females *i.e.* 82.35 % which clearly is the result of treatment given *i.e.* the 12.5 mg monosodium ethanoate dissolved in 1.75 ethanoic acid to make the solution to 250 ml. the sex ratio is definitely changed from the natural one and also from the one in the control group. So, it can be inferred that the above medication effects, in the form of alteration of sex ratios and produces female off-springs in horses. Hence, the efficacy of the above said material to produce preferential progenies of the female sex is well established.

Discussion

The drug is effective in altering the sex ratios in horses as is clear from the results in table given above. The results of animals in test category show female off-springs with a 82% success rate; where as the animals grouped under standard category show a mixed off-spring of female as well male colts in the near natural sex ratio. This shows that where as the drug altered the sex ratios and that too to a considerable modification of sex ratios; the animals under similar conditions produced mixed off-springs in the absence of application of drug. So, the inference clearly, is the efficacy of drug to alter the sex ratios in horses. A greater result that can be drawn from this experiment is that a mammalian sex fixer has arrived on the scene and manipulation of mammalian sex ratios is not an idealistic imagination but a reality in concrete terms and the possibility can not be ruled out that more sex fixers for mammals will be discovered in future.

References

Barrat, J. and Leger, D. 1979. Aveir des grossesses obstenues après stimulation de l' ovulation. A propos de 519 grossesses. *J. Gynecol. Obstet. Biol. Reprod. (Paris)* **8**: 332.

Beernik, F.J. and Ericsson, R.J. 1982. Male sex preselection through sperm isolation. *Fertil. Steril.* **38**: 493.

Bhattacharya, B.C. *et al.*, 1966. An attempt to determine the sex ratio of calves by artificial insemination with spermatozoa separated by sedimentation. *Nature* **211**: 863.

Corson, S.L. *et al.*, 1984. Sex selection by sperm separation and insemination. *Fertil. Steril.* **42**: 756.

Ericsson, R.J. *et al.*, 1973. Isolation of fractions rich in human Y sperm. *Nature* **246**: 421.

Gledhil, B.L. 1983. Control of mammalian sex ratio by sexing sperm. *Fertil. Steril.* **40(5)**: 572.

Gordon, M.J. 1958. The control of sex. *Scientific Am.* **199**: 87.

Lindahl, P.E. 1956. Counterstreaming centrifufation of bull spermatozoa. *Nature* **178**: 491.

Meistrich, M.L. 1982. Prspects of sexing mammalian sperm *in* R.P. Amann and G.E. Seidel, Jr. (*eds.*) Colorado University Associated Press.

Mitra and Chowdhury. 1990. Abstr. 2nd edn. *Animal Breed.* Abstrs. **58(4)**: 2354. etc.

Sampson, J.H. *et al.*, 1983. Gender after induction of ovulation and artificial insemination. *Fertil. Steril.* **40**: 481.

B. Sex assignment and fixing in sheep with monosodium ethanoate solution in ethanoic acid

Abstract

Sex fixing is a fascinating new concept in applied clinical pharmacology. Until now it was reported in certain ratites and lower reptiles. Of course, there have been numerous attempts to manipulate sex ratios in mammals but by and large, success has eluded them. Mammals are an important family and any success in producing desired sexed babies in them has considerable economic and social benefits. The present research is of a drug formula called 'aulprofem' which is a combination of sodium ethanoate and ethanoic acid in solution that when given orally to females of dairy animals, cows and buffaloes and other mammals such as sheep; yields female off-springs with near hundred percent success rate. The active constituents are therapeutically safe and pharmacologically compatible and do not exhibit any toxic or side effects. The various outcomes of this discovery may be a rapid increase in the numbers of females of high yielding milch varieties of dairy animals and thereby increasing milk production to unprecedented heights across the globe. Similarly, meat and wool productions can also be increased by increasing the overall numbers of meat and wool etc producing animals by producing more female off-springs of these animals with the help of this discovery.

Objectives

The objective of the study was to ascertain the sex fixing effect of monosodium ethanoate in ethanoic acid given by oral route in mammals particularly sheep. The sheep are an important member of family mammalian and they have a vital role in the economy of many countries of the world particularly Australia, Russia, China, India, and many European countries. Almost in all the nations of the world the sheep are an important source of wool and meat and they have a global role in the economy of the world. So, an attempt to alter the sex ratios in sheep has a lot to do with the world economy.

Introduction

Mammals are the most dominant animals on earth today. So the production of sexed offsprings in them has marked economic and social implications. Many scientists have worked on this since the first half of the last century. There were mainly three lines on which such research was conducted. First were the attempts to alter the –pH of the female genital tract, second were the attempts to produce sexed semen (a sexed semen is one containing the accentuated proportions of either X or Y bearing spermatozoa) and the third were the use of certain chemicals, hormones, sera etc to produce a shift in the sex ratio. Recently much emphasis is laid on the attempts to understand the exact mechanism of sperm-egg fusion. Candidate molecules have been identified that act as receptors on the ZP (zona pellucida) or the ovum membrane and bind with possible sperm ligands. Along with this the differential *in vitro* binding of X and Y containing sperms with different antigen-antibody media based on the assumption of differential binding sites on them and their successive elutriation in accentuated proportions has gained attraction. The present research is an *in vivo* attempt to produce female offsprings where sex ratios are manipulated probably by binding of receptor sites for Y sperm on the ovum by the action of certain moieties that are generated in the living system by the action of the oral administration of a solution of sodium ethanoate in ethanoic acid given to the female animals at the time of mating. The important consequences of this may be rapid increase in milk, meat and wool etc production in India and the world over to unfathomable heights by preferential production of females of dairy and related fields and the great social and economic benefits and changes associated with this. In a projection when only 10 percent of the total dairy animal wealth of India is brought under the application of this technique for a smaller time period of only 10 years; the considering the normal national levels of conception failure, calf mortality, adult mortality etc, nearly 160 million tonnes of increase in national milk production can be achieved. This is no short of a revolution. in case of sheep also, there can be rapid increase in the number of female animals resulting in much bigger flocks of these animals with the end result of bigger production of wool and meat also.

Materials and Methods

The present technology is an *in vivo* technique. It is based on the assumption of the production of certain moieties in the living system by the action of administration of certain chemicals that can block the binding or fusion of a particular X or Y sperm receptor on the ovum. The study was conducted in the form of *in vivo* trials on healthy animals between 3 to 7 years in age in the sheep farms selected for this purpose. All the 40 animals formed the test and control groups of 20 animals each. The other materials included the formulation that comprised a solution of monosodium ethanoate 12.5 mg dissolved in ethanoic acid 1.45 M q.s. to make the total solution to 225 ml. The solution was administered to the animal 30 minutes after natural service. The administration was by a single dose of 100 ml of above preparation via the oral route to the female animal. The results were recorded and calculated in simple percentage methods.

Observations

Study was conducted on 40 female animals, divided into two groups of control and test animals. The animals in the control group were not given any treatment of the test material and the animals in the test group were given the material by the oral route, 30 minutes after mating. The control group animals reported 6 animals as not conceived and none of the animals underwent miscarriage. The remaining 14 animals produced 26 off-springs, 12 males and 14 females. The animals in the test group reported 7 animals not conceived and no animal underwent miscarriage and remaining 13 animals produced 26 offspring, 06 males and 20 females.

	No. ofA animals	nimals not conceived	No. of miscarriages	Offspring	
				male	female
Test	20	07	0	06	20
Control	20	06	0	12	14

Results

The animals in the control group yielded mixed offspring, 12 males *i.e.* 46.15% and 15 females *i.e.* 53.85%. The animals in the test group yielded 26 off-springs, 06 males *i.e.* 23.07% and 20 females *i.e.* 76.93% which clearly is the result of treatment given *i.e.* the 5.0 mg monosodium ethanoate dissolved in 1.75 ethanoic acid to make the solution to 100 ml. So, it can be inferred that the above medication effects, in the form of alteration of sex ratios and produces female off-springs in sheep. Hence, the efficacy of the above said material to produce progeny o female sex is well established in sheep.

Discussion

The drug is effective in altering the sex ratios in sheep as is clear from the results in table given above. The results of animals in test category show female off-springs with a 76.93 % success rate; where as the animals grouped under standard category show a mixed off-spring of female as well male lambs. This shows that where as the drug altered the sex ratios and that too to a complete modification of sex ratios; the animals under similar conditions produced mixed off-springs in the absence of application of drug. So, the inference clearly, is the efficacy of drug to alter the sex ratios in sheep. A greater result that can be drawn from this experiment is that a mammalian sex fixer has arrived on the scene and manipulation of mammalian sex ratios is not an idealistic imagination but a reality in concrete terms and the possibility can not be ruled out that more sex fixers for mammals will be discovered in future.

Most of the animals that received the treatment delivered female offsprings. The failures can be attributed to manual error, spillage and faulty dose adjustment in case of heavier animals. Sex fixing in mammals has a long history. Many scientists have worked on this including Schroeder (1941), Bhattacharya *et al.* (1966), Ericsson *et al.* (1973), Gordon (1958), Lindahl (1956), Gledhill (1983), Corson *et al.*(1984), Beernik and Ericsson (1982), Barrat and Leger (1979), Sampson *et al.* (1983) etc. The major drawbacks of these techniques remained the lack of laboratory tests to evaluate the degree of sperm separation (Hafez, 1982) and the inability to know the precise mechanism of binding and fusion of mammalian sperm with ova. Overall, the manipulation of mammalian sex ratio has still remained a mirage on the horizon (Hunter, 1982). Recently there have been attempts to study the effects of certain compounds like tetraspanins like CD_9 (Hemler, 2003), CD_{81} (Cormier, 2004), glycosylphosphatidylinositol anchored proteins like CD_{55} (Coonrad, 1999), integrins (He *et al.*, 2003), disintegrins (Primakoff *et al.*, 2000), fertilins, cyritestin (Cho *et al.*, 1998) etc on the process of gamete fusion. These researches have clearly established the receptor sites on oocyte and their respective ligands on the surface of sperms. But the bigger questions are far from being answered. The 'tetraspanin-web' is still a big mystery (Boucheix *et al.*, 2001). The role of other candidate molecules the likes of which are mentioned above, needs a lot more investigating. Recently there have been attempts to differentially bind X and Y bearing spermatozoa using H-Y antigen antibody interactions using non protein substrata such as agarose beads and U.S. patents granted for them (Zavos *et al.*, 1991; Bryant *et al.*, 1984). These developments clearly demonstrate the existence of differential binding sites for X and Y sperms and this difference can be exploited in developing a viable technology for sex fixing in mammals, either *in vivo* or *in vitro*. The present technology is an *in vivo* technique and it can be hypothesized that there are differential receptors sites on the body of the ovum and their corresponding ligands on the bodies of the X and Y bearing sperms. A receptor site for X sperm does not receive the binding ligand of Y sperm as it does not correspond to its binding morphology and physiology and *vice versa*. The material of present research produces certain Y sperm binding ligand mimics (YSBLM) moieties in the living system and also in the genital system of the female animals undergoing this treatment in such a way that these YSBLM moieties differentially bind with receptor sites on the ovum thus occupying these sites and inhibit their subsequent binding and/or fusion with true binding ligands on the bodies of the Y sperms and this way over rule the chances of an XY zygote being formed. This leaves all the chances of an X sperm to come and combine freely with the X sperm specific receptor sites on the ovum and form an XX zygote. So, it can be inferred that the most appropriate candidate molecules that act as receptors on the oocyte like CD_9 and binding ligands on the sperm like fertilin beta are also X and Y sperm specific. So we can have tetraspanins like CD_9x and CD_9y and fertilins like fertilin beta-x and fertilin beta-y. The importance of this technology becomes even more marked with the view that it will not only provide an insight into the differential behaving and working of X and Y bearing sperms in the process of fertilization but will also definitely provide more valuable data and knowledge on the role played by the candidate molecules in understanding fully well the exact mechanism of gamete fusion in further research on this subject.

Summary

The drug formula or material involved in above study is effective in altering the sex ratios in sheep as is clear from the results obtained in this experiment. The rate of success in modification of sex ratios is tremendous. There is ample scope that this formulation when used on other mammals including humans will yield desired results only. This only leads to the derivation of one and only one conclusion that the first successful mammalian sex fixer of the world has arrived on the scene and is very much here to stay. The manipulation of mammalian sex ratio is not a dream or a mirage on the horizon (Short, 1982) now but a living reality and it cannot be ruled out that more and more of sex fixers meant for mammals will be discovered in the future to come.

References

Barrat, J. and Leger, D. 1979. Aveir des grossesses obstenues après stimulation de l' ovulation. A propos de 519 grossesses. *J. Gynecol. Obstet. Biol. Reprod. (Paris)* **8**: 332.

Beernik, F.J. and Ericsson, R.J. 1982. Male sex preselection through sperm isolation. *Fertil. Steril.* **38**: 493.

Bhattacharya, B.C. *et al.*, 1966. An attempt to determine the sex ratio of calves by artificial insemination with spermatozoa separated by sedimentation. *Nature* **211**: 863.

Boucheix, C. *et al.*, 2001. Tetraspanins. *Cell Mol. Life Sci.* **58**: 1189-1202.

Bryant, F. 1984. U.S. patent no.4, 448, 767.

Cho, C. *et al.*, 1998. Fertilization defects in sperm from mice lacking fertilins beta. *Science* **281**: 1857-59.

Coonrad, S.A. *et al.*, 1999. Treatment of mouse oocytes with PI-PPLC releases 70kDa (pl5) and 35 to 45 kDa (pl5.5) protein clusters from the egg surface and inhibits sperm oolema binding and fusion. *Dev. Biol.* **207**: 334-49.

Cormier, E.G. *et al.*, 2004. CD81is an entry coreceptor for hepatitis C virus. *Proc. Natl. Acad. USA*. **101**: 7270-74.

Corson, S.L. *et al.*, 1984. Sex selection by sperm separation and insemination. *Fertil. Steril.* **42**: 756.

Ericsson, R.J. *et al.*, 1973. Isolation of fractions rich in human Y sperm. *Nature* **246**: 421.

Gledhil, B.L. 1983. Control of mammalian sex ratio by sexing sperm. *Fertil. Steril.* **40(5)**: 572.

Gordon, M.J. 1958. The control of sex. *Scientific Am.* **199**: 87.

Hafez, E.S.E. 1982. Reproduction in farm animals. Reprint 5th edn. Lea and Fabiger, p. 499.

He, Z.Y. 2003. None of the integrins known to be present on the mouse egg orto be ADAM rreceptors are essential for sperm egg binding and fusion. *Dev. Biol.* **254**: 226-37.

Hemler, M.E. 2003. Tetraspanin proteins mediate cellular penetration, invasion and fusion events and define a novel type of membrane micro domain. Annu. Rev. Cell Dev. Biol. 19: 397-422.

Hunter, R.H.F. 1982. Reproduction of farm animals. Longman. London, pp. 138-39.

Lindahl, P.E. 1956. Counterstreaming centrifufation of bull spermatozoa. *Nature* **178**: 491.

Primakoff, P. *et al.*, 2000. The ADAM gene family: surface proteins with adhesion and protease activity. *Trends. Genet.* **16**: 83-7.

Sampson, J.H. *et al.*, 1983. Gender after induction of ovulation and artificial insemination. *Fertil. Steril.* **40**: 481.

Schroeder, V. 1941. Physica-chemical methods of sex regulation of progeny of mammals. Russian contributions. 1939 Genetics Congr. Am. Documentation Inst. Doc. 1565. *Abstr. In J. Hered.* **32**: 248.

Short, R.V. 1982. As quoted in reproduction of form animals by RHF Huntes. London. pp. 138-39.

Zavos, M and Dawson, K.A. 1991. Method for X and Y spermetozoa separation. U.S. Patent no. 4,999,283.

C. Sex fixing of progeny with monosodium ethanoate solution in ethanoic acid in goats

Objectives

The objective of the study was to ascertain the sex fixing effect of monosodium ethanoate in ethanoic acid given by oral route in mammals particularly goats. The goats are an important member of family mammalian and they have a vital role in the economy of many countries of the world particularly Asia and Africa, Russia, China, India, and many European countries. In many nations of the world the goats are an important source of milk, meat, leather and fur and they have a global role in the economy of the world. So, an attempt to alter the sex ratios in goats has a lot to do with the world economy.

Abstract

Sex fixing is a fascinating new concept in applied clinical pharmacology. Until now it was reported in certain ratites and lower reptiles. Of course, there have been numerous attempts to manipulate sex ratios in mammals but by and large, success has eluded them. Mammals are an important family and any success in producing desired sexed babies in them has considerable economic and social benefits. The present research is of a drug formula called 'aulprofem' which is a combination of sodium ethanoate and ethanoic acid in solution that when given orally to females of dairy animals, cows and buffaloes and other mammals such as goats; yields female off-springs with near hundred percent success rate. The active constituents are therapeutically safe and pharmacologically compatible and do not exhibit any toxic or side effects. The various outcomes of this discovery may be a rapid increase in the numbers of females of high yielding milch varieties of dairy animals and thereby increasing milk production to unprecedented heights across the globe. Similarly, meat and wool productions can also be increased by increasing the overall numbers of meat and wool etc producing animals by producing more female off-springs of these animals with the help of this discovery.

Introduction

Mammals are the most dominant animals on earth today. So the production of sexed offsprings in them has marked economic and social implications. Many scientists have worked on this since the first half of the last century. There were mainly three lines on which such research was conducted. First were the attempts to alter the –pH of the female genital tract, second were the attempts to produce sexed semen (a sexed semen is one containing the accentuated proportions of either X or Y bearing spermatozoa) and the third were the use of certain chemicals, hormones, sera etc to produce a shift in the sex ratio. Recently much emphasis is laid on the attempts to understand the exact mechanism of sperm-egg fusion. Candidate molecules have been identified that act as receptors on the ZP (zona pellucida) or the ovum membrane and bind with possible sperm ligands. Along with this the differential *in vitro* binding of X and Y containing sperms with different antigen-antibody media based on the assumption of differential binding sites on them and their successive elutriation in accentuated proportions has gained attraction. The present research is an *in vivo* attempt to produce female offsprings where sex ratios are manipulated probably by binding of receptor sites for Y sperm on the ovum by the action of certain moieties that are generated in the living system by the action of the oral administration of a solution of sodium ethanoate in ethanoic acid given to the female animals at the time of mating. The important consequences of this may be rapid increase in milk, meat and wool etc production in India and the world over to unfathomable heights by preferential production of females of dairy and related fields and the great social and economic benefits and changes associated with this. In a projection when only 10 percent of the total dairy animal wealth of India is brought under the application of this technique for a smaller time period of only 10 years; the considering the normal national levels of conception failure, calf mortality, adult mortality etc, nearly 160 million tonnes of increase in national milk production can be achieved. This is no short of a revolution in case of goats also, there can be rapid increase in the number of female animals resulting in much bigger flocks of these animals with the end result of bigger production of meat.

Materials and Methods

The present technology is an *in vivo* technique. It is based on the assumption of the production of certain moieties in the living system by the action of administration of certain chemicals that can block the binding or fusion of a particular X or Y sperm receptor on the ovum. The study was conducted in the form of *in vivo* trials on healthy animals between 3 to 7 years in age in the goat farms selected for this purpose. All the 50 animals formed the test and control groups of 25 animals each. The other materials included the formulation that comprised a solution of monosodium ethanoate 12.5 mg dissolved in ethanoic acid 1.45 M q.s. to make the total solution to 225 ml. The solution was administered to the animal 30 minutes after natural service. The administration was by a single dose of 100 ml of above preparation via the oral route to the female animal. The results were recorded and calculated in simple percentage methods.

Observations

Study was conducted on 50 female animals, divided into two groups of control and test animals. The animals in the control group were not given any treatment of the test material and the animals in the test group were given the material by the oral route, 30 minutes after mating. The control group animals reported 8 animals as not conceived and none of the animals underwent miscarriage. The remaining 17 animals produced 36 off-springs, 19 males and 17 females. The animals in the test group reported 8 animals not conceived and no animal underwent miscarriage and remaining 17 animals produced 34 offspring, 10 males and 24 females.

	No. ofA animals	nimals not conceived	No. of miscarriages	Offspring	
				male	female
Test	25	08	0	10	24
Control	25	08	0	19	17

Results

The animals in the control group yielded mixed offspring, 19 males *i.e.* 52.77% and 17 females *i.e.* 47.23%. The animals in the test group yielded 34 off-springs, 10 males *i.e.* 29.41% and 24 females *i.e.* 70.59% which clearly is the result of treatment given *i.e.* the 5.0 mg monosodium ethanoate dissolved in 1.75 ethanoic acid to make the solution to 100 ml. this is a considerable modification of sex ratio. So, it can be inferred that the above medication effects, in the form of alteration of sex ratios and produces female off-springs in goats. Hence, the efficacy of the above said material to produce progeny o female sex is well established in goats.

Discussion

The drug is effective in altering the sex ratios in goats as is clear from the results in table given above. The results of animals in test category show female off-springs with a 70% success rate; where as the animals grouped under standard category show a mixed off-spring of female as well male goat lambs. This shows that where as the drug altered the sex ratios and that too to a complete modification of sex ratios; the animals under similar conditions produced mixed off-springs in the absence of application of drug. So, the inference clearly, is the efficacy of drug to alter the sex ratios in goats. A greater result that can be drawn from this experiment is that a mammalian sex fixer has arrived on the scene and manipulation of mammalian sex ratios is not an idealistic imagination but a reality in concrete terms and the possibility can not be ruled out that more sex fixers for mammals will be discovered in future.

Most of the animals that received the treatment delivered female offsprings. The failures can be attributed to manual error, spillage and faulty dose adjustment in case of heavier animals. Sex fixing in mammals has a long history. Many scientists have worked on this including Schroeder (1941), Bhattacharya *et al.* (1966), Ericsson *et al.* (1973), Gordon (1958), Lindahl (1956), Gledhill (1983), Corson *et al.*(1984), Beernik and Ericsson (1982), Barrat and Leger (1979), Sampson *et al.* (1983) etc. The major drawbacks of these techniques remained the lack of laboratory tests to evaluate the degree of sperm separation (Hafez, 1982) and the inability to know the precise mechanism of binding and fusion of mammalian sperm with ova. Overall, the manipulation of mammalian sex ratio has still remained a mirage on the horizon (Hunter, 1982). Recently there have been attempts to study the effects of certain compounds like tetraspanins like CD_9 (Hemler, 2003), CD_{81} (Cormier, 2004), glycosylphosphatidylinositol anchored proteins like CD_{55} (Coonrad, 1999), integrins (He *et al.*, 2003), disintegrins (Primakoff *et al.*, 2000), fertilins, cyritestin (Cho *et al.*, 1998) etc on the process of gamete fusion. These researches have clearly established the receptor sites on oocyte and their respective ligands on the surface of sperms. But the bigger questions are far from being answered. The 'tetraspanin-web' is still a big mystery (Boucheix *et al.*, 2001). The role of other candidate molecules the likes of which are mentioned above, needs a lot more investigating. Recently there have been attempts to differentially bind X and Y bearing spermatozoa using H-Y antigen antibody interactions using non protein substrata such as agarose beads and U.S. patents granted for them (Zavos *et al.*, 1991; Bryant *et al.*, 1984). These developments clearly demonstrate the existence of differential binding sites for X and Y sperms and this difference can be exploited in developing a viable technology for sex fixing in mammals, either *in vivo* or *in vitro*. The present technology is an *in vivo* technique and it can be hypothesized that there are differential receptors sites on the body of the ovum and their corresponding ligands on the bodies of the X and Y bearing sperms. A receptor site for X sperm does not receive the binding ligand of Y sperm as it does not correspond to its binding morphology and physiology and *vice versa*. The material of present research produces certain Y sperm binding ligand mimics (YSBLM) moieties in the living system and also in the genital system of the female animals undergoing this treatment in such a way that these YSBLM moieties differentially bind with receptor sites on the ovum thus occupying these sites and inhibit their subsequent binding and/or fusion with true binding ligands on the bodies of the Y sperms and this way over rule the chances of an XY zygote being formed. This leaves all the chances of an X sperm to come and combine freely with the X sperm specific receptor sites on the ovum and form an XX zygote. So, it can be inferred that the most appropriate candidate molecules that act as receptors on the oocyte like CD_9 and binding ligands on the sperm like fertilin beta are also X and Y sperm specific. So we can have tetraspanins like CD_9x and CD_9y and fertilins like fertilin beta-x and fertilin beta-y. The importance of this technology becomes even more marked with the view that it will not only provide an insight into the differential behaving and working of X and Y bearing sperms in the process of fertilization but will also definitely provide more valuable data and knowledge on the role played by the candidate molecules in understanding fully well the exact mechanism of gamete fusion in further research on this subject.

Summary

The drug formula or material involved in above study is effective in altering the sex ratios in goats as is clear from the results obtained in this experiment. The rate of success in modification of sex ratios is tremendous. There is ample scope that this formulation when used on other mammals including humans will yield desired results only. This only leads to the derivation of one and only one conclusion that the first successful mammalian sex fixer of the world has arrived on the scene and is very much here to stay. The manipulation of mammalian sex ratio is not a dream or a mirage on the horizon (Short, 1982) now but a living reality and it cannot be ruled out that more and more of sex fixers meant for mammals will be discovered in the future to come.

References

Barrat, J. and Leger, D. 1979. Aveir des grossesses obstenues après stimulation de l' ovulation. A propos de 519 grossesses. *J. Gynecol. Obstet. Biol. Reprod. (Paris)* **8**: 332.

Beernik, F.J. and Ericsson, R.J. 1982. Male sex preselection through sperm isolation. *Fertil. Steril.* **38**: 493.

Bhattacharya, B.C. *et al.*, 1966. An attempt to determine the sex ratio of calves by artificial insemination with spermatozoa separated by sedimentation. *Nature* **211**: 863.

Boucheix, C. *et al.*, 2001. Tetraspanins. *Cell Mol. Life Sci.* **58**: 1189-1202.

Bryant, F. 1984. U.S. patent no.4, 448, 767.

Cho, C. *et al.*, 1998. Fertilization defects in sperm from mice lacking fertilins beta. *Science* **281**: 1857-59.

Coonrad, S.A. *et al.*, 1999. Treatment of mouse oocytes with PI-PPLC releases 70kDa (pl5) and 35 to 45 kDa (pl5.5) protein clusters from the egg surface and inhibits sperm oolema binding and fusion. *Dev. Biol.* **207**: 334-49.

Cormier, E.G. *et al.*, 2004. CD81is an entry coreceptor for hepatitis C virus. *Proc. Natl. Acad. USA.* **101**: 7270-74.

Corson, S.L. *et al.*, 1984. Sex selection by sperm separation and insemination. *Fertil. Steril.* **42**: 756.

Ericsson, R.J. *et al.*, 1973. Isolation of fractions rich in human Y sperm. *Nature* **246**: 421.

Gledhil, B.L. 1983. Control of mammalian sex ratio by sexing sperm. *Fertil. Steril.* **40(5)**: 572.

Gordon, M.J. 1958. The control of sex. *Scientific Am.* **199**: 87.

Hafez, E.S.E. 1982. Reproduction in farm animals. Reprint 5th edn. Lea and Fabiger, p. 499.

He, Z.Y. 2003. None of the integrins known to be present on the mouse egg orto be ADAM rreceptors are essential for sperm egg binding and fusion. *Dev. Biol.* **254**: 226-37.

Hemler, M.E. 2003. Tetraspanin proteins mediate cellular penetration, invasion and fusion events and define a novel type of membrane micro domain. Annu. Rev. Cell Dev. Biol. 19: 397-422.

Hunter, R.H.F. 1982. Reproduction of farm animals. Longman. London, pp. 138-39.

Lindahl, P.E. 1956. Counterstreaming centrifufation of bull spermatozoa. *Nature* **178**: 491.

Primakoff, P. *et al.*, 2000. The ADAM gene family: surface proteins with adhesion and protease activity. *Trends. Genet.* **16**: 83-7.

Sampson, J.H. *et al.*, 1983. Gender after induction of ovulation and artificial insemination. *Fertil. Steril.* **40**: 481.

Schroeder, V. 1941. Physica-chemical methods of sex regulation of progeny of mammals. Russian contributions. 1939 Genetics Congr. Am. Documentation Inst. Doc. 1565. *Abstr. In J. Hered.* **32**: 248.

Short, R.V. 1982. As quoted in reproduction of form animals by RHF Huntes. London. pp. 138-39.

Zavos, M and Dawson, K.A. 1991. Method for X and Y spermetozoa separation. U.S. Patent no. 4,999,283.

D. Sex fixing of progeny with monosodium ethanoate solution in ethanoic acid in dogs

Objectives

The objective of the study was to ascertain the sex fixing effect of monosodium ethanoate in ethanoic acid given by oral route in mammals particularly dogs. The dogs are an important member of family mammalian and they have a vital role in the social life of men. It is said that the dog is the oldest, first and most faithful friend of man and it has remained so since the times immemorial. Dog kennels and dog breeding is an important business also in many countries of the world particularly in Europe, America, Asia and Africa, Australia and almost all continents and places in the world. In many nations of the world the dogs are an important means of transport in the form of sledges, particularly in arctic countries. Dogs provide fur, leather and even their meat is eaten in some areas like north eastern India and they have a certain role in the socio aesthetics and socio economic status of men. So, an attempt to alter the sex ratios in dogs has a lot to do with the world of human sociological behavior and status.

Abstract

Sex fixing is a fascinating new concept in applied clinical pharmacology. Until now it was reported in certain ratites and lower reptiles. Of course, there have been numerous attempts to manipulate sex ratios in mammals but by and large, success has eluded them. Mammals are an important family and any success in producing desired sexed babies in them has considerable economic and social benefits. The present research is of a drug formula called 'aulprofem' which is a combination of sodium ethanoate and ethanoic acid in solution that when given orally to females of dairy animals, cows and buffaloes and other mammals such as dogs; yields female off-springs with near hundred percent success rate. The active constituents are therapeutically safe and pharmacologically compatible and do not exhibit any toxic or side effects. The various outcomes of this discovery may be a rapid increase in the numbers of females of high yielding milch varieties of dairy animals and thereby increasing milk production to unprecedented heights across the globe. Similarly, meat and wool productions can also be increased by increasing the overall numbers of meat and wool etc producing animals by producing more female off-springs of these animals with the help of this discovery.

Introduction

Mammals are the most dominant animals on earth today. So the production of sexed offsprings in them has marked economic and social implications. Many scientists have worked on this since the first half of the last century. There were mainly three lines on which such research was conducted. First were the attempts to alter the –pH of the female genital tract, second were the attempts to produce sexed semen (a sexed semen is one containing the accentuated proportions of either X or Y bearing spermatozoa) and the third were the use of certain chemicals, hormones, sera etc to produce a shift in the sex ratio. Recently much emphasis is laid on the attempts to understand the exact mechanism of sperm-egg fusion. Candidate molecules have been identified that act as receptors on the ZP (zona pellucida) or the ovum membrane and bind with possible sperm ligands. Along with this the differential *in vitro* binding of X and Y containing sperms with different antigen-antibody media based on the assumption of differential binding sites on them and their successive elutriation in accentuated proportions has gained attraction. The present research is an *in vivo* attempt to produce female offsprings where sex ratios are manipulated probably by binding of receptor sites for Y sperm on the ovum by the action of certain moieties that are generated in the living system by the action of the oral administration of a solution of sodium ethanoate in ethanoic acid given to the female animals at the time of mating. The important consequences of this may be rapid increase in milk, meat and wool etc production in India and the world over to unfathomable heights by preferential production of females of dairy and related fields and the great social and economic benefits and changes associated with this. In a projection when only 10 percent of the total dairy animal wealth of India is brought under the application of this technique for a smaller time period of only 10 years; the considering the normal national levels of conception failure, calf mortality, adult mortality etc, nearly 160 million tonnes of increase in national milk production can be achieved. These developments can be of good use in case of dogs which has by far the most faithful and dependable friend of humans from the prehistoric times and still continues to remain so.

Materials and Methods

The present technology is an *in vivo* technique. It is based on the assumption of the production of certain moieties in the living system by the action of administration of certain chemicals that can block the binding or fusion of a particular X or Y sperm receptor on the ovum. The study was conducted in the form of *in vivo* trials on healthy animals between 3 to 7 years in age in the dog kennels selected for this purpose. All the 12 animals formed the test and control groups of 06 animals each. The other materials included the formulation that comprised a solution of monosodium ethanoate 12.5 mg dissolved in ethanoic acid 1.45 M q.s. to make the total solution to 225 ml. The solution was administered to the animal 30 minutes after natural service. The administration was by a single dose of 100 ml of above preparation via the oral route to the female animal. The results were recorded and calculated in simple percentage methods.

Observations

Study was conducted on 12 female animals, divided into two groups of control and test animals. The animals in the control group were not given any treatment of the test material and the animals in the test group were given the material by the oral route, 30 minutes after mating. The control group animals reported no animals as not conceived and none of the animals underwent miscarriage. The 06 animals produced 26 off-springs, 13 males and 13 females. The animals in the test group also reported no animals as not conceived and no animal underwent miscarriage and 06 animals produced 24 offspring, 06 males and 18 females.

	No. ofA animals	nimals not conceived	No. of miscarriages	Offspring	
				male	female
Test	06	00	00	06	18
Control	06	00	00	13	13

Results

The animals in the control group yielded mixed offspring, 13 males *i.e.* 50.00% and 13 females *i.e.* 50.00%. The animals in the test group yielded 24 off-springs, 06 males *i.e.* 25.00% and 18 females *i.e.* 75.00% which clearly is the result of treatment given *i.e.* the 5.0 mg monosodium ethanoate dissolved in 1.75 ethanoic acid to make the solution to 100 ml. This is a considerable modification of sex ratio. So, it can be inferred that the above medication effects, in the form of alteration of sex ratios and produces female off-springs in dogs. Hence, the efficacy of the above said material to produce preferential progenies of female sex is well established in dogs.

Discussion

The drug is effective in altering the sex ratios in dogs as is clear from the results in table given above. The results of animals in test category show female off-springs with a 75 % success rate; where as the animals grouped under standard category show a mixed off-spring of female as well male puppies. This shows that where as the drug altered the sex ratios and that too to a considerable modification of sex ratios; the animals under similar conditions produced mixed off-springs in the absence of application of drug. So, the inference clearly, is the efficacy of drug to alter the sex ratios in dogs. A greater result that can be drawn from this experiment is that a mammalian sex fixer has arrived on the scene and manipulation of mammalian sex ratios is not an idealistic imagination but a reality in concrete terms and the possibility can not be ruled out that more sex fixers for mammals will be discovered in future.

Most of the animals that received the treatment delivered female offsprings. The failures can be attributed to manual error, spillage and faulty dose adjustment in case of heavier animals. Sex fixing in mammals has a long history. Many scientists have worked on this including Schroeder (1941), Bhattacharya *et al.* (1966), Ericsson *et al.* (1973), Gordon (1958), Lindahl (1956), Gledhill (1983), Corson *et al.*(1984), Beernik and Ericsson (1982), Barrat and Leger (1979), Sampson *et al.* (1983) etc. The major drawbacks of these techniques remained the lack of laboratory tests to evaluate the degree of sperm separation (Hafez, 1982) and the inability to know the precise mechanism of binding and fusion of mammalian sperm with ova. Overall, the manipulation of mammalian sex ratio has still remained a mirage on the horizon (Hunter, 1982). Recently there have been attempts to study the effects of certain compounds like tetraspanins like CD_9 (Hemler, 2003), CD_{81} (Cormier, 2004), glycosylphosphatidylinositol anchored proteins like CD_{55} (Coonrad, 1999), integrins (He *et al.*, 2003), disintegrins (Primakoff *et al.*, 2000), fertilins, cyritestin (Cho *et al.*, 1998) etc on the process of gamete fusion. These researches have clearly established the receptor sites on oocyte and their respective ligands on the surface of sperms. But the bigger questions are far from being answered. The 'tetraspanin-web' is still a big mystery (Boucheix *et al.*, 2001). The role of other candidate molecules the likes of which are mentioned above, needs a lot more investigating. Recently there have been attempts to differentially bind X and Y bearing spermatozoa using H-Y antigen antibody interactions using non protein substrata such as agarose beads and U.S. patents granted for them (Zavos *et al.*, 1991; Bryant *et al.*, 1984). These developments clearly demonstrate the existence of differential binding sites for X and Y sperms and this difference can be exploited in developing a viable technology for sex fixing in mammals, either *in vivo* or *in vitro*. The present technology is an *in vivo* technique and it can be hypothesized that there are differential receptors sites on the body of the ovum and their corresponding ligands on the bodies of the X and Y bearing sperms. A receptor site for X sperm does not receive the binding ligand of Y sperm as it does not correspond to its binding morphology and physiology and *vice versa*. The material of present research produces certain Y sperm binding ligand mimics (YSBLM) moieties in the living system and also in the genital system of the female animals undergoing this treatment in such a way that these YSBLM moieties differentially bind with receptor sites on the ovum thus occupying these sites and inhibit their subsequent binding and/or fusion with true binding ligands on the bodies of the Y sperms and this way over rule the chances of an XY zygote being formed. This leaves all the chances of an X sperm to come and combine freely with the X sperm specific receptor sites on the ovum and form an XX zygote. So, it can be inferred that the most appropriate candidate molecules that act as receptors on the oocyte like CD_9 and binding ligands on the sperm like fertilin beta are also X and Y sperm specific. So we can have tetraspanins like CD_9x and CD_9y and fertilins like fertilin beta-x and fertilin beta-y. The importance of this technology becomes even more marked with the view that it will not only provide an insight into the differential behaving and working of X and Y bearing sperms in the process of fertilization but will also definitely provide more valuable data and knowledge on the role played by the candidate molecules in understanding fully well the exact mechanism of gamete fusion in further research on this subject.

Summary

The drug formula or material involved in above study is effective in altering the sex ratios in dogs as is clear from the results obtained in this experiment. The rate of success in modification of sex ratios is tremendous. There is ample scope that this formulation when used on other mammals including humans will yield desired results only. This only leads to the derivation of one and only one conclusion that the first successful mammalian sex fixer of the world has arrived on the scene and is very much here to stay. The manipulation of mammalian sex ratio is not a dream or a mirage on the horizon (Short, 1982) now but a living reality and it cannot be ruled out that more and more of sex fixers meant for mammals will be discovered in the future to come.

References

Barrat, J. and Leger, D. 1979. Aveir des grossesses obstenues après stimulation de l' ovulation. A propos de 519 grossesses. *J. Gynecol. Obstet. Biol. Reprod. (Paris)* **8**: 332.

Beernik, F.J. and Ericsson, R.J. 1982. Male sex preselection through sperm isolation. *Fertil. Steril.* **38**: 493.

Bhattacharya, B.C. *et al.*, 1966. An attempt to determine the sex ratio of calves by artificial insemination with spermatozoa separated by sedimentation. *Nature* **211**: 863.

Boucheix, C. *et al.*, 2001. Tetraspanins. *Cell Mol. Life Sci.* **58**: 1189-1202.

Bryant, F. 1984. U.S. patent no.4, 448, 767.

Cho, C. *et al.*, 1998. Fertilization defects in sperm from mice lacking fertilins beta. *Science* **281**: 1857-59.

Coonrad, S.A. *et al.*, 1999. Treatment of mouse oocytes with PI-PPLC releases 70kDa (pl5) and 35 to 45 kDa (pl5.5) protein clusters from the egg surface and inhibits sperm oolema binding and fusion. *Dev. Biol.* **207**: 334-49.

Cormier, E.G. *et al.*, 2004. CD81is an entry coreceptor for hepatitis C virus. *Proc. Natl. Acad. USA.* **101**: 7270-74.

Corson, S.L. *et al.*, 1984. Sex selection by sperm separation and insemination. *Fertil. Steril.* **42**: 756.

Ericsson, R.J. *et al.*, 1973. Isolation of fractions rich in human Y sperm. *Nature* **246**: 421.

Gledhil, B.L. 1983. Control of mammalian sex ratio by sexing sperm. *Fertil. Steril.* **40(5)**: 572.

Gordon, M.J. 1958. The control of sex. *Scientific Am.* **199**: 87.

Hafez, E.S.E. 1982. Reproduction in farm animals. Reprint 5th edn. Lea and Fabiger, p. 499.

He, Z.Y. 2003. None of the integrins known to be present on the mouse egg orto be ADAM rreceptors are essential for sperm egg binding and fusion. *Dev. Biol.* **254**: 226-37.

Hemler, M.E. 2003. Tetraspanin proteins mediate cellular penetration, invasion and fusion events and define a novel type of membrane micro domain. Annu. Rev. Cell Dev. Biol. 19: 397-422.

Hunter, R.H.F. 1982. Reproduction of farm animals. Longman. London, pp. 138-39.

Lindahl, P.E. 1956. Counterstreaming centrifufation of bull spermatozoa. *Nature* **178**: 491.

Primakoff, P. *et al.*, 2000. The ADAM gene family: surface proteins with adhesion and protease activity. *Trends. Genet.* **16**: 83-7.

Sampson, J.H. *et al.*, 1983. Gender after induction of ovulation and artificial insemination. *Fertil. Steril.* **40**: 481.

Schroeder, V. 1941. Physica-chemical methods of sex regulation of progeny of mammals. Russian contributions. 1939 Genetics Congr. Am. Documentation Inst. Doc. 1565. *Abstr. In J. Hered.* **32**: 248.

Short, R.V. 1982. As quoted in reproduction of form animals by RHF Huntes. London. pp. 138-39.

Zavos, M and Dawson, K.A. 1991. Method for X and Y spermetozoa separation. U.S. Patent no. 4,999,283.

E. Sex fixing of progeny with monosodium ethanoate solution in ethanoic acid in pigs

Objectives

The objective of the study was to ascertain the sex fixing effect of monosodium ethanoate in ethanoic acid given by oral route in mammals particularly pigs. The pigs are an important member of family mammalian and they have a vital role in the economy of many countries of the world particularly Asia and Africa, Russia, China, India, and many European countries. In many nations of the world the pigs are an important source of meat, gelatin and fat and they have a dominant global role in food and pharma industry and the economy of the world. So, an attempt to alter the sex ratios in pigs has a lot to do with the world economy and health.

Abstract

Sex fixing is a fascinating new concept in applied clinical pharmacology. Until now it was reported in certain ratites and lower reptiles. Of course, there have been numerous attempts to manipulate sex ratios in mammals but by and large, success has eluded them. Mammals are an important family and any success in producing desired sexed babies in them has considerable economic and social benefits. The present research is of a drug formula called 'aulprofem' which is a combination of sodium ethanoate and ethanoic acid in solution that when given orally to females of dairy animals, cows and buffaloes and other mammals such as pigs; yields female off-springs with near hundred percent success rate. The active constituents are therapeutically safe and pharmacologically compatible and do not exhibit any toxic or side effects. The various outcomes of this discovery may be a rapid increase in the numbers of females of high yielding milch varieties of dairy animals and thereby increasing milk production to unprecedented heights across the globe. Similarly, meat and wool productions can also be increased by increasing the overall numbers of meat and wool etc producing animals by producing more female off-springs of these animals with the help of this discovery.

Introduction

Mammals are the most dominant animals on earth today. So the production of sexed offsprings in them has marked economic and social implications. Many scientists have worked on this since the first half of the last century. There were mainly three lines on which such research was conducted. First were the attempts to alter the –pH of the female genital tract, second were the attempts to produce sexed semen (a sexed semen is one containing the accentuated proportions of either X or Y bearing spermatozoa) and the third were the use of certain chemicals, hormones, sera etc to produce a shift in the sex ratio. Recently much emphasis is laid on the attempts to understand the exact mechanism of sperm-egg fusion. Candidate molecules have been identified that act as receptors on the ZP (zona pellucida) or the ovum membrane and bind with possible sperm ligands. Along with this the differential *in vitro* binding of X and Y containing sperms with different antigen-antibody media based on the assumption of differential binding sites on them and their successive elutriation in accentuated proportions has gained attraction. The present research is an *in vivo* attempt to produce female offsprings where sex ratios are manipulated probably by binding of receptor sites for Y sperm on the ovum by the action of certain moieties that are generated in the living system by the action of the oral administration of a solution of sodium ethanoate in ethanoic acid given to the female animals at the time of mating. The important consequences of this may be rapid increase in milk, meat and wool etc production in India and the world over to unfathomable heights by preferential production of females of dairy and related fields and the great social and economic benefits and changes associated with this. In a projection when only 10 percent of the total dairy animal wealth of India is brought under the application of this technique for a smaller time period of only 10 years; the considering the normal national levels of conception failure, calf mortality, adult mortality etc, nearly 160 million tonnes of increase in national milk production can be achieved. This is no short of a revolution in case of pigs also. There can be rapid increase in the number of female animals of this species resulting in much bigger herds of these animals with the end result of bigger production of meat, fat and gelatin for food and pharma industry.

Materials and Methods

The present technology is an *in vivo* technique. It is based on the assumption of the production of certain moieties in the living system by the action of administration of certain chemicals that can block the binding or fusion of a particular X or Y sperm receptor on the ovum. The study was conducted in the form of *in vivo* trials on healthy animals between 3 to 7 years in age in the pig farms selected for this purpose. All the 10 animals formed the test and control groups of 05 animals each. The other materials included the formulation that comprised a solution of monosodium ethanoate 12.5 mg dissolved in ethanoic acid 1.45 M q.s. to make the total solution to 225 ml. The solution was administered to the animal 30 minutes after natural service. The administration was by a single dose of

200 ml of above preparation via the oral route to the female animal. The results were recorded and calculated in simple percentage methods.

Observations

Study was conducted on 10 female animals, divided into two groups of control and test animals. The animals in the control group were not given any treatment of the test material and the animals in the test group were given the material by the oral route, 30 minutes after mating. The control group animals reported no animals as not conceived and none of the animals underwent miscarriage. The remaining 05 animals produced 40 off-springs, 19 males and 21 females. The animals in the test group reported none of the animals as not conceived and no animal underwent miscarriage and remaining 05 animals produced 38 offspring, 10 males and 28 females.

	No. ofA animals	nimals not conceived	No. of miscarriages	Offspring	
				male	female
Test	05	00	00	10	28
Control	05	00	00	19	21

Results

The animals in the control group yielded mixed offspring, 19 males *i.e.* 47.50% and 21 females *i.e.* 52.50%. The animals in the test group yielded 38 off-springs, 10 males *i.e.* 26.31% and 28 females *i.e.* 73.69% which clearly is the result of treatment given *i.e.* the 5.0 mg monosodium ethanoate dissolved in 1.75 ethanoic acid to make the solution to 200 ml. This is a considerable modification of sex ratio. So, it can be inferred that the above medication effects, in the form of alteration of sex ratios and produces female off-springs in pigs. Hence, the efficacy of the above said material to produce progeny o female sex is well established in pigs.

Discussion

The drug is effective in altering the sex ratios in pigs as is clear from the results in table given above. The results of animals in test category show female off-springs with a 73.69% success rate; where as the animals grouped under standard category show a mixed off-spring of female as well male piglets. This shows that where as the drug altered the sex ratios and that too to a complete modification of sex ratios; the animals under similar conditions produced mixed off-springs in the absence of application of drug. So, the inference clearly, is the efficacy of drug to alter the sex ratios in pigs. A greater result that can be drawn from this experiment is that a mammalian sex fixer has arrived on the scene and manipulation of mammalian sex ratios is not an idealistic imagination but a reality in concrete terms and the possibility can not be ruled out that more sex fixers for mammals will be discovered in future.

Most of the animals that received the treatment delivered female offsprings. The failures can be attributed to manual error, spillage and faulty dose adjustment in case of heavier animals. Sex fixing in mammals has a long history. Many scientists have worked on this including Schroeder (1941), Bhattacharya *et al.* (1966), Ericsson *et al.* (1973), Gordon (1958), Lindahl (1956), Gledhill (1983), Corson *et al.*(1984), Beernik and Ericsson (1982), Barrat and Leger (1979), Sampson *et al.* (1983) etc. The major drawbacks of these techniques remained the lack of laboratory tests to evaluate the degree of sperm separation (Hafez, 1982) and the inability to know the precise mechanism of binding and fusion of mammalian sperm with ova. Overall, the manipulation of mammalian sex ratio has still remained a mirage on the horizon (Hunter, 1982). Recently there have been attempts to study the effects of certain compounds like tetraspanins like CD_9 (Hemler, 2003), CD_{81} (Cormier, 2004), glycosylphosphatidylinositol anchored proteins like CD_{55} (Coonrad, 1999), integrins (He *et al.*, 2003), disintegrins (Primakoff *et al.*, 2000), fertilins, cyritestin (Cho *et al.*, 1998) etc on the process of gamete fusion. These researches have clearly established the receptor sites on oocyte and their respective ligands on the surface of sperms. But the bigger questions are far from being answered. The 'tetraspanin-web' is still a big mystery (Boucheix *et al.*, 2001). The role of other candidate molecules the likes of which are mentioned above, needs a lot more investigating. Recently there have been attempts to differentially bind X and Y bearing spermatozoa using H-Y antigen antibody interactions using non protein substrata such as agarose beads and U.S. patents granted for them (Zavos *et al.*, 1991; Bryant *et al.*, 1984). These developments clearly demonstrate the existence of differential binding sites for X and Y sperms and this difference can be exploited in developing a viable technology for sex fixing in mammals, either *in vivo* or *in vitro*. The present technology is an *in vivo* technique and it can be hypothesized that there are differential receptors sites on the body of the ovum and their corresponding ligands on the bodies of the X and Y bearing sperms. A receptor site for X sperm does not receive the binding ligand of Y sperm as it does not correspond to its binding morphology and physiology and *vice versa*. The material of present research produces certain Y sperm binding ligand mimics (YSBLM) moieties in the living system and also in the genital system of the female animals undergoing this treatment in such a way that these YSBLM moieties differentially bind with receptor sites on the ovum thus occupying these sites and inhibit their subsequent binding and/or fusion with true binding ligands on the bodies of the Y sperms and this way over rule the chances of an XY zygote being formed. This leaves all the chances of an X sperm to come and combine freely with the X sperm specific receptor sites on the ovum and form an XX zygote. So, it can be inferred that the most appropriate candidate molecules that act as receptors on the oocyte like CD_9 and binding ligands on the sperm like fertilin beta are also X and Y sperm specific. So we can have tetraspanins like CD_9x and CD_9y and fertilins like fertilin beta-x and fertilin beta-y. The importance of this technology becomes even more marked with the view that it will not only provide an insight into the differential behaving and working of X and Y bearing sperms in the process of fertilization but will also definitely provide more valuable data and knowledge on the role played by the candidate molecules in understanding fully well the exact mechanism of gamete fusion in further research on this subject.

Summary

The drug formula or material involved in above study is effective in altering the sex ratios in pigs as is clear from the results obtained in this experiment. The rate of success in modification of sex ratios is tremendous. There is ample scope that this formulation when used on other mammals including humans will yield desired results only. This only leads to the derivation of one and only one conclusion that the first successful mammalian sex fixer of the world has arrived on the scene and is very much here to stay. The manipulation of mammalian sex ratio is not a dream or a mirage on the horizon (Short, 1982) now but a living reality and it cannot be ruled out that more and more of sex fixers meant for mammals will be discovered in the future to come.

References

Barrat, J. and Leger, D. 1979. Aveir des grossesses obstenues après stimulation de l' ovulation. A propos de 519 grossesses. *J. Gynecol. Obstet. Biol. Reprod. (Paris)* **8**: 332.

Beernik, F.J. and Ericsson, R.J. 1982. Male sex preselection through sperm isolation. *Fertil. Steril.* **38**: 493.

Bhattacharya, B.C. *et al.*, 1966. An attempt to determine the sex ratio of calves by artificial insemination with spermatozoa separated by sedimentation. *Nature* **211**: 863.

Boucheix, C. *et al.*, 2001. Tetraspanins. *Cell Mol. Life Sci.* **58**: 1189-1202.

Bryant, F. 1984. U.S. patent no.4, 448, 767.

Cho, C. *et al.*, 1998. Fertilization defects in sperm from mice lacking fertilins beta. *Science* **281**: 1857-59.

Coonrad, S.A. *et al.*, 1999. Treatment of mouse oocytes with PI-PPLC releases 70kDa (pl5) and 35 to 45 kDa (pl5.5) protein clusters from the egg surface and inhibits sperm oolema binding and fusion. *Dev. Biol.* **207**: 334-49.

Cormier, E.G. *et al.*, 2004. CD81is an entry coreceptor for hepatitis C virus. *Proc. Natl. Acad. USA.* **101**: 7270-74.

Corson, S.L. *et al.*, 1984. Sex selection by sperm separation and insemination. *Fertil. Steril.* **42**: 756.

Ericsson, R.J. *et al.*, 1973. Isolation of fractions rich in human Y sperm. *Nature* **246**: 421.

Gledhil, B.L. 1983. Control of mammalian sex ratio by sexing sperm. *Fertil. Steril.* **40(5)**: 572.

Gordon, M.J. 1958. The control of sex. *Scientific Am.* **199**: 87.

Hafez, E.S.E. 1982. Reproduction in farm animals. Reprint 5th edn. Lea and Fabiger, p. 499.

He, Z.Y. 2003. None of the integrins known to be present on the mouse egg orto be ADAM rreceptors are essential for sperm egg binding and fusion. *Dev. Biol.* **254**: 226-37.

Hemler, M.E. 2003. Tetraspanin proteins mediate cellular penetration, invasion and fusion events and define a novel type of membrane micro domain. Annu. Rev. Cell Dev. Biol. 19: 397-422.

Hunter, R.H.F. 1982. Reproduction of farm animals. Longman. London, pp. 138-39.

Lindahl, P.E. 1956. Counterstreaming centrifufation of bull spermatozoa. *Nature* **178**: 491.

Primakoff, P. *et al.*, 2000. The ADAM gene family: surface proteins with adhesion and protease activity. *Trends. Genet.* **16**: 83-7.

Sampson, J.H. *et al.*, 1983. Gender after induction of ovulation and artificial insemination. *Fertil. Steril.* **40**: 481.

Schroeder, V. 1941. Physica-chemical methods of sex regulation of progeny of mammals. Russian contributions. 1939 Genetics Congr. Am. Documentation Inst. Doc. 1565. *Abstr. In J. Hered.* **32**: 248.

Short, R.V. 1982. As quoted in reproduction of form animals by RHF Huntes. London. pp. 138-39.

Zavos, M and Dawson, K.A. 1991. Method for X and Y spermetozoa separation. U.S. Patent no. 4,999,283.

PART - H

Test Dossiers

- Method of Analysis
- MSDS Data
- International Safety and Hazard Regulation Data

Method of Analysis

The product Aulprofem is analyzed as follows:

1. Description

A colorless, clear and completely transparent liquid solution is present.

2. Procedure no. I

PH value:

Calculate the pH value of the solution. It should range between 2.0 to 2.5.

3 Procedure no. II

Weight/ml of the preparation solution

Calculate the average weight/ml of the solution as per the following equation. Take 10 bottles selected on a random basis and measure the weight of the entire contents and the volume of the entire contents. Divide the gross weight of the contents by the volume of the contents as follows:

$$\text{Average weight/ml of Aulprofem solution} = \frac{\text{Weight of contents of 10 bottles}}{\text{Volume of contents of 10 bottles}}$$

$$= 0.95 \text{ to} 1.05 \text{ gm/ml} = (A)$$

4. Procedure no. III

Uniformity of the volume

Take 10 bottles and take and measure the contents of each and every bottle and then measure the total cumulative volume of all the bottles. Find out the bottles having the maximum volume and the bottles having the lowest volume in the above 10 bottles. Calculate the maximum deviation as follows. Note down the number of bottles deviating from the specifications mentioned in the monograph of the product.

$$\text{Maximum deviation (+Ve)} = \frac{(Vh - A) \times 100}{A}$$

$$\text{Minimum deviation (–Ve)} = \frac{(A - Vl) \times 100}{A}$$

Where,

A = Average volume of the bottles,

Vh = highest volume of the bottle in 10 bottles

Vl = Lowest volume of the bottle in 10 bottles

5. Procedure no. IV

Identification

The sample solution in the batch exhibits the same characteristics as that of the standard sodium ethanoate/ethanoic acid (Aulprofem) Reference Sample.

6. Procedure no. V

Assay

(90% to 110% of the label claim)

a. Assay for ethanoic acid

Take the Aulprofem solution in a conical flask and keep. Now take the standard 1N NaOH solution in the burette to the zero mark. Rinse a pipette with the Aulprofem solution and pipette out 25 ml of Aulprofem solution into a 250 ml conical flask. Add 2 drops of phenolphthalein indicator and stir. Titrate it with the standard NaOH solution as usual to the appearance of first permanent light pink coloration. Note down the reading of the burette and repeat the titration 3-4 times in order to get a concordant set of readings. Calculate the percentage of ethanoic acid in the sample according to the equality equation as given:

Each ml of 1 N NaOH = 0.06005 gm of ethanoic acid

V ml of 1 N NaOH = V × 0.06005 gm of ethanoic acid

According to normality equation: $N_1 \times V_1 = N_2 \times V_2$

$$N_2 = \frac{N_1 \times V_1}{V_2}$$

Where N_1 = Normality of NaOH,

V_1 = Volume of NaOH used

N_2 = Normality of Ethanoic acid,

V_2 = Volume of Ethanoic acid used

Strength of ethanoic acid in Aulprofem (S_1) = (Normality × Equivalent wt.) gms/liter

= (N_2 × 60.05) gms/liter

7. Procedure no. VI

b. Assay for Sodium ethanoate

Take the above solution neutralized by the NaOH solution and measure its volume and note it as Vs and titrate with 0.1 N sulphuric acid as told below. Take a standard solution of 0.1 N sulphuric acid in a burette previously rinsed with it to the zero mark. Then take the 25 ml Aulprofem solution obtained from the above mentioned titration and titrate with the standard sulphuric acid solution till the disappearance of pink color. Then calculate the normality of the sodium ethanoate solution (as the entire contents in the conical flask are now of sodium ethanoate only due to the prior reaction of Aulprofem solution with NaOH solution) according to the normality equation:

$$N_3 \times V_3 = N_4 \times V_4 \quad \text{or} \quad N_3 = \frac{N_4 \times V_4}{V_3}$$

Where, N_3 = Normality of sodium ethanoate,

V_3 = Volume of sodium ethanoate,

N_4 = Normality of sulphuric acid,

V_4 = Volume of sulphuric acid

Correlate this normality with the normality of the original 25 ml Aulprofem used in titration (a) as:

Correlated normality with respect to the original Aulprofem solution (N_5) = Resultant

$$\text{Normality } (N_3) \text{ after the titration} \times \frac{\text{Resultant voume after the titration (a) in 6}}{\text{Volume of Aulprofem taken } i.e. \text{ 25 ml}} = N_3 \times \frac{V_s}{25 \text{ ml}}$$

Strength of sodium ethanoate (S_2) = (Normality × equivalent weight) gms/liter = (N_5 × 41) gms/liter.

The strength of sodium ethanoate in original Aulprofem (S_4) = Strength of resultant sodium ethanoate S_2 – strength of sodium ethanoate equivalent to strength of ethanoic acid (S_3) in (a) of 6

$$S_2 - S_1 \times \frac{\text{Molecular weight of sodium ethanoate}}{\text{Molecular weight of ethanoic acid}} = S_2 - S_1 \times \frac{82}{60}$$

$= (S_2\text{-}S_3)$ gms/liter

$$\text{Where, } S_3 = S_1 \times \frac{82}{60}$$

Or Strength of sodium ethanoate in Aulprofem $(S_4) = (S_2\text{-}S_3)$ gms/liter

MSDS Data

Safety and Toxicity Data Sheet for Aulprofem

A brief data regarding safety and toxicity for Aulprofem is as given:

Flash Point: 324 Degree Celsius

Special Physical Criteria: The sodium ethanoate melts at about 100 degree Celsius to turn into a liquid. Being hygroscopic in nature, it forms a supersaturated solution of sodium ethanoate in water. This solution displays a particular property of cooling down its melting point to 58 degree Celsius in an exothermic reaction producing a sensation of hand warming type. At 120 degree Celsius the compound loses water of hygroscopic bonds to turn into anhydrous state. **Flash point is 324 degree Celsius.** The auto ignition temperature is 611 degree Celsius. This can only happen if it is spontaneously exposed into super thermo climates.

Toxicology

- **Toxicological information about sodium ethanoate is as given:**

  Toxicology study as available according to LD50/LC50 is as given:

  CAS# 127-09-3

  Draize test, rabbit, eye: 10 mg mild;

  Draize test, rabbit, skin: 500 mg/24 H mild

  Inhalation, rat: LC50 e" 30 gm/m /1H

  Oral, mouse: LD50 = 6891 mg/kg

  Oral, rat: LD50 = 3530 mg/kg

  CAS# 6131-90-4

- **Toxicological information about ethanoic acid is as given:**

  Acute oral lethal dose as calculated is:

  LD_{50} rat = 3310 mg/kg body weight

  LD_{50}, mouse = 4960 mg/ kg body weight

  Inhalation:

LC_{50}, mouse = 5620 ppm/1 hour

Dermal rabbit:

LD_{50}, rabbit = 1060 mg/kg body weight

Impression: 1 bottle of Aulprofem contains about 11.25 mg of sodium ethanoate and minus 10 % strong ethanoic acid. This comes out to be nearly 0.028 mg/kg body weight for sodium ethanoate and 59 mg/kg body weight for ethanoic acid for an animal of approximately 400 kilos. So, this is far, far safe as regards the safety criteria.

International Satety and Hazard Regulation Data

Additional Safety & Toxicity Data Sheet for Aulprofem

Chemical Name : Ethanoic acid, Sodium ethanoate
Use Type : Pharmaceutical aid, preservative, pH adjustment, sex fixer.

Detailed Toxicity Information about Aulprofem is as given

1. Acute Toxicity data is as given:

Acute Toxicity	**Ethanoic acid, Sodium ethanoate (combination)**	**Ethanoic acid (Parent chemical)**
WHO Acute Hazard	Not listed	Not listed
TRI Acute Hazard	Not listed	Not listed
MSDS	Not available	Not available
Acute rating from U.S., EPA product label	Not available	No consensus value
U.S. NTP Acute Toxicity Studies	No NTP Studies	Not Acutely Toxic to Moderately Toxic
Cholinesterase Inhibitor	No	No

2. Cancer Information data is as given:

Toxicity	**Ethanoic acid, Sodium ethanoate**	**Ethanoic acid**
IARC Carcinogens	Not listed	Not listed
U.S. NTP Carcinogens	Not listed	Not listed
California Prop 65 Known Carcinogens	Not listed	Not listed
U.S. EPA Carcinogens	Not listed	Not listed
TRI Carcinogens	Not listed	Not listed

3. Endocrine Toxicity Information is as given:

Toxicity	Ethanoic acid, Sodium ethanoate	Ethanoic acid
Illinois EPA list	Not listed	Not listed
Keith list	Not listed	Not listed
Colborn list	Not listed	Not listed
Benbrook list	Not listed	Not listed
EU list	Not listed	Not listed

4. Reproductive and Development Toxicity data is as given:

Toxicity	Ethanoic acid, Sodium ethanoate	Ethanoic acid
CA Prop 65 Development Toxin	Not listed	Not listed
U.S. TRI Development Toxin	Not listed	Not listed
CA Prop 65 Female Reprod. Toxin	Not listed	Not listed
CA Prop 65 Male Reprod. Toxin	Not listed	Not listed
U.S. TRI Reproductive Toxin	Not listed	Not listed

5. Toxicity data of special concern is as follows:

Toxicity	Ethanoic acid, Sodium ethanoate	Ethanoic acid
PAN Bad Actors list	Not listed	Not listed
PAN Dirty Dozen list	Not listed	Not listed

6. **Water Pollution Potential and Criteria** for Ethanoic acid, sodium ethanoate combination and ethanoic acid alone; both in the PAN Ground Water Contaminant Rating is insufficient and insignificant. This counts for the greater safety status and insignificant fears about its polluter capabilities

7. **Regulatory Information** for ethanoic acid, sodium ethanoate combination and ethanoic acid alone, both is that it is world wide registered for international regulatory status and there is not even a single country where this combination of ethanoic acid, sodium ethanoate both and ethanoic acid alone is banned.

8. U.S. and California Regulatory Status is as follows:

Registration	Ethanoic acid, Sodium ethanoate	Ethanoic acid
U.S. EPA Registered	Yes	Yes
U.S. EPA Hazardous Air Pollutant	Not listed	Not listed
U.S. EPA Min. Risk Pesticide (25b list)	No	No
CA Registered	No	Yes
CA Groundwater Contaminant	Not listed	Not listed
CA Toxic Air Contaminant	Not listed	Not listed

9. **Ecotoxicity data for Aulprofem is as follows:** As regards the aquatic and terrestrial toxicity of the ethanoic acid, sodium ethanoate combination and ethanoic acid alone; both is highly insignificant and that is the reason that no such data exists in the world about this.

10. **Ethanoic acid in its purest form** *i.e.* 100% strength can cause respiratory tract irritation and mild adverse effects on the central nervous system but certainly it is not lethal and in low concentrations, particularly when diluted to more than 5 times to its absolute strength; it does not cause such effects. In Aulprofem the dilution of ethanoic acid is plus 10 times to its absolute strength; so it can not cause such effects. Moreover it is in combination with sodium ethanoate. This gives the solution the nature of a buffer solution. This speaks in addition to the stability and safety of the solution.

11. **The auto-ignition point for sodium acetate is 611 degree Celsius** which is too high and more over this compound is in solution in water base. So, how can it catch fire? The compound is very stable under normal conditions of temperature and pressure *i.e.* 8 to 35 degree Celsius temperature and 1 atmospheric pressure.

Inference: The above data given is more than sufficient to prove the safety and stability and above all the zero hazard and risk levels connected with the solution of ethanoic acid, sodium ethanoate in combination and even the ethanoic acid as alone. The above facts have been taken from international standard sources and evidenced without partiality and prejudice and verified and given.

PART - I

Miscellaneous

- ✦ Publications
- ✦ Bits *n* Bytes

Publications

Aulakh, B.S. 2009. Procede de production *in vivo* de progeniture femelle de mammiferes. Brevet Canadien no. 2, 432, 172.

Aulakh, B.S. 2008. *In vivo* method for producing female offsprings in bovines. U.S. patent no. 7, 351, 581.

Aulakh, B.S. 2007. *In vivo* method for producing female offsprings in mammals. New Zealand patent no. 526489.

Aulakh, B.S. 2007. *In vivo* method for producing female offsprings in mammals. Australian patent no. 20011235973.

Aulakh, B.S. 2004. *In vivo* method for producing female offsprings in mammals. South African patent no. 203/4634.

Aulakh, B.S. 2008. Sex assignment and determination in dairy animals. *Pashudhan* **34(4)**: 7.

Aulakh, B.S. 2008. Preferential production of sexed progenies in cross-breeding to increase dairy productivity under commercial situations. Abstr. In compendium to national seminar on emerging opportunities for commercialization in dairying. NDRI Karnal, India.

Aulakh, B.S. 2008. A study in clinical pharmacology with world's first mammalian oral sex fixer drug formula 'aulprofem'. Abstr. In compendium to VIIIth annual conference of ISVPT. Mathura, India.

Aulakh, B.S. and Ali, A. 2008. The productivity enhancement through producing only female calf by binding Y sperm specific sites on ovum. Abstr. In compendium to international seminar on dairying. ILDEX 2008.

Bits *n* Bytes

Commercial applications of Aulprofem

Aulprofem is such a miraculous formula that is applicable in many animals and in this way various commercial advantages of it can be discussed as follows:

1. **Increased milk production:** As applicable to dairy animals Aulprofem can be instrumental in increasing milk production to unprecedented heights in India and worldwide. Just to discuss an example: India has an animal wealth of nearly 300 million dairy animals; cows and buffaloes (nearly 210 million cows and 90 million buffaloes). If a meager 1% of them are taken for the application of Aulprofem every year at a cost of Rs. 295 per dose, totaling to 88.5 crores of rupees. This will give rise to 9.0 lakh more female calves per year, keeping in view the nearly 35-40% national conception failure rate. Provided this phenomenon is continued for a minimum of 10 years, and keeping in mind 20% calf morality rate and 30% more general mortality; this will mean 50.40 lakh more female animals annually at the end of such an exercise. So, keeping in mind a conservative 8 liters of milk per day per animal on an average, this means 403.20 lakh liters of more milk per day. Calculated at a rate of rupees 20 per liter, this amounts to rupees 8064 lakhs (or 80.64 crores daily) or 29433.6 crore rupees annually, more in the GDP of India. This is an increase of 33258%. By no means, this is a smaller achievement! If this application is increased to say 10 percent, then the corresponding increases in GDP of India can be understood accordingly. On a simplistic estimate this means that we will be getting back Rs. 2, 94,336 crores back from Aulprofem in the form of increase in annual GDP of India by spending only a small amount of Rs. 885 crores per annum. This is not short of revolutionizing the entire agricultural economy of India!!
2. **Boost to meat production**: Similarly, on the lines of increasing milk production, there ca be effected a tremendous meat production

by selective, exotic breeding in case of meat producing species like pigs, goats, etc.

3. **Boost to wool production**: Aulprofem can effect major increase in wool production if it can be applied on sheep on a large scale nationally and internationally.
4. **Zygote labeling no more**: With Aulprofem in use, the expensive practices like zygote labeling and zygote discarding etc. will no more be needed.
5. **Developing exotic breeds of dogs, cats, horses etc**: The developing of exotic breeds of above mentioned and other related species of mammals can be made far more easier and practical with the help of Aulprofem.

❋ ❋

Market Scope and Potential

It is however not difficult to understand the market potential and reach of a product like Aulprofem but we will like to restrict ourselves to conservative estimates. To discuss a few points:

1. **Seiel and Amann estimate:** Seidel G.E. and Amann, R.P. in an article titled 'The impact of using sexed semen' ***(in Prospects for Sexing Mammalian Sperm. 1982. R.P. Amann and G.E. Seidel, Jr. (eds.). Boulder, Colorado. Colorado University Associated Press)*** has calculated the American market to be around $30 million a year for an economical discovery producing female sexed off-springs. The actual market can be a little more or less provided that these things can only be judged on the actual entry of such a product in the market.
2. **Indian Market:** India has a central dairy ministry and state dairy ministries in addition to the National Dairy Development Board (NDDB) which are well organized and massive organizations to manage its huge animal wealth of 280 million cows and buffaloes. They conduct about 1.5 to 2 crore AI's (Artificial Inseminations) every year. This simply means the existence of a market of a near 1.5 to 2 crore doses per year. Calculated at the rate of Rs. 500 per dose this amounts to a market of nearly 750 to 1000 crores of rupees per year. The usual rural and other dairy market is different and to add to this. So, the total market for a product like Aulprofem can be believed to be a considerable one. The markets for other animals like horses, sheep, goats, pigs, dogs, etc. is different. So, it can be inferred that the cumulative market for Aulprofem can

be considerable. The only question is to tap it efficiently and systematically.

3. **Off-shore Market:** Off-shore market that exists in Europe and other countries is huge and can be expected to run into hundreds of millions in dollars. The patent rights of Aulprofem are protected in U.S.A., Canada, Australia, New Zealand, South Africa and nearly 21 countries of Europe. So, a combined market for all of these countries can be expected to be a much bigger one.

❊❊

Aulprofem – Benefits to society other than commercial gains

Aulprofem is such a discovery that is bound to revolutionize human thought and imagination beyond consideration like any other great scientific discovery. Apart from providing commercial and economic benefits to society, it will have numerous other benefits as well:

1. **A new chapter in Biological Research**: Manipulation of mammalian sex ratios, behavior of chromosomes in sex determination in genotypic and phenotypic manifestations, advent and role of factors other than chromosomal ones in the processes involving sex determination and expression in embryos, fetuses etc., all being mega questions, will definitely be asking for newer explanations in the light of revelations of new type with the advent of discovery of Aulprofem on the scene. The scientific geniuses of tomorrow will be confronted with these questions and they will have to answer these queries. This will definitely open new vistas of scientific thinking, practice and applications and many new chapters in biological research will be opened.
2. **Victory of Femaleness**: Production of female sex in animals as well in humans with the help of Aulprofem definitely means nothing other than the victory of femaleness over male sex in human world dominated by masculine thought and practice. This will go a long way in establishing the long overdue respect for feminity and motherhood in the otherwise estranged world today where female ness is looked down upon and considered a drawback.
3. **Ensuring families of Choice-Man doing the Godly Act**: Aulprofem being a female sex fixer can be the best precursor to the production of a male sex fixer drug. This will enable almost every human inhabitor of the world to have a child of his choice

and as the new mindset and trend goes by almost across the globe, every human being will have off-springs of his choice; usually a boy and a girl and it will only be fools and sycophants who will produce more children or children of a particular sex and everybody will be happy having a boy and a girl – commonly called a complete family.

4. **Population Control**: As discussed in 3. if everybody can have a child of his choice *i.e.* a boy and a girl, then nobody will go on producing more children and population will be automatically controlled. Even after the discovery of a male sex fixer drug, which can be banned also; the male off-spring crazy people like the ones feared existing in India, Bangladesh or Pakistan etc. likely producing a lot of male children will automatically be acting as checks on population as males in turn cannot produce off-springs even if they are in great numbers.
5. **Revolutionizing Human thought and Mind**: Discovery of Aulprofem will definitely ignite the imagination and thinking of human mind worldwide like other great discoveries of last century like discovery of X-rays, discovery of Penicillin, Discovery of Radio-activity, discoveries of atomic spectrum etc.

Other Research Projects with us

Other than conducting research on Aulprofem we have been conducting research on certain other research projects at various degrees of progress and two of them need mention here as they have reached very advanced stages of development. They are as follows:

1. **Discovery of a GIT drug**: A drug very effective in gastrointestinal tract infections was discovered in our institute in 1996 and the series of experiments have proved its efficacy, consistency and accuracy. It is totally non-heavy metal and obviously does not produce bad metallic taste in mouth. The drug is very effective when administered alone and also in combination therapy with antibiotics like Norfloxacin, Ciprofloxacin etc. There is the residual spectroscopic analysis of this drug pending and we hope to complete it soon.
2. **Rabies Non-Vaccine Drug**: The experiments on a supper ambitious drug research project are underway with us. This research project gains special importance as it involves the development of a drug

therapy technique in virus related diseases as compared to traditional vaccine techniques which have proved not much effective against virus so far. This is definitely bound to open new fields of drug research against virus. Medical science of today has virtually proved a helpless non-performer against virus and this research will definitely open a new armored onslaught against virus. The world frightened to the core by diseases like AIDS, Hepatitis, Dengue and the likes, will be a relieved one when this research will be completed and humanity will then be very close to putting the ultimate war against virus to its most desirable finish *i.e.* the final victory over AIDS, Hepatitis and the likes. Definitely, this will be a true contribution to the spirit of medical profession, ethics and drugs research ranging throughout all the ages!

Cow Sibling Slaughter Prevention Campaign

World's biggest, beastardly and barbaric genocide is brewing just under our very nose. 10 to 15 million male calves are done to death in India every year because they are born in this world as males instead of being born as females and male calves are no longer needed in this era of farm mechanization. They are not liked by their dairy owner masters as they consume as their birth-right the commercial commodity milk from their mothers, the same milk that could be sold to fetch money. They are subjected to unbearable, torturous deaths, worse than the crucification in the following ways:

1. Luckiest amongst the killed calves are those who are suffocated to death just after birth to fetch more money for their nascent and un-infected skin to be used as comparatively costlier leather.
2. In the hot scorching winds of May-June, when the whole world rests in the cool environs of air-conditioners, coolers and fans, those new-born calves are pegged outsides in the burning sun, so that they are sun-stroked to death.
3. In the chilling cold of December-January, the calves are pegged out in the open throughout the long icy nights so that they are frozen to death by the before dawn.
4. Rests of the calves are subjected to the most heinous and un-tolerable death ever reported in history – the total starvation to death. Most of the calves are destined this tortuous end which sometimes prolongs to several weeks.

A society or a nation where such a cruel holocaust goes unabated does not deserve to be called a civilized society or a nation. The humans living in such a society definitely lose their precious right to be called humans.

So, it is one of the prime objects for us all involved in this magnificent project of Aulprofem to wage a relentless war against this un-natural, man's greed generated holocaust. A point here to note is that the metabolic steroid of oxytocin facilitates this holocaust as it becomes possible to milk the cow by injecting this biological poison to the animal at the time of milking. This hormone's metabolites even pass through the circulatory system of the animal to its milk and same milk is drunk by the humans, including children and adults, thereby poisoning their body systems to varying levels. This oxytocin injection was banned by the Indian government in the year 2002 but due to the all prevalent corruption in the drug control department and the drug inspectors occupied in the sole objective of minting money by corrupt means, that this nefarious injection is available in almost all drug and grocery stores throughout India. Hence, we all should strive to put this evil practice to end. This no doubt, will be a most pious cause if achieved.

❋ ❋

Nobel Prize for India

Nowhere in the entire history of mankind has there been a single anecdote that a great milestone applied research has been embedded with a sound commercial project. A research able to effect major changes in basic sciences and connected to a most noble cause of saving countless, innocent infantile cow siblings from dreadful clutches of death. Aulprofem is such a discovery and such a commercial concern! This can and will definitely rise to the level of a Nobel Prize, not because it has offered the most solid base for dethroning the mightiest Chromosomal Theory of Determination of Sex but also that it will lay the foundation for rapidest progress of nations that are poor and agro-based and have dairy and milk cattle to support their innumerable poverty ridden, half or total hungry masses. Let God bless them out of their poverty with the most nutritious diet on Earth; the milk, in the form of power of Aulprofem! And the same God bless India with its 2nd Nobel Prize in Science, yearning for which there already have passed a diamond jubilee of years, all in vain and thirst of the best glory of the world, *the Nobel Prize in Science.*

❋ ❋

About the Research–Aulprofem: Mending with difficulties

The discovery of the drug producing female off-springs in mammals was discovered in 1991 at the research farm of Gregor Mendel Institute for Research in Genetics at Gurdaspur, Punjab. Depending upon the highly remunerative and secretive nature of pharmaceutical research that today this field is, there were numerous hurdles in the way of this path breaking mega discovery. The main obstacle was Soviet style national scientific bureaucratic functioning that still prevails and dominates the scene in India and is the greatest hurdle in the way of genuine scientific research work that can be done in India. But fortunately, the mammoth project got underway and we were able to identify the active-constituents involved and patent was filed. This drug can increase the milk, meat and wool etc. production across the globe many times and revolutionize the agro based economies of most developing nations of the world. This will be the greatest achievement against poverty, hunger, mal-nutrition in a short span of time to come and a true contribution from a scientist from a country that is plagued with these problems.

❋ ❋

About us

Gregor Mendel Institute for Research in Genetics (GMIRG), is basically engaged in research in basic sciences including Genetics and Biology. Today being an era of Biotechnology, a special reference is laid on drugs research. So, it turned out that a set of experiments designed to evaluate some newer and simpler drug formulations yielded some wonderful results in the form of a drug formula that manipulated sex ratios in dairy animals with a near total success ratio. This miraculous achievement prompted us to carry out further schedules of experiments to verify this outcome and to ascertain certain other evaluations. This long series of experiments took nearly ten long and enthusiastic years to complete from 1991 to 2000 and by the end of year 2000, we were able to reach the levels of research up to ascertaining and establishing the active constituents involved. Hence, it became possible to file the drug patent on Dec. 15, 2000. The patent of this drug was filed at New York on Dec. 15, 2000 through the PCT route. The World Intellectual Property Organization, Geneva, Switzerland (WIPO) published this research on June 22, 2002, thereby limiting the international phase to Aug., 2003. Being a mega discovery, it was not easy to convince the patent offices in the world to get the final patents clear and granted. Fortunately enough, by the end of the year 2008; major patent offices in the world including USA, Australia, New Zealand, South Africa and Canada etc had granted the patents to this magnificent mile stone discovery in the field of reproductive physiology and applied biology. But fortunately enough, by the start of national phase, we were able to protect the patent rights in U.S.A., Australia, New Zealand, South Africa, Canada, India and Europe (nearly 21 countries including U.K., Germany, France, Italy, Spain, Portugal, Netherlands etc.). The drug is already in the market under the name, Aulprofem and steadily but firmly picking up sales.

PART - J

Charter of Fame

- Brevet Canadien
- U.S. Patent
- New Zealand Patent
- South African Patent
- Australian Patent

Office de la Propriété Intellectuelle du Canada
Un organisme d'Industrie Canada

Canadian Intellectual Property Office
An agency of Industry Canada

CA 2432172 C 2009/01/20

(11)(21) 2 432 172

(12) BREVET CANADIEN
CANADIAN PATENT

(13) C

(86) Date de dépôt PCT/PCT Filing Date: 2000/12/15
(87) Date publication PCT/PCT Publication Date: 2002/06/20
(45) Date de délivrance/Issue Date: 2009/01/20
(85) Entrée phase nationale/National Entry: 2003/06/16
(86) N° demande PCT/PCT Application No.: IN 2000/000125
(87) N° publication PCT/PCT Publication No.: 2002/047574

(51) Cl.Int./Int.Cl. *A61K 31/19* (2006.01), *A61D 19/02* (2006.01), *A61B 17/43* (2006.01), *A61K 9/00* (2006.01)

(72) Inventeur/Inventor:
AULAKH, BALWINDER SINGH, IN

(73) Propriétaire/Owner:
AULAKH, BALWINDER SINGH, IN

(74) Agent: BLAKE, CASSELS & GRAYDON LLP

(54) Titre : PROCEDE DE PRODUCTION IN VIVO DE PROGENITURE FEMELLE DE MAMMIFERES
(54) Title: AN IN VIVO METHOD FOR PRODUCING FEMALE OFFSPRINGS IN MAMMALS

(57) Abrégé/Abstract:
The invention provides an in vivo method whereby female offsprings can be produced in mammals, especially in bovine species, said method comprising the step of administering a therapeutically effective amount of a material comprising acetic acid, or its pharmaceutically acceptable derivatives to female mammals just after or before insemination

OPIC  CIPO

US007351581B2

(12) **United States Patent**
Aulakh

(10) **Patent No.: US 7,351,581 B2**
(45) **Date of Patent: Apr. 1, 2008**

(54) **IN VIVO METHOD FOR PRODUCING FEMALE OFFSPRINGS IN BOVINES**

(76) Inventor: **Balwinder Singh Aulakh**, House No. 2234/2C-29, Issanagari, Ludhiana (IN) 141 001

(*) Notice: Subject to any disclaimer, the term of this patent is extended or adjusted under 35 U.S.C. 154(b) by 611 days.

(21) Appl. No.: **10/450,353**

(22) PCT Filed: **Dec. 15, 2000**

(86) PCT No.: **PCT/IN00/00125**

§ 371 (c)(1),
(2), (4) Date: **Jul. 31, 2003**

(87) PCT Pub. No.: **WO02/47574**

PCT Pub. Date: **Jun. 20, 2002**

(65) **Prior Publication Data**

US 2004/0048806 A1 Mar. 11, 2004

(51) **Int. Cl.**
C12N 5/00 (2006.01)
C12P 7/54 (2006.01)

(52) U.S. Cl. **435/325**; 435/140; 424/400; 514/23

(58) **Field of Classification Search** 435/140, 435/325; 424/400
See application file for complete search history.

(56) **References Cited**

U.S. PATENT DOCUMENTS

3,219,525	A	11/1965	Berkow	
3,749,089	A	7/1973	Derr	
4,159,718	A *	7/1979	Bower	.... 604/212
4,336,801	A	6/1982	Sentell et al.	
4,999,283	A	3/1991	Zavos et al.	

OTHER PUBLICATIONS

The Vinegar Institute—Frequently Asked Questions at http://www.versatilevinegar.org/faqs.html, 2005.*
To et al., Gender Preselection, J. HOng Kong Med. Assoc., 46(4):331-336 (Dec. 1994).*

* cited by examiner

Primary Examiner—Jon Weber
Assistant Examiner—Tiffany M. Gough
(74) *Attorney, Agent, or Firm*—Rankin, Hill, Porter & Clark LLP

(57) **ABSTRACT**

The invention provides an in'vivo method whereby female offsprings can be produced in bovines said method comprising the step of administering a therapeutically effective amount of a material comprising acetic acid, or its pharmaceutically acceptable derivatives to female bovines just after or before insemination.

10 Claims, No Drawings

The Patent Act 1953

Patent Renewal Certificate

Letters Patent No: 526489

An in vivo method for producing female offsprings in mammals

in the name of

Balwinder Singh Aulakh

has been renewed to 15 December 2010

Payment of $340.00 was requested by:

John Service, M. J. Service & Associates, Suite 145, 63 Remuera Road, Newmarket, Auckland 1050

on 29 February 2008

Neville Harris
Commissioner of Patents, Trade Marks and Designs

205 Victoria Street, Wellington
PO Box 9241, Marion Square, Wellington 6141, New Zealand or DX SX 11129, Wellington
Freephone: 0508 4 IPONZ (0508 447 669), International callers dial: +64 3 962 2607, Fax: +64 4 978 3691, www.iponz.govt.nz

REPUBLIC OF SOUTH AFRICA

REPUBLIEK VAN SUID AFRIKA

PATENTS ACT, 1978

CERTIFICATE

In accordance with section 44 (1) of the Patents Act, No. 57 of 1978, it is hereby certified that

AULAKH, Balwinder; Singh

has been granted a patent in respect of an invention described and claimed in complete specification deposited at the Patent Office under the number

2003/4634

A copy of the complete specification is annexed, together with the relevant Form P2.

In testimony thereof, the seal of the Patent Office has been affixed at Pretoria with effect from the **24th** day of **November 2004**

Registrar of Patents

Australian Government

IP Australia

LETTERS PATENT

STANDARD PATENT

2001235973

I, Philip Spann, the Acting Commissioner of Patents, grant a Standard Patent with the following particulars:

Name and Address of Patentee(s):

Balwinder Singh Aulakh
House No.2234/2C-29, Issanagari, Ludhiana, 141 001, India

Name of Actual Inventor(s):

Aulakh, Balwinder Singh.

Title of Invention:

An in vivo method for producing female offsprings in mammals

Term of Letters Patent:

Twenty years from 15 December 2000

Dated this 2nd day of August 2007

Philip Spann
Acting Commissioner of Patents

PATENTS ACT 1990

Appendix – I

Persons

Abdul Kalam (1931): The scientist and president of India from 2002 to 2007. He was famous for his high principled and austere life style. He was a known advocate of genetically modified crops.

Adam Smith (1723-1790): The great economist, pioneer and founder of the theory of *'lassez faire'* in the national and international trade. His theory came under criticism by later great economists like Karl Marx and J. M. Keynes etc.

Ajim Premji (1945): The Indian version of Bill Gates. He was the richest person of India for many years and he, like Bill Gates earned fortunes from the business in software making in computers.

Akbar Jalaluddin Mohammed (1542-1605): He was the greatest of Mughal rulers in India and his quest for a son led him to seek the blessings of famous *Sufi* saint Khwaja Salimmuddin Chisti at Fatehpur Sikri

Aladdin: The famous character of Arabian nights. He was symbolized as a crusader for humanity against the evil forces of world.

Attila, the Hun (406-453): The famous general and ruler of Huns in the Eastern Europe and central Asia. He invaded Europe and north Asia and conquered and defeated many nations and countries. He invaded Rome also and was one of the forces, resulting in the ultimate fall of the great Roman Empire and its control passing into the temporal authority of church.

Augustus Caesar (63 BC-14 AD): The nephew of Julius Caesar who became king of Rome after him and made Rome the most powerful kingdom and monarchy.

Bal Thackray: A right wing Maratha Hindu regional separatist violence based political ideologue and leader who pursued an equivalent to Nazi ideology in the Maharashtra province of India. He was responsible for riots in Bombay in 1992 after the demolition of Babri mosque at Ayodhya which killed nearly 2000 people.

Becquerel Antoine Henry (1852-1908): The discoverer of radioactivity. He worked with Curies; Marie and Pierre in the discovery of Radium. He shared the Nobel Prize in physics with them in 1903.

Bhima: The famous strongman and friend and disciple of lord Krishna in the story of *Mahabharata*.

Bhindranwale Jarnail Singh: A right wing Sikh extremist ideologue and leader who pursued an equivalent to Nazi ideology in the Punjab province of India. He was killed in a military operation in 1984.

Bin Laden Osama (1957): The Saudi national and a billionaire businessman who funded and founded many a terrorist organizations in the world including the dreaded *Al-Queda* which bombed the World Trade Centre in New York in September, 2001.

Bismarck Chancellor (1815-1898): He is also known as 'the iron man of Germany'. He was the chancellor of Prussia and played an important role in the unification of German states under the leadership of king Wilhelm of Prussia.

Borlaug Norman (1914): The American scientist and the founder of green revolutions in the world. He was awarded a Nobel Prize for his contribution aimed at alleviating the humanity from horrors of hunger.

Bush George junior (1946-): The U.S. president who took the fight against terrorism up to foreign lands after the *Al-Queda* terrorists destroyed world trade twin towers in New York in September, 2001.

Chaplin Charlie (889-1977): The world famous comedian and intellectual. He was a very close friend of Joseph Stalin, the ruler of Soviet Union.

Chidambaram P (1945-): The finance minister under Manmohan Singh. He appeared helpless in front of the onslaught of inflation and economic recession that set in the world markets and of course in India in 2007 afterwards.

Curies: Marie Curie (1867-1934) and Pierre Curie (1859-1906) who along with Henry Becquerel worked on radioactivity and discovered Radium and shared the Nobel Prize in Physics with him in 1903. Marie Curie later got another Nobel Prize, this time in chemistry in 1911.

Darwin Charles (1809-1882): The famous natural history scientist and biologist. His evolution theories are still considered the greatest contribution that a single man can make towards the cause of scientific development and the liberation of human thought and mind from forces of conservatism and religious backwardness.

Dickens Charles (1812-1870): The great writer and literary genius. His writings gave a true picture of life in the pro and post revolution France.

Dracula Vladislav (IIIrd) (1431-1476): The ruler of Wallachia state in Romania from 1448 to 1452 and then from 1456 until his death. He was famous for the torment that he inflicted upon his people. Also known as Vlad, the impaler.

Edison Thomas Alva (847-1931): The greatest inventor in human history. It is said that he made one thousand inventions including the ones of electric bulb, cinema, gramophone, rubber, telephone bell and many others. He had poor background but he grew to be one of the most successful and richest persons of his times and he started and owned the mega multinational General Electric Company, popularly known as GE.

Einstein Albert (1879-1955): The great scientist and a scientific genius. He discovered the famous 'Theory of Relativity' that virtually laid the foundations of modern day Quantum Science, a concept that is very much different in nature and away from the conventional science particularly physics. He was awarded Nobel Prize in Physics in 1921.

Farid Baba (1173-1266): A great *Sufi* saint born in present day Faridkot in Pakistan which is named after him. He preached religious tolerance and utmost meditation and obedience to God.

Fast Howard (1914-2003): The celebrated writer and thinker who is one of the most influential and successful literary geniuses of the modern times. He wrote 'The Eternal Rebel' a novel that made him one of the heavy weight writers of all times.

Fleming Alexander (1881-1955): He was a laboratory technician in a government hospital in London. He discovered penicillin and founded the entire new branch of antibiotics. This is still considered as one of the greatest discoveries ever reported in the entire history of medical sciences. He was awarded Nobel Prize in Medicine in 1945 for this.

Freud Sigmund (1856-1939): The world famous philosopher and psycho-analyst. He is considered as one of the all time great men in human history. He put forward his famous theory that sex is at the root cause of all our thinking and actions and this is the most basic instinct of human personality.

Galileo (1564-1642): The great scientist and mathematician who believed and propagated in Copernicus' theory about earth's revolving around the Sun. He discovered telescope and used it to make valuable new discoveries about astronomy. He also discovered famous 'pendulum's principle'.

Garibaldi (1807-1882): A famous statesman and leader of Italy who along with Cavour fought and succeeded in unifying Italy and making it a nation state.

Gates Bill (1955) The international computer expert and entrepreneur who developed useful and easy to use software for the computers called Windows . He made billions in business and for many years was the number one rich person of the world.

Goebeles: He was friend and accomplice of Adolph Hitler.

Gorky Maxim (1868-1936): The famous author of the novel 'Mother' whose book is considered as one of the greatest literary achievement and master piece. His writings influenced the Russian youth in days of Soviet revolution.

Hegel (1770-1831): The famous philosopher and thinker whose famous principle on the materialistic approach to history and philosophy and his famous thesis of dialectical materialism attracted him immortal name and fame in the history of mankind.

Henry VI (1421-1471): He was the king of England who married six times but could not get a male heir to the throne.

Himmler: He was friend and accomplice of Adolph Hitler.

Hitler Adolph (1889-1945): The dictator of Germany who improved upon the political philosophy of fascism and raised it to the next level of Nazism which put the famous indoctrination regarding imperialism of Lenin into a permanent grave.

Iacocca Lea (1924-): The famous business and management icon and the general manager of Chrysler motor company in 1992 who led to the revival of the company's trade and reputation after the slowdown of 1991-92.

Janak: The king of Mithila state in ancient India who was famous for his principled and pious rule. He was father of Sita, the wife of lord Ram.

Kennedy John F. (1917-1963): The popular and progressive president of the United States who waged a relentless war against the apartheid KKK in U. S. and the international drug cartels and mafia. He was killed in 1966.

Kennedy Robert F. Jr. (1925-1968): He was the advocate general of United States under his elder brother John F. Kennedy who was then the president. Actually, it was Robert who spearheaded a relentless war against drug mafia. He was also killed when he tried to contest for the post of president of United States after the murder of his brother.

Keynes J.M (1883-1946): The great economist and founder of the strategy of middle path better known as the theory of mixed economy in economics. He wanted to wed free markets with limited state control for better welfare and upliftment of the common masses.

Khan Genghis (1167-1227): The great Mongolian general and ruler who invaded and conquered most parts of Asia, Europe and North Africa. He is supposed to have ruled the vastest tracts of land in the recorded history of mankind. He was known for his blood thirst, savagery and barbarism but he had a very good and efficient administrative skill and insight and he developed a very efficient and working administrative methodology and trade structure running throughout his vast empires.

Khomeini Ayatollah (1902-1989): The Shiite Muslim religious leader who became head of the state in Iran. He is supposed to have started the intolerant jihadist & extremist version of Islam to carry out political crusades.

Laski Harold J (1893-1950): Professor Harold J Laski was one of the leading figures in the development of modern thought and philosophy. He was an active political crusader and a central personality in the Labor Party of England. He influenced the thinking of young men in years before and after the World War II.

Lincoln Abraham (1809-1865): The famous statesman and president of the United States of America who played an important role to unify his country during the turbulent days of civil war in America. He fought to free the black Negroes from slavery and was shot by a racial maniac.

Louis Pasteur (1822-1895): The famous French scientist who invented rabies vaccine.

Mandela Nelson Rohilahla (1918-): The great freedom fighter and president of the republic of South Africa from 1994-1999. His name has come to symbolize the synonym of anti-apartheid. He led a legendary movement against the apartheid racial government in South Africa and freed the country from the yoke of racial white rule.

Maninder Pandher: The master of Surinder Kohli who was a partner in his crimes and was also arrested in 2006 and sentenced to gallows on February, 13, 2009 by a Central Bureau of Investigation Special Court in Ghaziabad. A full report appeared in The Tribune at Chandigarh on February, 14, 2009, Saturday.

Manmohan Singh (1932): The prime minister of India from 2004 to 2009. He was a highly principled and honest person. In spite to that he was a world class economist but he could not control the rising prices of daily need commodities.

Marx Karl (1818-1883): The famous German economist and political thinker. He is well known for his philosophy called Marxism.

Mendel Gregor (1822-1884): The father of genetics. He founded his theory about laws of inheritance. He was a poor person and a part time teacher in a church run children's school where he experimented on green peas that led him to discover the first and fundamental principles in heredity.

Milosevic Slobodan (1941): The president of Serbia from 1989 to 1997. He was responsible for the mass genocide of Muslim and other minorities in Serbia for which he was arrested in April, 2001 and handed over to the International Tribunal for Justice in Hague.

Modi Narendra Kumar: A right wing Hindu communalism based political leader who pursued an equivalent to Nazi ideology in the Gujarat province of India. He was the then chief minister and responsible for riots in Gujarat in 2002 after the train burning incident at Godhara railway station which killed nearly 2000 people.

Mor-Mukatdhari, Murali Manohar: These are just other names of lord Krishna.

Mukhtaran Mai: She was a teen aged girl which was gang raped at the behest of local tribal leaders in Pakistan. Later this episode got highlighted and brought tremendous shame and bad name for the government of General Pervez Musharref who was then dictator of Pakistan.

Munshi Prem Chand (1880-1936): The master story teller and father of Hindi novel. He wrote many a master piece short stories that touch the human heart in a unique moving way even today. He has carved for himself an immortal place in the history of Hindi literature.

Mussolini Benito (1883-1945): The Italian dictator who founded the principle of fascism and put the Lenin's principle regarding the imperialism being the final stage of capitalism, into a permanent eclipse.

Napoleon Bonaparte (1769-1821): The emperor of France from 1804 to 1815. He loved his wife Josephine very much but in order to get a male heir to the throne, he divorced her and married the Austrian princess.

Nasreen Taslima (1962): The famous writer of Bangladesh who suffered the wrath of fundamentalist conservative Islamic clergy due to her novel the 'Lajja'.

Neruda Pablo (1904-1973): The famous poet of Chille whose poetry inspired generations of men in entire Latin American countries. He was awarded Nobel Prize in Literature for his contribution in the field of literature.

Osho Rajnish Chander Mohan Jain (1931-1990): A pseudo-empiricist philosopher who also was a follower of Sigmund Freud. He made a great name, following and fortune in India and abroad. He loved to have the costliest of the cars, diamonds and other things of rare luxury. He also had the company of most beautiful and graceful women and he never made this fact a hiding. He was a brilliant writer, genius and a great orator and he attracted people towards him from all walks of life.

Peter the great (1672-1725): The Russian emperor who married four times in his old age to get a male heir after his two sons were killed in tragic deaths.

Raman C V (1888-1970): The great Indian physicist who discovered Raman Rays and Raman Effect for which he was awarded Nobel Prize in Physics in 1930.

Roentgen Wilhelm Conrad (1845-1923): The discoverer of X-rays. He was awarded the first Nobel Prize in Physics in 1901 for his discovery.

Ross Ronald (1857-1932): The Royal Indian Army officer and surgeon. He discovered the full life history of malarial parasite, the *Plasmodium vivax* and paved the further way for research in this regard. He was awarded Nobel Prize in Medicine for this in 1902.

Rhone: He was friend and accomplice of Hitler.

Rousseau jean Jacques (1712-1778): The famous French thinker and social and political ideologue who wrote the famous book 'Social Contract' which became one of the guiding forces and spirit of the French revolutionary struggle.

Russell Bertrand (1872-1970): A well known genius and thinker of modern times. He was awarded Nobel Prize in Literature in 1950.

Stevenson George (1781-1848): Son of a poor worker in a coal mine and a mechanic who developed the first railway engine and locomotive. He is the brain and inspiration behind the automobile revolution that overtook the world in later years at the end of nineteenth century and at the start of twentieth century.

Sultana Razia (1205-1240): The daughter of king Iltumish of Delhi who became Queen after him in 1236. Known as an able ruler and a good lover. She fell in love with a Negro slave and this led to her fall and death.

Surinder Kohli: The famous villain of Nithari episode in India. He along with his master Maninder Pandher, killed small children and young girls, raped and killed them and then ate their flesh. He was arrested in June, 2006 and tried and sentenced to gallows on February, 13, 2009 by the special court of Central Bureau of Investigation in Ghaziabad. A report appeared in the Tribune on 14 February, 2009.

Tagore Rabindranath (1861-1941): The famous Indian poet and writer. He was awarded Nobel Prize in literature for his book Gitanjali.

Thatcher Margaret (1925): The famous iron lady and British prime minister who crushed the Irish national movement with iron hand.

Voltaire (1694-1778): The celebrated French thinker and political ideologue whose writings influenced the French revolutionary struggle along with other writers such as Rousseau etc.

Watson and Crick: James Watson (1928), Francis Crick (1916-2004) who worked the complicacies of the double helix DNA model along with Frederick Wilkins (1916-2004) and shared the Nobel Prize in Medicine along with him in 1962.

Watt James (1736-1819): An U.S. engineer and scientist who made valuable discoveries and improvements on steam engines.

Wright brothers: Oliver Wright (1871-1948) and Wilbur Wright (1867-1912) who created history by making the first aero plane. They started their careers as ordinary bicycle mechanics and engineers but later became interested in making models for flying machines and succeeded.

Appendix – II

Places

Bukhari Sharief: A place in holy Gospels of Islam where the most reverend Prophet Hazrat Mohammed delivered the sermon on female hood.

Fatehpur, Sikri: A city in Uttar Pradesh province of India.

Kanya Kumari: The last corner of Indian shore where three oceans; the Indian Ocean, the Arabian Sea and the Bay of Bengal meet.

Kohima: A city in India in the heights of Himalayas beyond Imphal in Arunanchal Pradesh province.

Paatran: A small town in Patiala district of Punjab province in India where in August, 2006, the infamous human female infanticide episode took place.

Silicon Valley: The hub of international IT industry near Los Angeles in USA

Soviet, Union, U.S.S.R.: These were the names of the federation of the Russian states before the end of communist rule in 1991.

States of India: Haryana, Andhra Pradesh, Arunanchal, Maharashtra, Bihar, Karnataka, Madhya Pradesh, Uttar Pradesh, Himachal Pradesh, Punjab, Gujarat etc. These are all provinces in the republic of India.

Appendix – III

Names

Adam's breed: Here it means the entire humanity.

Al-Queda: The extremist outfit founded by the billionaire industrialist and Saudi Arabian national Osama-bin Laden. This is a dreaded terrorist organization that is supposed to have trained and financed the Taliban misadventure in Afghanistan. He was a close friend of Taliban leader and ruler of Afghanistan, Mullah Mohammed Omar. This organization carried the operation of destruction of twin towers of the World Trade Centre in the United States. After the war of Afghanistan, he fled to tribal areas of Pakistan-Afghanistan border and is still hiding and evading arrest.

BMW: The German car and automobile company

General Motors, Ford and Chrysler: They are the car maker automobile companies of USA.

Global Research. February, 2, 2008. Potential health hazards of genetically engineered foods. An article by Stephen Lyndman.

Hezbollah: Also known as the Mujahideen, they were right wing Islamic terrorists. They were patronized by various Muslim ruled governments like Colonel Nasser of Egypt, Col. Gadaffi of Libya, the government in Algeria, the Idi Amin of Uganda etc. Their modern day avatars are the Hamas of Palestine and Taliban of Afghanistan and Pakistan.

Hyundai: The Korean automobile company.

Indian Express. September 15, 2006, Friday. Health department officials seized at least 50 aborted fetuses from Sahib Hospital in Patran in Patiala district of Punjab province in India. On August 19, Anjali Bhawara, commissioner, Patiala range, was appointed the inquiry chief to conduct a full enquiry into this episode.

Khmer Rouge: This was a Norodnik anarchist movement which under the garb of Maoism and communism took control of the government in Cambodia. They let loose one of the bloodiest genocides of the entire history in Cambodia. Their leader Pol Pot is one of the most hated persons across the world.

KKK: The right wing anti Negro apartheid movement in the United States. After the killing of President Kennedy, this was suppressed by the U.S. government.

Legion, the commando regiment: this was the world's first professional elite commando unit under the command of Francisco Franco (1892-1975) and his predecessor Millan Astray also known as the Spanish Foreign Legion. This was so successful in African war in Morocco that it prompted other governments to raise commando regiments in their countries as well. Under Franco, its strength rose to an enormous of 8,000 troops from the earlier 2,000 troops.

Monsanto Imagine, Nov. 2005. Evolve new GM varieties: Kalam. There are other articles like the one in Financial Express of Nov. 10, 2006, titled, 'Tech-savvy Kalam advocates nano use, favours tronsgenic route for farm research'. Internet edition. And another article in The Hindu of Oct. 16, 2008, titled, 'He shows no signs of slowing down'. Internet edition.

Raymond's: The internationally reputed cloth and suiting brand.

Suzuki, Toyota, Honda: The Japanese car and automobile manufacturers.

Taliban: The modern day version of Hitler's philosophy of Nazism. The Taliban means religious Islamic students but in practice they are proponents of the hardliner militant and intolerant version of Islam and they brought countless misery on women, children and general masses in Afghanistan and Pakistan.

Tribune. August 6, 2003. There was a big story of the presence of pesticide and insecticide residues in the cold drinks and aerated waters of multinational companies like Coke and Pepsi.

Appendix – IV

Conversions used in the Book

1 million	=	10 lakhs
10 millions	=	1 crore
1 U.S. Dollar	=	50 rupees

Subject Index

Symbols

–pH 78
    values 79

A

Aberrations 104
Abject poverty 174
Abortions 54
Abstinence 112
Abstract 279
Acetate
    assimilation enzyme 274
    thiokinase 274
Acetic acid 266
Acetyl 262
    phosphate 274
Acetyl_P 274
Acetylation 239
    deacetylation 274
Acetyl-coA 262
    deacetylase 273
    synthetase suspension 301, 304
Acid
    condition 111
    white vinegar 112
Acidic contents 260
Acidifiers 259
Acidifying agents 258
Acrosomal vesicle 69
Acrosome 62
    reaction 63
Action
    reversal 88
    threshold 88
Active
    acetyl radical 262
    constituents 257
Acumen 239
Acute
    abdominal disorders 261
    injury 291
Adam 182
ADAM 61
Adam and eve 239
ADAM family 68, 85
Adam's breed 189
ADAM-1 68
ADAM-2 68
ADAM-3 68
Adaptation 128
Addict 237
Addiction 42, 237
Adenosine diphosphate 274
Adenylyl cyclase enzyme 62
Adolescents 129
Adrenaline 93
Adult mortality 281
Aesthetic finishes 252
Aesthetics 237
Agama agama 75
Agarose beads 80, 86
Ageing 111, 274
Agglutination 292
Aggravation 260
Agrarian 226
    country 182
    economy 181
Agricultural 182
    chemicals 186
    economy 133
    land 130
    produce 36
    universities 193
Agriculturists 187
Agro based 161, 197
AI 258
AIDS 139
    virus 139
Ailment 139, 142, 285
Albumin column separation 78
Alcohol 237
Alimentary tract 264
Alkaline
    baking soda 112
    condition 111
    phosphatase 285
        activity 285
        apparatus 285
Aluminium hydroxide 258
Amenities 208, 240
American 97
    negroes 97
Amino acids 273
Ammonium chloride 188
Amphiphilic 83

Ample scope 338
Amylase concentrations 308
Anatomical 142
Anchor 67
Ancillary 121
Anecdote 361
Anemia 147
Angiosperms 73
Anhydrous state 347
Animal
  houses 225
  husbandry 189
  keepers 197
  populations 108
  species 30
  wealth 176, 323
Animalia 73
Annals 203
Annelids 72
Annihilation duo 176
Annual
  employment 198
  epidemic 140
Anode 78
Anorectics 259
Antacids 258
Anti-acidity agents 260
Anti-biotic 18, 88, 261
  properties 139
Anti-body 79, 90
Anti-cancer drugs 274
Anti-carcinogenic 88
Anti-depressants 259
Antigen 266
  anti-body
    interactions 282
    media 280
Anti-histamines 147
Anti-malarial 88
Anti-septics 264
Anti-sera 79, 111
Appetite 119, 259
Apple 189
Applied
  clinical pharmacology 84
  sciences 151
Aquatic 102
  animals 183
Arable 151
  land 122
Arctic countries 328
Arid zones 102
Arthropods 73
Artificial 261
  insemination 79, 261
Aspiration 55
Assets 162
Asthmas 129
Athlete's breakfast 266
Atmospheric air 21, 192
Atomic
  organizations 84
  spectrum 359
ATP freeze dried powder 301
Auctions 221
Auto-ignition 347
  point 351
Avatar 207

B

Bacterial diseases 19
Balanced diet 185
Bane 127
Barbiturates 139
Barren land 239
Barrier 63
Basic
  diet 183, 184
  knowledge 150
  necessities 180
  sciences 149
Beast 244
Beef 183
Bees 147
Behavior 90
Benefactors 45
Beverages 259
Bicarbonate ions 62
Bimodal 80
Binding ligands 83, 85, 86
Bio-products 136
Bio-accumulation 146
  property 175
Bio-availability 257
Biochemical 102, 139
  chemistry 84
  equilibrium 90
  processes 268
Biochemistry 77, 84
  analyzer 287
Bio-film development 274
Biological
  amines 147
  exercise 134

existence 20, 192
expense 101
mutations 147
periods 201
phenomenon 191
poison 135, 361
research 191, 192
sciences 43, 51
systems 55
truth 89
Biologists 76
Biology 117
Bio-physical 53
Bio-synthesis 67
Bird flues 139
Birth 77, 79
ratio 80
right 360
Bisexual 73
Bisons 118
Black Friday 222
Blank
performer 87
sample 304
Blisters 260
Blockage 262
Blockers 268
Blood 84, 119
concentrations 272
pressure 129, 267
Body
fluid 84
immunities 196
syndromes 188
weight 257
Boom 221
time 209
Bovine species 21, 203
Brain 119
Brawl 155
Bread 75, 125, 159
improvement 162
grounds 129
programs 159
Breakthrough 50
Brethren 233
Brokerage 219
Brothels 244
Bt cotton 123
Bucket science 84
Buffaloes 157, 280
Buffer stocks 187
Buffering 89
reagents 258
Bull
semen 78
spermatozoa 78
Burette 344
By-product 119, 231
progenies 188

C

C. reactive proteins 290
Caenorhabditis elegans 67
Calcium 69, 128, 266
ion 63
oscillations 66
Calf 134
mortality 281
Caliber 41
Calibration 33
Camphor 188
Cancer 35, 123
Cancerous 37
Candidate molecules 64
Canonical 83
Capacitation 62
Capillaries 111
Capital 219
Capsule biosynthesis 274
Carbohydrate 67
Carbon dioxide 268
Carbonate reacted sperms 245
Carcinogenic 263
Cardio- 123
vascular 135
function 288
Carnivores 184
Cartels 237
Casino economics 210
Cathode 79
CD_{55} 68, 85, 282, 325
CD_{81} 85, 282, 325, 331, 337
CD_{9} 61, 85, 282, 325, 331, 337
Cell division 60, 65
Cell-cell 82
fusion 60
Cellsoft 80
Cellular functions 68
Cellulose 184
Cementing 241
Centrally acting drugs 259
Centrifugation 78
sedimentation 79

Cereals 184
Chemical 53, 54, 102
 death 127
 fertilizers 126, 132
 poisoning 182
 pollution 196
 reactions 231
 sulphates 126
 toxicities 127
Chemistry 44
Chemo-attractant 62, 93
Chicken 183
Children 129
Chills 121
Chlamydomonas 67, 73
Cholesterol
 levels 288
 oxidase enzymes 289
 standard reagent 289
 sterase 289
Chordates 75, 150
Chromatin
 matter 102
 structure 274
Chromosomal
 factors 102, 103
 part 138
 theory 97, 103, 106, 138, 149
Chromosomes 52, 65
Chronic 237
 disease 242
Ciprofloxacin 359
Cirrhosis 257
Citrate 262, 303
Citric acid 262
Civilization 122
Clinical 84, 112, 258
 applications 151
 co-relation 88
 invention 149
 pharmacology 279
 studies 78, 258
Clomiphene citrate 80
Cloning discovery 207
Clothing 125, 177
Cluster 64, 92
CNS 262
 function 290
Cocaine 42, 237
Cockroaches 73
Coelenterates 72
Coenzyme A 266
Cohesion 241
Coital facilitators 102
Coitus 62
Cold 299
Color 233
 blindness 139
Colorful 238
Colossal 203
 standing 44
Columns 79
Commercial activity 218
Commodities 21
Common cold 140
Communes 118
Complete family 359
Complexity 111
Composition patterns 84
Compounds 111
Computers 205
Conceived 26
Concentration 258
 threshold 87
 time curve 258
Conception 77
 failure 281
 rate 54, 111
 time 263
Concise 22
Configuration 77
Conformational change 82
Congenital 54, 149
 deformities 54
Conical flask 344
Conjugation 266
Connectivity 207
Conspicuous 238
Contaminant rating 350
Contamination 188
Contentment 24
Continents 206
Convection countercurrent galvanization 79
Copulation 73
Core 41
Cortical granule 66
 exocytosis 69
 reaction 69
Cough 299
Covalently 80
Cow 280
 herds 253
 progenies 162

Creatinine 287
- clearance 257
- concentration 287
- standard 287

Creation 59, 236
Creators 203
Creature 118, 233
Criterion 178
Crocodiles 118
Crops 131
- breeding 158
  - programs 158

Cross
- breeding programs 157
- fertilization 72
- over
  - action 89
  - attempt 89
  - window 89

CRP 290
- latex
  - antigen 291
  - reagent 291

Crusades 119
Cryo sleep 261
- preservation 54

Cubans 97
Cultivation 182
Culture
- media 54, 55
- mediums 129

Cumulus 63
- oophorus 62

Cyclic AMP 62
Cyclobenzaprine 259
Cyclones 177
Cyritestin 68, 85, 325, 331, 337
Cytological 78
Cytoplasm 62
Cytoplasmic
- continuity 82
- digesting 82

Cytosol 62, 63

D

Dagger toothed tiger 118
Dairy 41, 162, 193, 197
- animals 280
- based industry 193
- bovines 192
- cows 281
- diets 196
- farming 186
- industry 225
- meals 195
- outputs 250
- production 198
- profession 163
- sectors 225

Dance language 147
Data 206
- centers 206
- organizations 206

DDT 45, 147, 175
Death 129, 176
Deep penile penetration 112
Defects 54
Deformities 54
Degree 53
Deletion 68, 102
Delicate organs 260
Deliveries 258
Deluge 182
Demographic
- records 98
- surveys 106

Density gradients 79
Depolarization 65
Derivative technique 80
Designated territory 197
Designer 252
- baby 252
- ordering 85

Desired sexed offspring 106
Destitute 177
Deteriorated 141
Developing country 129
Deviations 102
Diabetes 129
Diabetic epidemics 188
Dialysis 257
Dictation 202
Diet 78
Dietary
- needs 186
- requirements 183

Diethanolamine buffer 285
Digestible 218
Digestion 102
Digestive
- system 148
- tract 184

Dihydrotestosterone 76, 110

Dinosaurs 34
Diploid 70
  cells 53, 111
Discharges 102
Discomfort 258
Discovery 43, 46, 122, 139, 280
Disease 122, 129
  free society 146
Disintegrins 85
  loop 68
Dismounted 261
Disposable
  applicator sticks 291
  plastic droppers 291
Distal leaflets 82
DNA content 79
Dog 329
  breeding 328
Domain 44, 54
Dose regimen 257, 258
Douche-nozzles 112
Draize test 347
Draughts 177
Drinking water 195
Drosophila 83
  myoblast fusion 67
  pseudo-obscura 102
Drug
  action 140
    modules 88
  cartels 42, 240
  chemical 140
  dependence 262
  formulation 55, 362
  mafia trade 241
  preparations 106
  research 140
  therapies 85
Dry
  fruits 239
  rations 180
Duf 67, 83
Dwarf 128
Dynasty lines 157

E

Earthquakes 30
Earthworms 73
EC_1 69
EC_2 69
Echinodermates 72
Eco system 32
Ecological
  balance 151
  disaster 151
Economic
  activity 227
  benefits 195, 335
  recession 222
Economical discovery 357
Eden 176
EGF 67
Egg 59
  yolk 79
Ejaculated 261
Electrical 53
Electrolytes equilibria 89
Electronics 44
Electrophoretic 79
Elimination 257
Elucidation 84
Elutriation 280
Emancipator 42
Embryo 54
  labeling 43
  transfer 43
Embryonic
  death rates 100
  states 99
Embryos 111
Embryotoxic 263
Employment guarantee plan 198
Endocrinology 84
Endogenous 273
Endurance 128
Energy 227
Enrichment 79
Enterprise 220
Entity 127
Entrant 49, 141
Entrepreneurs 208
Environment 30, 129, 183, 186
  damage 188
Environmental 102
  damage 146, 186
Environmentalists 189
Environs 55
Enzymes studies 307
Epidermal growth factor 67
Epilepsy 129
Equatorial 64
Equilibrium 15, 89

Equipments 231
Era 117
Erythrocytes 273
Ester 262
Estradiol-17 beta 76, 110
Eternal 23
Ethanoate 87
Ethanoic acid 279
Eukaryotes 274
Euphoria 237
Eureka 23
Evil creature 30
Evolution 13, 43
Evolutionary 20, 91, 128
Excretion 102, 257
Exhaustion quotient 54
Exocytosis 60, 63
Exogenous 273
Exotic 158
  breed 357
    stock 55
  breeding 157, 357
    programs 158
Expensiveness 111
Expiry criteria 87
Extinct 33, 35
Extra
  cellular
    domain 64
    glycoprotein 67
    loops 69
    matrix 62
    space 69
  mitotic spindles 65
Eye 125
  disorders 129
Eyesight 125

F

Fabric 117
Factors 106
Fair gender 200
Famines 51, 177
Farm 118
  community 193
  lands 151
  scientists 187
  sector 187
Fascism 230
Fat 334
Fatal 258
Fatality 264
Fatigue 54
Fatty acids 274
Fauna and *flora* 151, 196
F-body 78
Feed productions 225
Felines 184
Female
  calf 134
  calves 162
  folks 200, 203
  genital tract 112
  off-spring 77
  orgasm 112
  sexed off-spring 150
  world 201
Femaleness 194, 202
Feminine 202, 252
Feminity 358
Femur 45
Ferns 34
Fertile
  land 239
  soils 186
Fertilin 64
  beta 68, 85, 86
Fertilins 61, 85, 325, 331, 337
Fertility 68
  enhancers 102
Fertilization 66, 338
Fertilizers 36
Fetal
  formation 263
  stages 100
Fetotoxic 263
Fetuses 100
Fiasco 124, 222
Field trials 112
Filtrate 79
Financial institutions 212
Fish 183
Flagellar biosynthesis 274
Flesh 184
Flocks 323
Flow
  cytometry 79
  sorting 79
Fluctuations 98, 103, 104
Fluid 79
Fodder 132
  crops 37, 189
Foliage 157

Follicle cells 62
Food 125, 272
 crisis 133, 187
 front 187
 grain 184
  production 186, 187
 scarcity 187, 189
 security 132
Forced natural death 162
Forefathers 84
Foreign exchange 194
Forest trees 157
Formulative
 factors 106
 theory 106
Fortified 105
Fractions 14
Fragment 236
Fruits 123, 183
Function
 activity 111
 defect 67
 environment 111
Fundamental discovery 149
Fungi 73
Fur 328
Fus1 67
Fusion 43, 59
 agents 68
 event 82
 peptide 83
Fusogenic 64

G

G.M. crops 131
Gallery 90
Gamete 63
 fusion 67, 85, 283, 337
Gamma globulin 297
Garb 232
Gas exchange 148
Gastric tube 260
Gastrointestinal tract 359
GDP 19, 191, 226
Gelatin 334
Gene
 knockout 68
 manipulations 123
 mutation 147
 research 143
Generation 24, 59, 159
Generator 119
Genes 102
Genetic 102, 123, 143
 balance 191
 code 149
 constitution 53, 111
 disorders 123
 dwarfs 182
 engineering 123
 factors 102
 fuel 130
 manipulation 123
 modification 131
 mutant 132
 mutations 132
 research 147
 tempering 111, 188
 transfer 102
Genetically
 modified 131
 mutant varieties 186
Genital 133
 instrument 251
 organs 55, 102, 252
 parts 252
 system 282
 tract 54, 102
Genocide 234
Genre 202
Geographical 21, 232
 habitats 151
Germination 133
Germplasm 102
Gigantic 21
Girl child 200
GIT 261
Glands 102
Global warmings 201
Globe 31, 98
Globulins 111
Gloom 222
Glucose 307
Glues 231
Glut 228
Glycerol 79
Glycine 79
Glycogen 266
Glycoprotein 62
Glycosylphosphatidylinositol 61
 anchored proteins 85, 282
Goats 192, 322
Gold 239

Golden triangle 238
Gonadotrophins 80
Gory 121
Gospel 24
GPI 67
Grains 183
  productions 190
Gram
  negative bacteria 147
  positive bacteria 147
Great nation 174
Green
  missiles 130
  revolution 45, 51, 122
Gross domestic produce 229
Groundwater 126
Gymnosperms 34

H

Habit forming 262
Habitat 107
Habitual 206
Half lives 257
Haphazardness 139
Haploid cells 52, 110
Haploid pronuclei 70
Harmone 84
Harmony 233
Harvests 210
Hbs 67, 83
Health 334
  hazards 132
  prophylaxis 179
Heart 119
  diseases 129
  rate 267
Hectares 122
Hemophilia 139
Hepatic
  function 262
  impairment 259
  malfunction 262
Hepatitis C virus 68
Herbicide 37, 123, 128
Herbs 128
Hereditary 149
Hermaphrodites 72
Heroin 237
Hetero nature 93
High yielding
  quotient 162
  traits 162
Hinge like motion 82
Histological 147
Holocaust 135, 361
Homo
  nature 93
Homo sapiens 118, 137
  septens 178
Homologous 274
Horses 192, 313
Hot temperate zones 102
Housing 177
Hues 233
Human
  body 119
  data 106
  demography 172
  flesh 132
  lives 176
  psychology 241
  race 185
  species 106
  subjects 297
  volunteers 295
Humanity 128
Humerous 45
Humid 102
Hunger 51, 172, 174
  masses 250
H-Y antigen 110
  antibody 85
    interactions 331
Hyaluronidase 62
Hybrid dwarf varieties 127
Hydrogen bonds 83
Hydrolytic enzymes 64
Hydrophobic region 64
Hygienic 22
Hygienically potable grades 179
Hygroscopic 347
  bonds 347
Hyper
  osmotic medium 78
  polarized 62
Hypersensitivity 259
Hypnotic 139
Hypo
  osmotic medium 78
  price
    effect 213
    syndrome 213
Hypothetical 268

I

ICAR 129
Ice ages 201
Ig 83
Ig super family 68
IgA 296
IgE 296
IgG 296
IgM 296
Immobilized 79
Immune response 89
Immunity 179
Immuno
  absorbent 80
  separation 79
Immunoglobulin 67
Immunological reaction 87
Immunology 77
Impaired renal 262
Impairment 68
In vitro
  binding 280, 329
  environment 54
  technologies 52, 55
In vivo 13, 83
  attempt 317, 329
  methodology 55
  technology 52
Inca 40
Incinerated 123
Incinerators 120
Income levels 250
Incubated 25
Industrial
  revolution 222
  slowdown 208
Infantile cow siblings 361
Infants 79, 120
Infections 147
Infinite 27
Inflammation 291
Influx 63
Information technology 205
Ingredients 152, 158
Inhabitations 178
Inherent 45, 54
Inorganic and organic domains 143
Insect
  transmitted diseases 175
  vector diseases 147
Insecticide 21, 123, 147
Insects 34
Insemination 55, 110
Instincts 117
Institutes 127
Instruments 252
Insulin 147
Integrins 85, 331, 337
Intellectuals 222
Intelligentia 189
Intensive
  agriculture 189
  farming 189
Intercourse 112, 264
Internet 206
Intersexes 14
Intestines 184
Intoxication 42
Intra cytoplasmic 69
Intracellular exocytosis 63
Intranet 206
Introduction 279
Investigating research 70, 84
Ions 89
Irradiation 111
Isoelectric focusing 79
Isotonic saline 307
IT
  boom 207
  revolution 205
IVF (*in vitro* fertilization) 54
IVF techniques 143

J

Jeopardy 139
Juices 184

K

Kennels 328
Ketone bodies 273
Kidney failures 188
Killer diseases 188
Killing powers 176
Kinetic 81
  motion 24
Kingdom *animalia* 202
Kiwis 76
Knees 232
Krieg 39, 119

L

L malate dehydrogenase/citrate
    suspension 301
Label claim 344
Labor 263
    time 263
Laboratory
    glass pipettes 287
    preparations 258
    tests 53
Lactating 157, 264
Laissez faire 119
Land 122
Latex slide 290
Lay-offs 227
LC_{50} 347, 348
Leakage 264
Leather 328
Lesions 299
Lethal 141
    dose 347
Lethargy 92
Leucoderma 129
LGBS 37, 123, 139
Life 91
    calendar 129
    conditions 175, 177
    giving 27
    spans 175
    styles 107
Ligands 85, 331
Light treatment 147
Lipid 62, 64
    anchor 67
    mixing 82
Liposomes 82
Lips 260
Liquidity 227
Livable 22
Live
    births 80
    system 91
Livelihood 223
Liver 36, 257
    cirrhosis 259
    function 285
    malfunctions 188
    problems 196
    therapy 147
Livestock
    circles 281
    disciplines 260
    industry 225
    sectors 225
    world 253
Living
    conditions 174
    organism 90, 91
    pastures 35
    systems 90, 179
Lizards 75
Localized adventures 189
Lock and key mechanism 82
Lungs 148
    function 299
Lupus vulgaris 147
Lush green weeds 238

M

Macro levels 175
Macrophage 68
Mad cows 139
Mafiosi 244
Magnesium 128
    chloride 285
    hydroxide 258
Magnetic resonance imaging 147
Magnitude 120
Malaria 45
Male
    calf 134
    calves 161
    mortality 108
    off-spring 77
Malnutrition 45, 250
Malonate 274
Mamba 235
Mammalian
    fertilization 62
    myoblast fusion 68
    off-spring 113
    sex
        fixer 133, 138, 332
        manipulation 138
        ratio 55, 326, 337
    vaginas 112
    world 203
Mammals 106, 150
Mammothness 21
Mammoths 118
Maneuver 54, 157
Manipulation 55, 80, 337

Mankind 120, 175
Manual 54
  error 282
Manufactory 231
Marginal farmer 161
Marginalized farmers 197
Masculine 252
Mass information 208
Materials and methods 279
Mating 258, 263
Maya 40
MDR malaria 139, 142
MDRTB 139, 142
Meat 31, 50, 179
  diet 184
  industry 108
  processing
    industry 194
    species 193
  production 193
Mechanism 59, 84, 105
Mechanized development 162
Media 54, 111
Medical
  field 146
  health 179
  help 178
  history 258
  profession 360
  science 146
Medicinal
  research 142
  science 44
Mega crash 222
Meiosis 62
Meiotic spindle 64
Membrane
  bilayers 83
  potential 62, 66
Menace 42, 234
Meningitis 291
Mental
  bias 180
  health 90
Merger 59
Meshings 184
Mesmerizing 35
Messiah 189
Metabolic
  cycles 262
  harmone 134, 135
  pathways 135
Metabolism 62, 135
Metabolites 135, 257
Metallic taste 359
Metamorphosis 120
Meteoric 234
Methodology 14, 54, 55, 79, 84, 110, 215
Methyl-iso-cyanide 120
Mice 132
Micro liters 84
Microbes 122
Micropipettes 287
Microvillar
  free region 64
  rich region 64
Microvilli 64
Midas touch 239
Migraines 129
Migration 111
Milch
  animals 198
  cattle 156
  females 159
  varieties 20, 280
Milk 50, 126
  machines 157
  production 21, 46, 134, 189
  signs 258
  yields 225
Mimics 86
Mineral 128
  absorption 267
  wealth 31
Ministries 237
Minuscule 41
Mirage 14, 338
Mire 52
Miscarriage 54, 281, 314
Misery 231
Missionary position 112
Mitochondria 93
  enzyme 273
  signaling 274
Mitotic spindle 70
Mobile telephony 207, 222
Mobility 104
Model 258
  DT60 307
  DTE 307
  DTSC 307
Modified 105
  theory 105
Modus operandi 161

Moiety 68, 85, 280
Molecular science 84
Monoclonal antibody 80
Monograph 343
Monophage 68
Monosodium ethanoate 281
Morphine 237
Morphology 77, 79, 112
Mortality 98
  data 98
  equation 175
  rates 98
  sex ratio 100
Mother
  earth 188
  nature 143, 203
Motility 62, 64, 78
Mouth
  burns 260
  cavity 265
Mucous membrane 260
Mud 178
Multifaceted 231
Multiple births 80
Multipolar 65
Murites 128
Muscle tones 102
Mutagenic 263
Mutant 182
Mutating virus 141
Mutation 35, 102
Mutton 183

N

N and C termini 69
NADH 300
NaOH solution 345
Narcotic
  mafia
    syndicates 241
    trade 245
  production 240
  trade 237, 240
National
  asset 226
  developments 225
  economy 157, 218
  investment 226
  milk production 281
  money 226
  prestige 194
  saving 226
Nations 225
Native traits 157
Natural 261
  ally 244
  environments 129
  equilibrium 213
  farming 190
  forces 128
  habitats 35
  insects 128
  job 136
  selection 33, 35, 74, 101, 128
  service 335
  strength 128
  system 128
  weeds 128
Naturalists 189
Nausea 258
NDDB 357
Nectar 136
Negative control 291
Neo-borns 129
Neo-natal 183
  septicemia 291, 292
Neo-plasm 263
Neuro 123
Neuroleptics 259
Neurological 142, 147
  disorders 37
Niche 32, 155
Niche 32
Nicotinamide-adenine-dinucleotide 300, 303
Nitrogen 128, 183
  assimilation 274
Nitrous oxide 147
Nobel prize 127
Non-chordates 75
Non-diploid cells 65
Non-protein
  diluents 266
  substratum 79
Non-vegetative foods 184
Norfloxacin 359
Nostrils 24, 121, 260
Nourishers 203
NSAIDs 88
N-terminus 83
Nutrition 134, 179
Nutritional
  diet 179
  food 189

ingredients 183
qualities 158
values 157
Nutritionists 185
Nutritious diet 179

O

Obstacles 62
Oceans 27
Odorants 92
Off-springs 50, 258
Oilseeds 123
O-linked oligosaccharides 63
Omnipresent 52, 162
Onions 123
Oocyte 62, 81, 85, 149
Oolema 83, 85
OPD testing 299
Opioids 259
Opium 237
Optic 37, 124
Optical system 295
Optimum 55
Oral 55
dose 257, 258
estrogen pill 137
route 260
Ordering 252
Organic 55, 84
evolution 35
farming 186
live matter 34
species 34
Organism 33, 72
Organization 82
Organized 76
chromatin 110
Organs 135
Orientation 69, 84
Oscillations 15
Osmo-regulation 274
Ostriches 76
Ova 55, 97
Overdosage 264
Overdose 264
Oviducts 245
Ovulated egg 62
Ovulation 112
monitors 112
Ovum 43, 62, 83
Oxaloacetate 300, 303
Oxidative dismutation reaction 274
Oxygen 205
content 148
Oxytocin 134
injection 134

P

Pain threshold 88
Pangs 250
Papaya 75
Parameters 53, 87, 142
Parasympathetic nervous systems 262
Pastures 141
Patent 76, 362
office 76, 363
rights 363
Pathogenicity 274
Pathway 273
Patriotism 194
Peak 257
levels 272
Pedestal 44
Penicillin 122, 147
Peptides 68
Percolate 46
Performance 252
Performing health 54
Peri-buccal areas 260
Peripheral tissues 266
Peri-urban areas 197
Perivitelline
cavity 69
space 64
Perspective 157
Perspiration 102
Pesticide 21, 123
Pests 128
Pets 80
pH adjustment 349
pH value 343
Pharma
industry 334
sectors 157
Pharmaceutical
aid 258
compounds 139
environment 268
formulations 106
Pharmacologically compatible 280
Pharmacology 279
Phenolphthalein indicator 344
Phenomenon 73, 117

Phenotypically 252
Phosphate 37, 128, 274
Phosphatidylinositol specific
  phospholipase 67
Phyllogenetic 33
Phylogenic history 91
Physical 102
  health 183
  merger 81
Physics 44
Physiological 102, 111
  research 143
Physiology 77, 85
Picric acid 287
Pig farms 253
Pig-A 67
Pigs 192, 334
Pilus assembly 274
Pivotal 174
Planet 118
Plant kingdom 72
Plantations 188
Plasma 84
  clearance 257
  concentrations 257
  elimination 274
  level 287
  membrane 64
  protein 272
  separation heparin 285
Pneumonia 299
p-nitro phenyl phosphate 285
Pogrom 121
Poison 21, 126
  agents 186
  residues 126
  revolution 129
Polio 139
Pollination 73
Pollutants 188
Pollution 30
Polygamy 49
Polyspermy 65
Poor-o-phobia 179
Poppy 42, 238
  cultivation 240
Populace 45
Populations, 31
Positive control 291
Post
  disease tetanus 139
  market 260
Postulates 128
Potable water 178
Potatoes 123
Potential role 68
Poverty 172
  fighter 175
  line 195
  ridden 361
Power center 93
Pox 139
Practical manifestation 198
Predator 117
Predetermination 107
Predicted shift 106
Preference 49
Pregnancies 132
Pregnant women 196
Preponderance 111
Preservation 54
Preservative 349
Prey 117
Primary
  block 66
  effect 89
  sex ratio 74, 94, 98
Product 343
Progenies 56, 150, 157
Progeny
  delivering 104
  testing 14
Programmed 28
Prokaryotic 33
Pro-male 201
Promethazine 259
Prontosil 147
Prophylactic 122
Prophylaxis 179
Prosperity 124
Prostate cancer 147
Protein 64
  carbohydrate 81, 85
  content 196
  free substrate 80
  kinase C 69
  lipid 82, 85
  protein 82, 85
  substratum 79
  yields 128
Prototype 274
Prowlness 176
Proximal 82
Psychotropic drugs 259
Puddings 184

Pulmonary
 diseases 299
 infections 300
Pulps 184
Pulses 123, 183
Putative 83
Pyruvate dehydrogenase 273
Pythons 118

Q

Quaternary 218

R

Rabbits 263
Racecourses 313
Radicals 128
Radio-activity 18
Radiographic study 300
Radiological examination 299
Rations 224
Ratites 75, 150
Real estate 208
Receptor 63
 domains 86
 points 92
 sites 85
Recession 208
Recovery 258
Reindeers 32
Remuneration 198
Renal 36, 123
 function 257, 288
 impairment 259
 malfunctions 196
Reproduce 28
Reproduction 55
Reproductive
 organs 62
 physiologists 76, 110
 physiology 112, 143
 tract 62, 112
Reptiles 150
Repulsive 189
Research
 domain 42
 institutes 193
Reserve forests 157
Residues 193
Resources 163
Respiration 102
Respiratory tract irritation 351
Retinopathies 188
Revival 205
Revolution 24, 281
RF latex
 antigen 297
 reagent 297
Rheumatoid
 arthritis 296
 factor test 296
Rice 183
Risk marketing 215
Roasted chickens 120
Rst 67, 83
Rubber teats 291
Rumen 260
Ruptures 262
Rural
 areas 195, 197
 concern 197
 debtless-ness 197
 development 197
 economy 197
 localities 226

S

S.S.R. 97
Sabotage 234
Safety criteria 348
Salads 123
Sanitation 178
Sati 200
Savagery 236
Scars 299
Scenario 31
Schedules 258
Science 28
Scientific
 breakthrough 191
 community 112, 150
 developments 182
 disciplines 150
 discovery 50
 environment 139
 exploration 150
Scientists 185
Scope 54
Sea food 183
Season 78
Second green revolution 122, 130

Secondary
- block 66
- grip 64
- response 89
- sex ratio 97

Secretion 285
Sedimentation 79
See saw mechanism 94
Segregation 65
Seizures 259
Self fertilization 72
Semen 54, 79
Semi
- employed 198
- feudal 201
- urban 197

Seminal plasma 78
Sense 237
Separation 53
Sephadex column separation 79
Serotonon reuptake inhibitors 259
Serum 84, 257
Sex
- assignment 191
- chromosomes 83
- control technology 15
- expression 94
- fixed 28, 258
  - drug 31
- fixing 41, 75
  - act 289
  - activity 259
  - perspectives 48
  - technology 42, 158
  - trials 258
- pre-assignment 314
- pre-determination 15
- pre-selection 72
- ratio 91, 100
  - effect 259
  - manipulation 258

Sexed
- babies 280
- embryos 54
- off-spring 54, 75, 280
- progenies 87
- semen 52, 54, 108, 111, 280
  - technologies 52
- zygotes 54

Sexual
- manifestation 98, 104
- partners 102
- preponderance 103
- reproduction 72
- tract 111

Sexuality organs 252
Shallow penetration 112
Shariat laws 240
Sheds 225
Sheep 192, 316
Shelter 125, 178
Side effects 184
Skeletal 263
Skimmed milk powder 79
Slaughter house 162
Sledges 328
Sleeping sickness 139
Sluggish markets 226
Smack 237
Small babies 196
Snakes 118
Snaring 24
Sns 67, 83
Social
- benefits 280
- implications 280
- security 178, 242

Societal mindset 200
Sodium
- bicarbonate 262
- citrate 79
- ethanoate 113, 279
- fluoride 286
- hydroxide 262
- ions 268
- salts 265
- toxicity 262

Soil 129, 182
- strata 182

Solace 238
Solitary inventor 140
Sorting rate 79
Souls 238
Soups 184
Span 274
SPE-9 67
Species 53
Specifications 343
Spectrophotometer 285, 289
Spectroscopic 84
- analysis 359

Spects 125
Speculators 218
Sperm 43, 59
- binding 85

mass 78
motility 111
separation 53, 79, 108, 331
velocities 80
Spermatozoa 52
Sperm 54
egg fusion 67, 317
oocyte 85
fusion 68
interactions 69
Spillage 260, 265
Spinal 262
Spines 121
Spirogyra 73
Split genes 148
Standard creatinine solution 287
Starvation 45, 51, 127
State subsidies 224
Statistics 98
Stereochemistry outlooks 90
Sterilization procedures 157
Stimulant 92
Stomach 260
Strains 139
Strata 40
Stratum 46
Straw 178
Streptomycin 147
Stress 78
Strong
acids 262, 265
alkalies 262
Stud farms 313
Sub-prime crisis 221
Success rate 258
Sulphate 37, 128
Sulphonamide 88, 139, 147
Sulphur 183
Sulphuric acid 345
Sunlight 261
Super-monsoons 177
Supersaturated solution 347
Surface
biochemistry 78
characteristics 53
Surgical procedures 147
Survival 54, 128, 129
capacity 98
Swine 253
Sycophants 359
Sympathetic 262
Syndicates 237
Synthetic reactions 273
Syphilis 188

T

Target
areas 107
specificity 81
Tartarates 262
Tartaric acid 262
Taste 237
TCA cycle 273
TDSD 75
technique 75
Technical experts 208
Technique 79
Technocrats 129
Technology 28, 76
Teratogenic effects 263
Terminal 238
diseases 124
half-life 274
Terrestrial 183
Terrorism 234
Tertiary 218
Test tube baby 43
Testis
enriched 68
specific 68
Tetraspanin 61, 68, 85, 282, 337
enriched domains 68
family 68
web 68, 85, 325
Thalassemia 139
Thallophytes 73
Thatch 178
Theoreticians 232
Theorists 35
Therapeutic 251
action 89
application 272
dose 267, 268
effects 88
Therapeutically safe 280
Therapy 141
Thermo climates 347
Thermostatic cuvette 286
Throats 142
Thumb rule 138
Time
calendar 35
curve 257

Topological 64
Tornadoes 30
Toxic 280
    poisons 151
    residues 190
    residuism 182
Toxicity 21, 132
    data 349
        sheet 349
Toxicology 279
    criteria 285
    studies 279
Toxification 132
Toxins 187
Tracts 102
Trait 139, 251
Tranquilizers 88
Transmembrane domain 83
Treatment 106
Trials 81
Tricarboxylic acid cycle 266
Tsunamis 30, 201
Tubectomised 157
Tuberculosis 147, 299
Tumors 147, 263
Tundra 102
Twilight 129
Twins 54
Typhus 45

U

Un prejudicial 203
Under ground water tables 186
Underfed 177
Underground water 182
    beds 186
Uneasiness 261
Unhaemolyzed serum 287
Unidirectional mass clusters 104
Unisexual 73
Unitary 103
    sex ratio 104
    values 104
United nations
    agencies 107
    States 107
    statistical demographic books 107
Universal
    happenings 201
    truth 140
Universe 118, 140
Unorganized
    chromatin 76
    sector 195
Unpalatable 189
Urban areas 195
Urea 126, 128
    content 287
    fertilizers 132
Urine 304
Uterine contractor 134

V

Vaccination therapies 147
Vaccine 19
    development 139
Vagina 62, 111
Vaginal 102, 113
    douching 112
    media 267
    penetration 112
    –pH 111
    techniques 267
    tract 111
Valleys 102
Vapors 88
Veal industry 135
Vector 73
Vegetables 123, 180, 183
Vegetative 184
Veil 134
Velocity sedimentation 79
Venomous propaganda 222
Ventricular function 267
Vesicle fusion 60
Veterinary 260
    industry 225
Viability 54
    loss 79
Viable
    methodology 106
    off-springs 281
Vicinity 84, 218
Vicious 189
Victors 128
Vigil 244
Vigor 33
Viral diseases 139
Virus 360
    cell 82
        fusion 60
    diseases 139
    receptor 82

Vistas 140
Vital fronts 180
Vitality 54
Vitamins 183
Volcanic belts 102
Voyage 27

W

Weak acids 262
Weedicide 37, 128
Wheat 183
  cultivation 123
WHO 175
Womanhood 203
Womenfolk 200
Wool 31, 50
  industry 194
  production 46, 193, 316
World
  economy 313
  history 203
Wreckful 234

X

X bearing sperms 90
X chromosome 77
X fractions 79
X-rays 359
XX 77
XX zygote 86, 283
XY chromosomal 77
XY zygote 86, 104, 332

Y

Y bearing spermatozoa 337
Y chromosome 77
Y fractions 79
Y linked histo-compatibility antigen 79
Y specific 86
Y sperm binding ligand mimics 337
Yak 32
Yellow fever 147
Yoga 154
YSBLM 86, 282
  moieties 90, 325, 332
Y-sorting 79
Y-specific antibodies 266

Z

Zinc 128
Zona 62
  pellucida 62, 84, 85
ZP 329
ZP_1 62
ZP_2 62
ZP_3 62
Zulu 40
Zygote 54, 70, 263
  discarding 357
  labeling 357